Management of Migraine Pain

Shvetank Bhatt • C. R. Patil
Neeraj Mahindroo
Editors

Management of Migraine Pain

Emerging Opportunities and Challenges

Springer

Editors
Shvetank Bhatt
School of Health Sciences and Technology
Dr. Vishwanath Karad MIT World Peace
University
Pune, Maharashtra, India

C. R. Patil
Department of Pharmacology
R. C Patel Institute of Pharmaceutical
Education & Research
Shirpur, Maharashtra, India

Neeraj Mahindroo
School of Health Sciences and Technology
Dr. Vishwanath Karad MIT World Peace
University
Pune, Maharashtra, India

ISBN 978-981-97-4528-9 ISBN 978-981-97-4529-6 (eBook)
https://doi.org/10.1007/978-981-97-4529-6

This Springer imprint is published by the registered company Springer Nature Singapore Pte Ltd.
The registered company address is: 152 Beach Road, #21-01/04 Gateway East, Singapore 189721, Singapore

Preface

Migraine is a major headache disorder characterized by moderate to severe, unilateral, pulsating headache worsening with physical activities that may be associated with sensory disturbances classically named as aura. Its episodes may last from a few hours to a couple of days. It is associated with nauseating feeling. Migraine affects the most productive age group ranging from puberty to middle age and may last for the entire lifespan of a patient. Women are more prone to migraine.

There is a need to determine the economic consequences of migraine associated with loss of productivity, worsening of quality of life, and need of prolonged treatment. The etio-pathogenesis of migraine is being progressively understood. Apart from the conventional understanding of the role of monoamine neurotransmitters and vasoactive mediators affecting neuro-vasculature of the brain, the advent of 'omics' sciences have added to the knowledge on pathogenesis of migraine. These details need to be presented to the audience as consolidated narrative.

Further, the recent advances in the contemporary medicines as well as controlled trials on the complementary therapies may provide the physicians with better modalities to treat their patients. There is a need to have a systematic and updated appraisal of such developments in the clinical testing of different treatment options.

We hope the book shall be a useful compilation for undergraduate, postgraduate, doctoral students, and researchers working in cancer and drug delivery research, research and development, and national research institutes. We hope to receive feedback, suggestions, and inputs from researchers and students that will help improve the next edition of the book.

Pune, Maharashtra, India — Shvetank Bhatt
Shirpur, Maharashtra, India — C. R. Patil
Pune, Maharashtra, India — Neeraj Mahindroo

Acknowledgment

We would like to express our sincere gratitude to the Honorable Founder-President, Dr. Vishwanath Karad, and Executive President, Mr. Rahul V. Karad, Dr. Vishwanath Karad, MIT World Peace University, Pune, India, for their motivation. We would also like to extend our gratitude to the faculty members of the School of Health Sciences and Technology at Dr. Vishwanath Karad, MIT World Peace University, Pune, India.

Contents

Editors and Contributors

About the Editors

Shvetank Bhatt is currently working as an associate professor in the School of Health Sciences and Technology, Dr. Vishwanath Karad MIT World Peace University, Pune, Maharashtra, India. He has done M. Pharm in Pharmacology from Manipal College of Pharmaceutical Sciences, MAHE, Manipal, Karnataka, and Ph.D. in Neuropharmacology from Birla Institute of Technology and Science (BITS) Pilani, Pilani Campus, Rajasthan. He has total 16 years of industrial and academic research experience. His area of specialization is CNS disorders, pain, inflammation, and immunooncology. He is recipient of Prof. Duggirala Visweswaram and Prof. Sreemantalu Satyanarayana Award (Best Paper in Pharmacology-IJPER) in 2012 and 2014. He has published more than 80 papers and 20 book chapters. Dr. Bhatt is having total citation 1445; h-index 18 and i10-index 42. He is life member of the Association of Pharmaceutical Teachers of India (APTI) and the Indian Pharmacological Society (IPS).

C. R. Patil is working as a professor and head, Dept. of Pharmacology, R.C. Patel Institute of Pharmaceutical Education and Research, Shirpur, Maharashtra. He has done M. Pharm in Pharmacology from Gujarat University, Ahmedabad, and Ph.D. from Shivaji University Kolhapur. He has more than 24 years of teaching and research experience. His area of expertise is CVS pharmacology, pain, and inflammation. He is recipient of various awards including Innovator Teacher Award (2017), MHRD, Govt. of India, New Delhi, and Young Scientist Award (2017), CCRH-AYUSH, New Delhi. He has published more than 100 papers and book chapters. Dr. Patil is having total citation 4200; h-index 34 and i10-index 66.

Neeraj Mahindroo is medicinal chemist with an experience of over 25 years in academics and research. Currently, he is working as professor and dean at the School of Health Sciences and Technology, MIT World Peace University, Pune, India. Earlier, he was founder Dean of the School of Health Sciences and Technology, UPES, Dehradun. He was also the Dean of the Faculty of Pharmaceutical Sciences and Dean of Research and Development at Shoolini University. The School of

Pharmaceutical Sciences was ranked in the top 40 under the leadership for four consecutive years (2016 (19), 2017 (39), 2018 (30), 2019 (39)) by the National Institutional Ranking Framework (NIRF), Ministry of HRD, Government of India. He is Member of University NIRF, NAAC, and NBA core team. He has experience in PCI and AICTE fillings and inspections. His research interests are drug discovery for neglected diseases, neurological disorders, and cancer, screening of fractionated extracts, and isolated compounds from plants from the North-Western Himalayas. He has published 45 papers with more than 2194 citations (h-index = 25; i10-index =36), 5 patents, and 10 patent applications. He is Member of the Editorial Advisory Board of ACS Medicinal Chemistry Letters and member of CII National Committee on R&D and Innovation 2020–21, part of the sub-group on Healthcare and Chemicals.

Contributors

Gita Sanjay Aher Department of Pharmacology, RC Patel Institute of Pharmaceutical Education and Research, Shirpur, Maharashtra, India

Senthil Visaga Ambi Biopharmaceutical Research Lab, Anusandhan Kendra-1, School of Chemical and Biotechnology, SASTRA Deemed-to-be-University, Thanjavur, Tamil Nadu, India

Department of Bioengineering, School of Chemical and Biotechnology, SASTRA Deemed-to-be-University, Thanjavur, Tamil Nadu, India

Moushumi Baidya Bharat Pharmaceutical Technology, Agartala, West Tripura, India

Shvetank Bhatt School of Health Sciences and Technology, Dr. Vishwanath Karad MIT World Peace University, Pune, Maharashtra, India

Bappaditya Chatterjee School of Pharmacy, GITAM University, Hyderabad, India

Sakshi Arun Chaudhari Department of Regulatory Toxicology, National Institute of Pharmaceutical Education and Research (NIPER), Raebareli, Lucknow, Uttar Pradesh, India

Apoorva Chittoda Department of Pharmacology and Toxicology, National Institute of Pharmaceutical Education and Research (NIPER), Raebareli, Lucknow, Uttar Pradesh, India

Thangaraj Devadoss Department of Pharmaceutical Chemistry, Faculty of Pharmacy, Karpagam Academy of Higher Education, Coimbatore, Tamil Nadu, India

Arghya Kusum Dhar School of Pharmacy, The Neotia University, Sarisha, West Bengal, India

Gayathri Gopal Biopharmaceutical Research Lab, Anusandhan Kendra-1, School of Chemical and Biotechnology, SASTRA Deemed-to-be-University, Thanjavur, Tamil Nadu, India

Department of Bioengineering, School of Chemical and Biotechnology, SASTRA Deemed-to-be-University, Thanjavur, Tamil Nadu, India

Sidhartha Sankar Kar Faculty of Pharmacy, C.V. Raman Global University, Bhubaneswar, Odisha, India

Anurag Kuhad Pharmacology Research Laboratory, University Institute of Pharmaceutical Sciences, UGC-Centre of Advanced Study, Panjab University, Chandigarh, India

Neha Mattoo Pharmacology Research Laboratory, University Institute of Pharmaceutical Sciences, UGC-Centre of Advanced Study, Panjab University, Chandigarh, India

Yashdeep Mukheja Pharmacology Research Laboratory, University Institute of Pharmaceutical Sciences, UGC-Centre of Advanced Study, Panjab University, Chandigarh, India

Aryaa Nigade School of Health Sciences and Technology, Dr. Vishwanath Karad MIT World Peace University, Pune, Maharashtra, India

Priyanka Pandey Department of Pharmaceutical Engineering and Technology, Indian Institute of Technology (B.H.U), Varanasi, Uttar Pradesh, India

Shivam Kumar Pandey Department of Pharmacology and Toxicology, National Institute of Pharmaceutical Education and Research (NIPER), Raebareli, Lucknow, Uttar Pradesh, India

Viral Patel Department of Pharmaceutics, Ramanbhai Patel College of Pharmacy, Charotar University of Science and Technology (CHARUSAT), Anand, Gujarat, India

Gauri Pathak School of Health Sciences and Technology, Dr. Vishwanath Karad MIT World Peace University, Pune, Maharashtra, India

Kalpesh Ramdas Patil Department of Pharmacology, R. C. Patel Institute of Pharmaceutical Education and Research, Dhule, Maharashtra, India

Khusbu Sunil Patil Department of Pharmacology, RC Patel Institute of Pharmaceutical Education and Research, Shirpur, Maharashtra, India

Omkar C. Patil Prathima Institute of Medical Sciences, Karimnagar, Telangana, India

Riya Patil Department of Pharmaceutics, Shobhaben Pratapbhai Patel School of Pharmacy & Technology Management, SVKMS NMIMS University, Mumbai, India

Surabhi Patil Department of Pharmacology, Institute of Pharmacy, Nirma University, Ahmedabad, Gujarat, India

Ravindra Babu Pingili School of Pharmacy and Technology Management, SVKM's NMIMS (Deemed to be University), Shirpur, Maharashtra, India

Rohini Pujari School of Health Sciences and Technology, Dr. Vishwanath Karad MIT World Peace University, Pune, Maharashtra, India

Tania Purswani Department of Pharmaceutics, Shobhaben Pratapbhai Patel School of Pharmacy & Technology Management, SVKMS NMIMS University, Mumbai, India

Mithun Singh Rajput Department of Pharmacology, Institute of Pharmacy, Nirma University, Ahmedabad, Gujarat, India

Jean Sophy Roy Biopharmaceutical Research Lab, Anusandhan Kendra-1, School of Chemical and Biotechnology, SASTRA Deemed-to-be-University, Thanjavur, Tamil Nadu, India

Harish Selvaraj Biopharmaceutical Research Lab, Anusandhan Kendra-1, School of Chemical and Biotechnology, SASTRA Deemed-to-be-University, Thanjavur, Tamil Nadu, India

Jigna Shah Department of Pharmacology, Institute of Pharmacy, Nirma University, Ahmedabad, Gujarat, India

Suhashini Shanmuganathan Biopharmaceutical Research Lab, Anusandhan Kendra-1, School of Chemical and Biotechnology, SASTRA Deemed-to-be-University, Thanjavur, Tamil Nadu, India

Saritha Shetty Department of Pharmaceutics, Shobhaben Pratapbhai Patel School of Pharmacy & Technology Management, SVKMS NMIMS University, Mumbai, India

Rakesh Kumar Singh Department of Pharmacology and Toxicology, National Institute of Pharmaceutical Education and Research (NIPER), Raebareli, Lucknow, Uttar Pradesh, India

Shreya Singh Department of Pharmacology and Toxicology, National Institute of Pharmaceutical Education and Research (NIPER), Raebareli, Lucknow, Uttar Pradesh, India

Urvashi Soni School of Health Sciences and Technology, Dr. Vishwanath Karad MIT World Peace University, Pune, Maharashtra, India

Aashka Thakur Department of Pharmaceutics, Shobhaben Pratapbhai Patel School of Pharmacy & Technology Management, SVKMS NMIMS University, Mumbai, India

Vinod Tiwari Department of Pharmaceutical Engineering and Technology, Indian Institute of Technology (B.H.U), Varanasi, Uttar Pradesh, India

Shiv Kumar Yadav Department of Pharmacy Practice, R. C. Patel Institute of Pharmaceutical Education and Research, Dhule, Maharashtra, India

1 Introduction, Epidemiology, Etiology, and Pathophysiology of Migraine Pain

Khusbu Sunil Patil, Gita Sanjay Aher, Omkar C. Patil, and Shvetank Bhatt

1.1 Introduction

Migraine is disorder of pulsatile, throbbing headache mostly present unilaterally. Attacks might range from a few hours to days. Often associated with nausea, vomiting, visual, and auditory symptoms (Goadsby et al. 2017). Mostly, the symptoms are visual and neurological. The symptoms occurring before the actual attack which may extend into the attack are known as aura (Cutrer and Huerter 2007). Migraine attacks can be with or without association of aura.

The term "migraine" originates from the Greek word *hemikrania*, which was later adapted into Latin as *hemigranea*. The French translation of such a term is "migraine." Migraine frequently leads to disability and absenteeism from work, representing a common cause. Migraine is a headache of primary type and involves the neurovascular and inflammatory aetiology. Migraine attacks are intricate neurological events that develop over a span of hours to days, occurring in a repetitive manner. The most common type of migraine is without aura (Ruschel and De Jesus 2023). Migraine is a form of headache distinguished by repeated episodes of attack, pulsating pain on one side of the head, ranging from moderate to severe. The pain arises from the stimulation of nerve fibers within the walls of the blood vessels in the brain, which travel through the protective membranes known as meninges.

K. S. Patil · G. S. Aher
Department of Pharmacology, RC Patel Institute of Pharmaceutical Education and Research, Shirpur, Maharashtra, India

O. C. Patil
Prathima Institute of Medical Sciences, Karimnagar, Telangana, India

S. Bhatt (✉)
School of Health Sciences and Technology, Dr. Vishwanath Karad MIT World Peace University, Pune, Maharashtra, India

S. Bhatt et al. (eds.), *Management of Migraine Pain*,
https://doi.org/10.1007/978-981-97-4529-6_1

Migraines consist of four phases, and all these phases occurs during the attack as shown in Fig. 1.1:

1. Prodrome (hours to days prior to headache)
2. Aura (5–60 min prior to headache)
3. Headache (4–72 h)
4. Postdrome (24–42 h following headache)

Several distinct factors can elevate the likelihood of experiencing a migraine. The factors that initiate the headache process, triggering migraines, differ among individuals and encompass a range of elements such as sudden changes in weather or environment, insufficient or excessive sleep, exposure to strong odors or fumes, emotional factors, stress, overexertion, loud or abrupt noises, motion sickness, low blood sugar, skipped meals, tobacco use, depression, anxiety, head trauma, and the aftermath of a hangover (Khan et al. 2021). As the role of these triggers, food habits, altered circadian rhythm, and so forth is well established, removal of such triggers provide a significant benefit to the migraine patients. Hence, in treatment of migraine, the alterations in food habits, sleep patterns, physical activity and other non-pharmacological interventions play a vital therapeutic role. Moreover, maintaining a consistent pattern of food intake may prove to be significantly relieving for migraine pain (Currò et al. 2021). Basic treatment used in migraine are nonsteroidal anti-inflammatory drugs (NSAID), Triptans, antiemetic, calcitonin gene-related peptides (CGRP) antagonist, selective serotonin receptor agonist, ergots, and so forth. Alternative treatments employed for migraines encompass various approaches, including regular exercise, yoga, relaxation training, cognitive-behavioral therapy, biofeedback, minimizing triggers, detoxification, and the use of supplements such as butterbur and melatonin (Gazerani 2020). Based on our finding, the hypothesis suggesting a connection between eating habits and migraine headaches has been substantiated (Nazari and Eghbali 2012). However, changing the food patterns and other triggers have been indicated to provide a great relief to the patients in addition to the conventional therapeutic interventions.

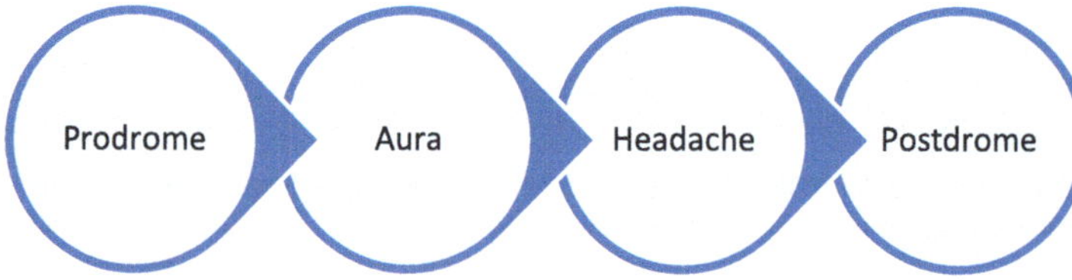

Fig. 1.1 Phases of migraine

1.2 Epidemiology and Connections with Predominance/Lower Occurrence

Migraines are typically differentiated from regular headaches by their severity. According to a recent estimate, the global prevalence of migraine is 14–15%. Migraines contribute to 4.9% of the worldwide burden of illness, measured in terms of years lived with disability (YLDs) (Steiner and Stovner 2023). The global occurrence of migraine is progressively increasing across decades. In 2019, the prevalence of migraines rose to 87.6 million, marking an increase from 62.6 million in the 1990s (Fan et al. 2023). Migraine occurs at a significantly higher rate in countries such as India, China, Indonesia, and the USA, collectively representing over 40% of the global incidence (Fan et al. 2023). Certain systematic reviews and even meta-analysis have indicated positive correlations between the occurrence of migraine and the dietary habits, eating patterns, beverages, and food components (de Oliveira et al. 2023; Ceren Akgor et al. 2023; Hindiyeh et al. 2020). These habits and dietary components may act as triggers or suppressors of the migraine.

1.3 Etiology of Migraine

Various causes or risk factors have been found to initiate the attack. Although vasoconstriction is associated in initiation of migraine, the main pathogenesis is due to inflammation and vasodilation of major cerebral vessels, thus leading to dural stretch causing pain. Its symptoms encompass severe headache, nausea, vomiting, anorexia, diplopia, dizziness, poor muscle coordination, slurred speech, tinnitus, fainting, droopy eyelid, large pupil, vertigo, pricking or stabbing sensation, and so forth. Migraines tend to occur with greater frequency in the morning, especially upon awakening. Certain individuals may experience migraines at anticipated intervals, such as before menstruation or during weekends following a stressful work-week. After experiencing a migraine, individuals often feel fatigued or weakened, but typically, they are free of symptoms during the periods between attacks (Gil-Gouveia and Martins 2018). Furthermore, the potential pathways linking dietary habits to migraines have not been thoroughly investigated. Current evidence from Mendelian Randomization (MR) studies suggests a potential causal link between various risk factors, such as systolic blood pressure (SBP), diastolic blood pressure (DBP), serum total calcium, neuroticism, difficulty awakening, insomnia, anxiety, migraine, and major depressive disorder (MDD) with headaches. Furthermore, prior MR studies have provided evidence indicating that certain dietary habits, such as alcohol intake, are linked to these migraine risk factors, including blood pressure. As a result, one or more of these identified migraine risk factors may serve as potential mediators in the relationship between specific dietary habits and the occurrence of migraines (Liu et al. 2023). The important risk factors involved in the pathological progression of migraine are discussed below.

Sex: Migraine attacks are more common in women than men.

Age: Most of the people affected with migraine attack are the age between 10 and 40 years. However, some women may be affected with the migraine attack after age of 50 years, that is, just after menopause.

Family history: Genetic link of migraine attack is also established well. If one parent has a history of migraine, their child has a 50% chance of getting them. If both parents have them, the risk increases up to 75%.

Comorbid disorders: Comorbid disorders such as depression, anxiety, Alzheimer's disease, epilepsy, sleep disorder, post-traumatic stress disorders, diabetes, hypertension, cancer, and so forth can increase the chance of migraine attack.

Hormonal changes: Hormonal changes in women during menstrual cycle or postmenopause is a major trigger for migraine attack. During pregnancy, women may also be affected with migraine pain, which may or may not be associated with hypertension or high sugar levels in the blood. The attack may be reversible after childbirth.

Stress: Stress is another important factor that leads to progression of migraine attack.

Food and Drinking Habits: This is also an important factor. Skipping meal is one of the important factors that leads to migraine pain. Certain nutrients hold potential in the management of migraines. For example, magnesium displays vasodilator effects and has been linked to a decrease in the frequency of migraines. Micronutrients like Coenzyme Q10, riboflavin (vitamin B2), and omega-3 fatty acids have shown effectiveness in migraine prophylaxis (Fig. 1.2). Additionally, understanding the influence of specific food additives and preservatives on migraine susceptibility is crucial for individuals considering dietary modifications.

1.4 Pathogenesis

The main pathogenesis that leads to migraine is improper functioning of neurons in the brain, particularly the neurons of trigeminal nucleus. This causes release of various neurotransmitters like CGRP, 5-hydroxy tryptamine (5-HT), dopamine, and so forth, which in turn causes increased inflammation and vasodilation of major cerebral blood vessels. This causes a stretch of dura mater, which results to headache (Arulmozhi et al. 2005).

Evidence for release of CGRP, 5-HT, and dopamine association in pathogenesis can be seen by effectiveness of various medications like CGRP antagonists (gepants) and monoclonal antibodies against CGRP. Similarly, for serotonin, we have various receptors like 5-HT 1B, 5-HT 1D, 5-HT 1F which are present, presynaptically acting as presynaptic break for release of neurotransmitters. Thus, its agonists (like triptans and ditans) are used in therapeutics. Also, dopamine antagonists are used for the treatment of migraine pain. All these empower the idea of neurotransmitter-mediated inflammation and vasodilatation (Noseda et al. 2017).

In case of genetic disorders of migraine like familial hemiplegic migraine, the neuronal dysfunction is attributed to mutation of various ion channels, thus

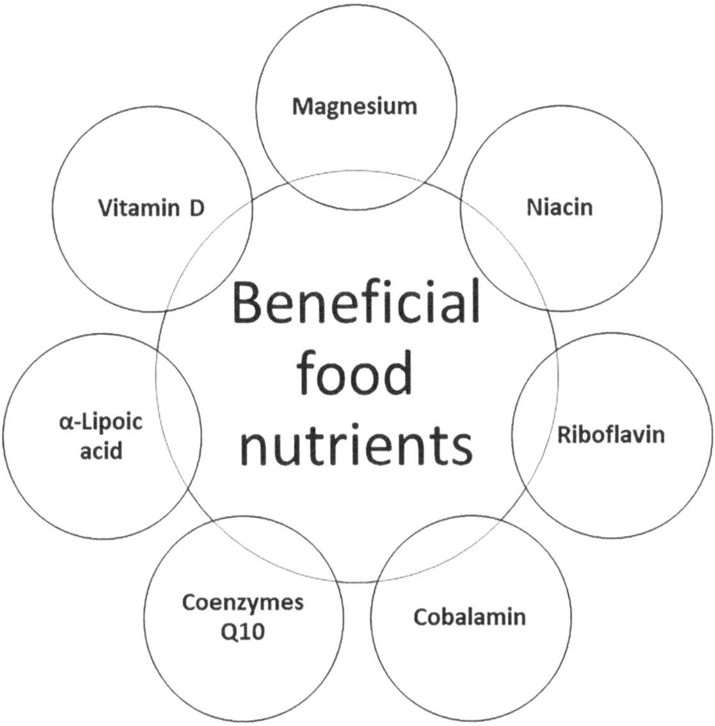

Fig. 1.2 Beneficial food nutrients for migraine pain

Table 1.1 Ion channel and mutations are seen in various diseases

Disease	Ion channel	Mutation involved
FHM 1	Voltage-gated calcium channel	CACANA1A
FMH 2	Na—K ATPase	ATP1A2
FMH 3	Voltage-gated sodium channels	SCN1A

causing increased excitability. Following ion channel and mutations are seen in various diseases (Table 1.1). This increased excitability is again related with release of various neurotransmitters, thereby causing inflammation and vasodilation (Pietrobon 2018).

In addition, the role of various food items, probiotics, and prebiotics is studied well in the pathogenesis of migraine pain. In addition to above factors, food habits, micronutrients, and presynbiotics, prosynbiotics, and synbiotics also have great influence on migraine pain. Modification in their levels in our body may also be responsible for migraine attack. How these factors affect migraine are discussed in the next section of this chapter.

1.5 Connections with Food Habits, Micronutrients, Pre-, Pro-, and Synbiotics

Food Habit: Some of the factors that drive the association between diet and migraine are gut microbiota, inflammatory mediators, stress hormones, the serotonin pathway, neuropeptides, and nutritional substances. Potential novel therapeutic targets for migraine could be developed if we gain a better understanding of the gut–brain–microbiota (GBM) axis. The potential impact of food on migraines is attributed to the direct link between the central nervous system and the enteric nervous system.

Micronutrients: The deficiency of many nutrients including magnesium, niacin, riboflavin, cobalamin, coenzymes Q10, carnitine, α-lipoic acid, and vitamin D is linked with migraine (Nattagh-Eshtivani et al. 2018).

Probiotic: Probiotics are living microorganisms that provide good health. These are good microorganisms that are helpful to maintain the integrity of gastrointestinal tract and blood brain barrier. The proper use of probiotics also improves immunity, reduces inflammation, and is involved in synthesis of certain neurotransmitters. They can be found in yogurt, dietary supplements, and other beauty products.

Gastrointestinal disturbances are common and often accompanied with migraine. Previous research suggested a higher incidence of headaches in persons that are affected with gastrointestinal disorders. The concept of a gut–brain axis is important as this axis is having bidirectional relationship and influence working of gut and brain significantly. Microbiome can be a third component of this axis. It is found that these microbiotas have significant influence on pain and inflammation. Specifically, certain probiotics, predominantly strains of lactobacilli and bifidobacteria, have the ability to improve the integrity of the gut epithelial cell barrier, thereby preserving integrity of GI tract. In general, the evidence indicates that modifying the gut microbiota through dietary changes and potentially reducing inflammation can exert a notable influence on migraines (Gazerani 2020).

1. **Bifidobacteria**: Bifidobacteria support immune system and restrict the growth of unfavorable bacteria in the intestine. These bacteria support the breakdown of lactose into nutrients that can be used by our body. They contribute to restructuring of the balance of gut microbiome by creating an environment less appreciable for pathogenic bacteria.
2. **Lactobacillus**: Lactobacillus bacteria produce lactase, which breaks down lactose, and lactic acid, which restricts the growth of harmful bacteria. They also enhance utilization of muscle fuel, mineral absorption, and immune responses. Lactobacillus species are commonly found in fermented food like yogurt, miso, sauerkraut, and pickles.
3. **Saccharomyces boulardii**: This yeast-based probiotic helps combat diarrhea and other digestive issues. It is especially proficient in promoting gastrointestinal well-being and addressing inflammation within the intestines. Saccharomyces boulardii is commonly found in fermented food like kimchi, natto, fish sauces, juices, raw fruits, and vegetables (Hill et al. 2014).

Prebiotic: Prebiotics are substances in fiber that feed probiotics. The gastrointestinal microbiota, a diverse assembly of microorganisms that are present in the gut, is implicated in the physiology and integrity of various intestinal and extra-intestinal segments, such as stomach, brain, nerves, large intestine immune components, and so forth. These prebiotics are utilized by microbiome, and these microbiomes have various physiological roles in the body (Naghibi et al. 2019).

1. **Fructooligosaccharides (FOS)**: FOS are a type of prebiotic that can precisely favor the growth and activity of beneficial gut bacteria. Studies suggest that FOS supplementation may promote gut eubiosis and help in the reduction of migraine attacks by supporting maintain the healthy microbes inside the gatrointestinal tract.
2. **Galactooligosaccharides (GOS)**: GOS are another type of prebiotic that can serve as diet for beneficial gut bacteria. By promoting the growth of these useful bacteria, GOS supplementation may attribute to maintain integrity of gut and potentially impact migraine symptoms indirectly through gut microbiota modulation.
3. **Inulin**: Inulin is a prebiotic fiber that can enhance the growth of beneficial microbe in the gut. Research suggests that inulin supplementation may help improve integrity of gut and significantly produce beneficial effects on migraine patients (Di Lauro et al. 2023).
4. **Butterbur**: Butterbur extracted from the leaves of *Petasites hybridus*, is a supplement of herbal origin that has demonstrated its potential in the prophylaxis of migraines in some studies in adults. The drug inhibits working of L-type voltage-gated calcium channels, leading to a reduction in tone of blood vessels and the excitation of nerve cells. The active componets of herb include Sesquiterpenes (Petasin and Isopetasin), which have been observed to produce anti-inflammatory effects through inhibition of cyclooxygenase-2 enzyme. The overall effect will be reduction of synthesis of leukotriene and reduction in the release of prostaglandins (Borlak et al. 2022).
5. **Feverfew**: Feverfew (*Tanacetum parthenium*) is a medicinal herb commonly used for the treatment of migraine and arthritis. The effects of feverfew are primarily due to the presence of bioactive compound, parthenolide, which is a sesquiterpene lactone. Parthenolides inhibit production of prostaglandin, disrupt the flexibility of blood vessels, and potentially affect the secretion of serotonin (5-HT). Parthenolides and lactones are spasmolytic that make smooth muscles nonresponsive to endogenous substances such as noradrenaline, acetylcholine, bradykinin, prostaglandins, histamine, and 5-HT in a noncompetitive manner. These antagonistic properties are steady with an antimigraine effect through inhibition of the influx of extracellular calcium into smooth muscle cells of blood vasculature (Buckle et al. 1985).

Food relieving symptoms: Indeed, a notable correlation was observed between headaches and the intake of certain food items within the primary categories of proteins, carbohydrates, fat, and fruits and vegetables (Nazari and Eghbali 2012).

Diet significantly influences pathogenesis of migraine. According to the American Migraine Foundation, some dietary recommendations for migraine prevention include consuming oral magnesium citrate, limiting saturated and trans fats, increasing seafood consumption for omega-3 fats, and avoiding known triggers such as alcohol, aged cheeses, caffeine, and chocolate. Furthermore, studies indicate that adopting a ketogenic diet, incorporating food high in omega-3 fatty acids, and consuming food rich in magnesium may contribute to the prevention of migraine attacks. It is also recommended to eat small, frequent meals, avoid skipping meals, and stay hydrated by drinking water instead of sugary drinks. However, it is important to note that individual responses to specific food can vary, so it's essential to identify and avoid personal trigger food. Additional research is necessary to establish evidence-based dietary recommendations for managing migraines (Gazerani 2020).

Mechanisms of action of some important food: These include reducing inflammation, enhancing serotonin regulation, improving vascular tone, relaxing blood vessels, and maintaining nerve function to help relieve migraine symptoms (de Almeida Soares et al. 2018).

The literature does not provide the clear picture of the mechanisms through which food triggers migraine attacks. The available mechanisms mainly involve the liberation of biogenic amine such as tyramine, resulting in the release of nitric oxide due to the presence of sulfur compounds in food. The food containing flavonoids has capability of changing neurotransmitter levels such as dopamine and histamine. Some substances in migraine triggers inhibit the enzyme group "sulfotransferases (SULTs)," which plays a role in neurotransmitter metabolism; for example, quercetin and catechin in chocolate, hesperidin in oranges and lemons, quinic and caffeic acid in coffee, and epicatechin in tea and red wine are known to inhibit SULTs, specifically the enzymes SULT1A1 and SULT1A3 (Ceren Akgor et al. 2023).

Though multiple evidences appear to underscore the effects of diet on the migraine frequency and severity, there are very few actual studies where the causality has been substantiated in the form of cause and effect. There may be an overlap between migraine-induced craving for certain type of food that is not actually responsible for the subsequent attack. Hence, there is a need of more studies to consolidate how the food habits and food components exactly contribute to the occurrence of migraine (Lisicki and Schoenen 2021).

1.6 Conclusion

Various evidences and factors affecting pathophysiology suggested the role of serotonin, dopamine, adrenergic system, autacoids, inflammatory markers, and CGRP in the progression of migraine attack. In addition, various unusual food habits such as skipping of meal also lead to the development of migraine pain.

References

Arulmozhi DK, Veeranjaneyulu A, Bodhankar SL (2005) Migraine: current concepts and emerging therapies. Vasc Pharmacol 43(3):176–187

Borlak J, Diener HC, Kleeberg-Hartmann J, Messlinger K, Silberstein S (2022) Petasites for migraine prevention: new data on mode of action, pharmacology and safety. A narrative review. Front Neurol 13:864689

Buckle D, Livingstone A, Simmons K (1985) Efficacy of feverfew as prophylactic treatment of migraine. Br Med J (Clin Res Ed) 291(6507):1508

Ceren Akgor M, Vuralli D, Sucu DH, Gokce S, Tasdelen B, Gultekin F et al (2023) Distinct food triggers for migraine, medication overuse headache and irritable bowel syndrome. J Clin Med 12(20):6488

Currò CT, Ciacciarelli A, Vitale C, Vinci ES, Toscano A, Vita G et al (2021) Chronic migraine in the first COVID-19 lockdown: the impact of sleep, remote working, and other life/psychological changes. Neurol Sci 42(11):4403–4418

Cutrer FM, Huerter K (2007) Migraine aura. Neurologist 13(3):118–125

de Almeida Soares A, Louçana PMC, Nasi EP, de Holanda Sousa KM, de Sousa Sá OM, Silva-Néto RP (2018) A double- blind, randomized, and placebo-controlled clinical trial with omega-3 polyunsaturated fatty acids (OPFA ω-3) for the prevention of migraine in chronic migraine patients using amitriptyline. Nutr Neurosci 21(3):219–223

de Oliveira LL, de Barros Lôbo IM, de Sousa NH, Capistano PS, Santos da Silva HM, Pinho LS et al (2023) Association between dietary habits, lifestyle and migraine attacks during social isolation in the COVID-19 pandemic: a systematic review of observational studies. J Biomed Res Environ Sci 4(4):760–772

Di Lauro M, Guerriero C, Cornali K, Albanese M, Costacurta M, Mercuri NB et al (2023) Linking migraine to gut dysbiosis and chronic non-communicable diseases. Nutrients 15(20):1–23

Fan L, Wu Y, Wei J, Xia F, Cai Y, Zhang S et al (2023) Global, regional, and national time trends in incidence for migraine, from 1990 to 2019: an age-period-cohort analysis for the GBD 2019. J Headache Pain 24(1):79

Gazerani P (2020) Migraine and diet. Nutrients 12(6):1658

Gil-Gouveia R, Martins IP (2018) Clinical description of attack-related cognitive symptoms in migraine: a systematic review. Cephalalgia 38(7):1335–1350

Goadsby PJ, Holland PR, Martins-Oliveira M, Hoffmann J, Schankin C, Akerman S (2017) Pathophysiology of migraine: a disorder of sensory processing. Physiol Rev 97(2):553–622

Hill C, Guarner F, Reid G, Gibson GR, Merenstein DJ, Pot B et al (2014) Expert consensus document: the international scientific association for probiotics and prebiotics consensus statement on the scope and appropriate use of the term probiotic. Nat Rev Gastroenterol Hepatol 11(8):506–514

Hindiyeh NA, Zhang N, Farrar M, Banerjee P, Lombard L, Aurora SK (2020) The role of diet and nutrition in migraine triggers and treatment: a systematic literature review. Headache J Head Face Pain 60(7):1300–1316

Khan J, Al Asoom LI, Al Sunni A, Rafique N, Latif R, Al Saif S et al (2021) Genetics, pathophysiology, diagnosis, treatment, management, and prevention of migraine. Biomed Pharmacother 139:111557

Lisicki M, Schoenen J (2021) Old habits die hard: dietary habits of migraine patients challenge our understanding of dietary triggers. Front Neurol 12:748419

Liu X, Yu Y, Hou L, Yu Y, Wu Y, Wu S et al (2023) Association between dietary habits and the risk of migraine: a Mendelian randomization study. Front Nutr 10:1123657

Naghibi MM, Day R, Stone S, Harper A (2019) Probiotics for the prophylaxis of migraine: a systematic review of randomized placebo controlled trials. J Clin Med 8(9):1441

Nattagh-Eshtivani E, Sani MA, Dahri M, Ghalichi F, Ghavami A, Arjang P et al (2018) The role of nutrients in the pathogenesis and treatment of migraine headaches: review. Biomed Pharmacother 102:317–325
Nazari F, Eghbali M (2012) Migraine and its relationship with dietary habits in women. Iran J Nurs Midwifery Res 17(2 Suppl 1):S65–S71
Noseda R, Borsook D, Burstein R (2017) Neuropeptides and neurotransmitters that modulate thalamo-cortical pathways relevant to migraine headache. Headache J Head Face Pain 57:97–111
Pietrobon D (2018) Ion channels in migraine disorders. Curr Opin Physiol 2:98–108
Ruschel MAP, De Jesus O (2023) Migraine headache. In: StatPearls [Internet]. StatPearls Publishing
Steiner TJ, Stovner LJ (2023) Global epidemiology of migraine and its implications for public health and health policy. Nat Rev Neurol 19(2):109–117

The Neurobiology of Migraine and Overview of Migraine Management

2

Yashdeep Mukheja and Anurag Kuhad

2.1 Introduction

Migraine, a pervasive and often incapacitating condition, is marked by unilateral or bilateral headache pain alongside various associated symptoms. Migraines are sometimes preceded with an aura, encompassing sensations like colored lights and extremity numbness, may accompany the pain. This precursor is often followed by full-blown headache along with symptoms such as nausea, vomiting, photophobia, and phono phobia. Migraineurs are classified based on frequency: those experiencing 14 or fewer headache days per month are considered to have episodic migraine, while chronic migraine involves 15 or more headache days monthly. In females, migraine is frequently linked to changes in the menstrual cycle. The condition is biologically predisposed, likely involving genetic mechanisms that sensitize individuals to environmental triggers. On the other hand, the pathophysiology of migraine is increasingly well characterized (Eigenbrodt et al. 2021). Historically, researchers believed that migraine was caused by vascular abnormalities and that migraine aura and headache pain stemmed primarily from intracranial and extracranial arterial vasospasm/dilation. The brain stem is often identified as the primary site of this neurological disruption although some researchers argue for the integral role of neurovascular abnormalities (Asghar et al. 2011; Sánchez del Rio and Linera 2004; Welch et al. 2001). However, it is becoming evident that migraine encompasses fundamental central nervous system malfunction, trigeminovascular system engagement, central neuron sensitization, and increased sensitivity to pain.

Y. Mukheja · A. Kuhad (✉)
Pharmacology Research Laboratory, University Institute of Pharmaceutical Sciences, UGC-Centre of Advanced Study, Panjab University, Chandigarh, India
e-mail: anurag.kuhad@pu.ac.in

S. Bhatt et al. (eds.), *Management of Migraine Pain*,
https://doi.org/10.1007/978-981-97-4529-6_2

2.2 International Classification of Headache Disorders Guidelines for Migraine

Migraine diagnosis poses a unique challenge as there are no definitive gold standard tests available for its diagnosis. Its diagnosis solely depends upon a comprehensive evaluation of patient's medical history, symptoms, and clinical presentation. The International Headache Society classification committee introduced the initial International Classification of Headache Disorders (ICHD) in 1988. These guidelines were refined it in 2004 with minor adjustments. In 2006, additional revisions were proposed, particularly addressing chronic migraine (CM), resulting in the ICHD-2R criteria. Despite the unveiling of the third edition, ICHD-3, in 2018, the present update relies on the ICHD-2 criteria, owing to a scarcity of studies utilizing the ICHD-3 criteria, according to ICHD-3 guidelines (Diener et al. 2019).

According to ICHD, migraine headache is classified into two categories (Table 2.1):

2.3 Factors Affecting Migraine

2.3.1 Genetic Factors

Certain individuals are more susceptible to migraine attacks because of genetic predisposition, suggesting a genetic link. Studies estimate a heritability of 40–50%, suggesting a substantial genetic component in this debilitating disorder (D'Amico 1991; Honkasalo 1995; Stewart 1997). Genetic research, on familial subtype, Familial Hemiplegic Migraine (FHM), a rare subtype exhibiting a Mendelian

Table 2.1 Migraine classification: ICHD criteria

Migraine without aura	Migraine with aura
A headache persisting for 4–72 h (when not treated or treatment is ineffective)	The emergence of neurological symptoms within a 10- to 60-min window preceding the onset of the actual headache, typically lasting no more than an hour, characterizes the aura phase
Headache exhibits at least two of the ensuing features: • Location on one side • Pulsating nature • Intensity ranging from moderate to severe • Impact on physical activity (e.g., walking, stairs)	Aura symptoms encompass: • Paraesthesia (abnormal sensations or numbness) • Hemiparesis (unilateral weakness) • Pins and needles • Aphasia (difficulty speaking) • Disorientation
During the headache, one or more of the following symptoms may occur: • Nausea and/or vomiting • Sensitivity to light and/or sound	The headache may be preceded by nausea, vomiting, light sensitivity, and sound sensitivity during this phase

inheritance pattern has provided insights into the genetic causes for migraine. However, other studies have identified potential genetic underpinnings of common migraine.

Several large-scale genetic analyses have pinpointed regions of the genome loci associated with common migraine in general population (Chasman et al. 2011; Esserlind et al. 2013). The LRP1 gene, transcribed in brain and vasculature tissue, and the TRPM8 gene have emerged as a potential candidate for migraine (Anttila 2010). LRP1 has been implicated in glutamatergic signaling, whereas TRPM8 codes for an ion channel, which is a sensor of temperature and temperature-induced pain; the gene has been a target in neuropathic pain models (Chasman et al. 2011). Research has conclusively shown the association of these two genes with migraine as well as identified another gene, the TSPAN-2 gene (Esserlind et al. 2013). TSPAN-2 has been implicated in oligodendrogenesis. Research also suggests a genetic basis of impaired glutamate signaling in migraine pathophysiology, showing an association between two candidate genes involved in glutamate homeostasis, PGCP and MTDH, with migraine, particularly migraine with aura (Anttila 2010).

Though rarer, FHM's clear pattern of inheritance simplifies gene hunting, potentially unlocking causes and treatments for more common forms of migraine (Goadsby and Kullmann 2005). FHM is a rare, autosomal dominant kind of aura producing migraine that causes temporary unilateral paralysis of the limbs (Joutel et al. 1993). Three forms of FHM have been identified (although up to six are believed to exist), each classified according to mutations in specific ion channel genes that likely increase cortical excitability, a pivotal phenomenon in migraine pathophysiology (Di Lorenzo et al. 2012). FHM type I occurs due to mutations in calcium channel, particularly in brain cells of cerebellum and brain stem (Ducros et al. 2001; Martínez et al. 2016). FHM type II is connected to gene mutation of ATP1A2, which results in abnormal functioning of a $Na+/K+$ ATPase expressed in astrocytes (Castro et al. 2008; De Fusco et al. 2003; Marconi et al. 2003; Vanmolkot et al. 2006). This mutation causes excessive release of glutamate into neuronal synapses. Lastly, FMH type III is linked to a mutation in voltage-gated sodium channels expressed in cortical neurons leading to a sustained influx of sodium producing hyperexcitability or inexcitability of neurons (Dichgans et al. 2005; Kahlig et al. 2008).

These mutations extend beyond FHM; patients with sporadic hemiplegic migraine without neurological features consistent with FHM and family members of those patients carried mutations similar to those found in FHM patients (de Vries et al. 2007; Russell 2008). These findings suggest a potential shared genetic cause across various forms of migraine with aura. Additionally, the shared pathophysiology of cortical hyperexcitability and certain channelopathies overlapping with epilepsy have led some researchers to speculate the efficacy of antiepileptic drugs as migraine prophylactic treatments (Bianchin et al. 2010; Huang et al. 2017). Despite the genetic connections to FHM and mutations identified in the broader migraine population, this knowledge has not translated into novel migraine therapies. Although an FHM type 1 mutation has been linked to alteration in calcitonin gene-related peptide (CGRP) expression in the trigeminal ganglion of mice (Mathew et al. 2011), no relation is established between CGRP and migraine in FHM patients (Hansen 2008, 2011). However, other studies have established a link between CGRP and migraines in other types of migraine headaches (Hansen et al. 2010; Lassen et al. 2002), leading to advances of CGRP-targeted therapies for migraine.

2.3.2 Environmental Factors: Migraine Triggers

Experts speculate that individuals susceptible to migraines may experience an onset triggered by typically non-harmful environmental factors. The precise mechanism by which these triggers initiate a migraine attack remains unclear. Migraine sufferers can often pinpoint specific events or substances, as depicted in Fig. 2.1, that serve as catalysts for their migraines. While the predictability of these triggers varies among individuals, the factors triggering migraine attacks offer valuable insights into the condition's origin. Numerous migraine triggers, such as inadequate sleep, hunger, alcohol, and stress, exert direct influence on the brain and adds more credibility to theories emphasizing neurological origins rather than vascular causes. Recognition of such triggers helped inspire the study of possible treatments such as the melatonin agonist ramelteon, an insomnia medication. Furthermore, medications that influence serotonin and nitric oxide have been noted to provoke migraines and have been employed to induce headaches in clinical investigations.

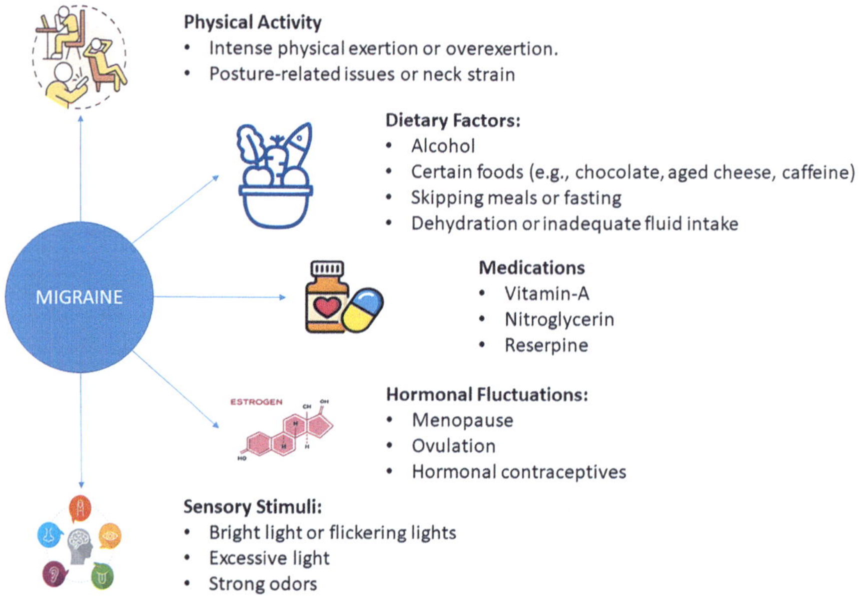

Fig. 2.1 Myriad of factors such as physical activity, diet, hormones, medications that play a vital role in triggering migraine

2.4 Key Pathways and Drug Targets

During 1930s, a leading neurologist named Harold Wolff suggested a theory for migraine headaches. He believed that cause of migraine is sudden changes in blood vessel size in the brain. His idea was that migraine aura happened due to temporary lack of blood flow caused by narrowed blood vessels inside the skull. The throbbing headache itself, he thought, resulted from these vessels then widening outside the skull, which irritated nerves around them. Interestingly, Wolff noticed that administration of ergotamine tartrate narrows blood vessels and helped ease the headache pain (Pardutz and Schoenen 2010).

Evolution in techniques for measuring blood flow, developed and refined since Wolff's era, has corroborated the association between the migraine aura phSAF21a-qase and a significant reduction in cerebral blood flow, nearly reaching ischemic levels. Detailed examinations of cerebral blood flow during a migraine episode reveal a gradual propagation of oligemia (diminished blood volume) across the cortical surface at a velocity of 2–3 mm per min, initiating predominantly in the occipital cortex and progressing towards the temporal and parietal lobes. This reduction in blood flow typically aligns with the side of headache pain and is opposite to the manifestation of aura symptoms. A study described that reduced blood flow (called spreading oligemia) might also happen in migraines without aura. This suggests that the blood flow reduction might be too subtle to cause noticeable symptoms in some cases. These findings support the idea that migraines with and without aura might be different ways the brain shows the same underlying problem (Woods et al. 1994).

Several molecules have been implicated in migraine. One particular molecule, CGRP, has generated a great deal of interest among physicians and drug developers as a target for comprehensive migraine treatments.

2.4.1 Serotonin

5-HT, a chemical messenger in brain also acts on blood vessels, has been linked to migraine development (Lance 1991). Triptans, the most efficacious group of drugs for treating acute migraine work by agonizing the 5-HT1 receptor. The key role of serotonin in migraine pathophysiology is well established, but the exact mechanism is unclear (Drummond 2006). The antihypertensive medication reserpine, known for releasing biogenic amines such as dopamine, norepinephrine, and serotonin from synaptic vesicles in the central nervous system, triggers migraine headaches. These headaches can often be relieved by giving medications that increase serotonin levels in the brain which include serotonin, substances the body converts into serotonin (precursors), or drugs that mimic serotonin's effects (5-HT agonists). During a migraine episode, the plasma serotonin levels decrease, and there is a subsequent increase in urinary levels of serotonin and its major metabolite, 5-hydroxyindoleacetic acid, indicating the importance of metabolism of serotonin in the pathophysiology of the disorder (Drummond 2006).

2.4.2 Dopamine

Dopamine is known to play a role in pathophysiology of migraine (Charbit et al. 2010; D'Andrea and Leon 2010). Substantia nigra, region in brain stem which is rich in dopaminergic innervation, seems to get activated during the aura phase (Blin 1991; Cerbo 1997; Del 1994). Dopamine has the potential to trigger feelings of nausea and decrease gastric motility, and its direct correlation with the vomiting observed during a migraine attack is evident.

2.4.3 Glutamate

Glutamate had been identified as an important molecule in migraine pathophysiology (Hamberger 1993). Glutamate binds on various postsynaptic receptors including NMDA, AMPA, and kainate receptors—all of which are ionotropic—and metabotropic glutamate receptors. Binding of glutamate to these receptors activates the receptors and increases the excitability of the neurons. It is believed that cortical hyperexcitability observed in migraineurs is due to overstimulation of the NMDA receptors (Rogawski 2008).

2.4.4 Gamma Amino Butyric Acid

Gamma amino butyric acid (GABA) emerges as a significant player in the initial stages of a migraine onset. Antiepileptic drugs showcasing efficacy in migraine prophylaxis exert GABA-enhancing effects, contributing to their therapeutic effectiveness by limiting the neuronal hyperexcitability. While the specific mechanisms behind GABA-mediated migraine prevention are not yet well-defined or explicitly known, GABA's inhibition of the serotonin-containing brain stem raphe nucleus has the potential to lead to a decrease in the overall levels of serotonin systemically.

2.4.5 CGRP

Emerging evidences suggest the critical role of CGRP in migraine pathophysiology. CGRP is the most powerful vasodilator neuropeptide, which causes site-specific vasodilation and the extravasation of plasma and plasma proteins into surrounding tissue on its release. The finding that intravenous CGRP injection causes migraine episodes in people with both migraine with aura and migraine without aura lends more credence to the involvement of CGRP in migraine (Hansen et al. 2010; Lassen et al. 2002). Furthermore, studies indicate that CGRP baseline levels may be higher in those with chronic migraine than in those with episodic migraine and healthy controls (Carod-Artal et al. 2011). Positive results from large clinical trials using CGRP receptor antagonists pave the way for CGRP-based therapies in migraine management. (Connor et al. 2009; Ho et al. 2008). Research also indicates towards the ability of triptans in controlling the release of neuropeptides such as CGRP, which may contribute to their clinical efficacy (Eltorp et al. 2000); furthermore, CGRP levels are normalized following sumatriptan treatment (Deen et al. 2017).

2.4.6 Pituitary Adenylate Cyclase-Activating Peptide-38

Pituitary adenylate cyclase-activating peptide (PACAP)-38 is a vasodilator peptide involved in a number of functions, including neurogenic inflammation and nociception (Kaiser and Russo 2013). PACAP-38 receptors are found throughout the trigeminal system (Sundrum and Walker 2018), and research has shown that, in rats, PACAP-38 induces CGRP release within the trigeminal nucleus caudalis, a brain region of importance in the transmission of migraine (Jansen-Olesen et al. 2014). Studies provide compelling evidence for PACAP-38's involvement in migraines. Infusions of PACAP-38 triggered migraine-like attacks in migraineurs without aura and headaches in healthy individuals (Schytz et al. 2009). Additionally, migraine patients exhibited elevated PACAP-38 levels during attacks, but these levels dropped below those of healthy subjects between attacks (Tuka et al. 2013).

2.4.7 Nitric Oxide

Compounds that increase nitric oxide (NO) levels have been shown to trigger migraines without aura in individuals susceptible to these headaches. This observation has led researchers to explore NO as a potentially key biochemical factor in migraine development (Olesen 2008; van der Kuy and Lohman 2003). Emerging evidence suggests that NO might initiate localized inflammation and blood vessel dilation through the release of neuropeptides like CGRP, which contributes to migraine pain (Prakash et al. 2014).

2.4.8 Hormones

Menstrual migraines are often due to hormonal fluctuations (Newman 2007). These fluctuations may trigger abnormal responses in brain neurotransmitter like serotonin and opioids, potentially leading to migraine attacks (Newman 2007). Furthermore, the estrogen levels during menstruation and menopause decreases, which impact serotonin regulation leading to migraine attacks (Vetvik and MacGregor 2021).

2.4.9 Magnesium

Magnesium deficiency, potentially linked to mitochondrial dysfunction (Lodi et al. 2001), may contribute to migraine development, including menstrual migraine by increasing NMDA receptor activity and cortical hyperexcitability. A link has been established between magnesium deficiencies in women with menstrual migraine (Mauskop et al. 2002), with some women experiencing relief through magnesium supplementation.

2.5 Neurogenic Inflammations in Migraine

2.5.1 Spreading Nerve Cell Depression in Migraine Aura

It is believed that reduced neuronal activity during migraine attack leads to the spreading oligemia. In the 1940s, A.A.P. Leão and colleagues, while investigating experimental epilepsy models, discovered that external stimuli could trigger a brief phase of neuronal excitation followed by prolonged nerve cell depression and a disruption of brain ion balance (Leao 1944). The diminished neuronal activity initiated from the stimulus site in one hemisphere and gradually propagated in all directions at a pace ranging from 2 to 5 mm per minute. This wave of decreased neuronal activity resulted in alterations in cerebral blood flow, closely resembling the distribution of the spreading neuronal depression. This pattern mirrors the occurrences of oligemia and visual aura observed in human migraine attacks. Accumulating evidence suggests an association between spreading depression and migraines (Eikermann-Haerter et al. 2012). The uncertainty revolves around whether spreading depression and spreading oligemia are exclusive to individuals experiencing migraine with aura or are common occurrences across all migraine patients. Spreading depression may occur in migraine without aura but not to an extent that provokes the neurological disturbances characterizing migraine with aura; however, study data have challenged this notion (Hauge et al. 2009).

2.5.2 Role of the Trigeminovascular System

It is believed that neurogenic inflammation of the meninges occurs within the trigeminovascular system, where an increase in the release of neuropeptides (e.g., CGRP) occurs from trigeminal neuron depolarization and consequently increases vascular permeability or dilation of cerebral blood vessels and causes plasma protein extravasation (Ramagopalan 2007). The precise etiology of migraine pain is unknown yet, although current research suggests involvement of the trigeminal nerve. While the exact source of irritation to this nerve is yet to be identified, the brainstem, specifically the trigeminal nucleus caudalis, is increasingly suspected to be the initiation site. This activation may coincide with or follow a cortical event such as spreading depolarization (Doods 2007).

2.6 Other Processes Involved in Migraine

2.6.1 Central Sensitization

Central sensitization is crucial in pathogenesis and progression of episodic migraine to chronic migraine (Dodick 2006). Central sensitization is a process where increased synaptic transmission in the central nervous system leads to a lowered pain threshold, an intensified pain response, and the extension of pain sensitivity to unaffected areas. In the context of migraines, neurogenic inflammation reduces the threshold for activating cranial nociceptors. Repeated stimulation of the trigeminal nerve, resulting from peripheral sensitization, induces increased excitability of second-order neurons within the trigeminal nucleus caudalis. Clinically, this results in hyperalgesia (an increased pain response) and cutaneous allodynia (a painful response to non-noxious stimuli), both serving as indicators of central sensitization in the context of migraines.

2.6.2 Brain Stem as Migraine Generator

According to a study conducted in 1995 by Weiller et al., utilizing Positron Emission Tomography (PET), those with migraine without aura exhibited higher blood flow in the brain stem and cortical areas during and after episodes. Subcutaneous injections of 6 mL of sumatriptan were shown to be beneficial in treating headaches caused by migraines, phonophobia, and photophobia. PET scans revealed normal blood flow in cortical locations, but the locus coeruleus and raphe exhibited elevated blood volume long after the headache and associated migraine symptoms went away. Additionally, a second functional magnetic resonance imaging study revealed activity in the red nucleus of the brain stem and the dopamine-producing substantia nigra in individuals with migraines without aura (Weiller et al. 1995).

2.6.3 Chronification of Migraine

Migraine is increasingly viewed as a progressive disorder in which headache frequency can evolve from an episodic (i.e., intermittent) to a chronic (i.e., daily or near daily) state in some patients. The progression from EM to CM can last from months to years and occurs gradually in approximately 80% of cases that progress and abruptly in 20% of cases (Bigal et al. 2004; Mathew et al. 1982). Preliminary data also indicate that episodic migraineurs progress to CM at the rate of 2.5% per year, and some CM patients also remit from CM to EM (Bigal and Lipton 2008). Although several modifiable risk factors for migraine chronification have been identified, little is understood about its actual causes. Studies indicate that migraine progresses together with a variety of anatomical changes, including reductions in cortical gray matter volume (Kim et al. 2008; Valfrè et al. 2008), iron accumulation in the periaqueductal gray matter (Kruit et al. 2009), and subcortical white matter lesions. Functional changes also occur including reversible and increased cortical hyperexcitability (Aurora et al. 2007).

2.7 Treatment Overview

Migraine patients often start treatment on their own by administering over-the-counter (OTC) nonsteroidal anti-inflammatory drugs (NSAIDs) for headache relief. If proper relief is not achieved, then refer to a primary care physician or a neurologist, who then carefully diagnoses and determines the type and severity of migraine. For mild pain, NSAIDs with varying characteristics are prescribed. However, if headaches persist or are moderate in intensity, triptans come into play, administered orally or non-orally based on severity and patient factors. Figure 2.2 describes in detail about the patient journey of a migraine patient and is necessary for us to understand so as to effectively manage migraine.

Approved medications for acute migraine attacks, whether with or without aura, lack specific approvals for treating headaches separately, with or without aura. There's also no specific approval for either episodic migraine or chronic migraine despite most being approved based on clinical trials involving episodic migraine. Doctors generally follow consistent prescribing patterns for acute treatment in both groups as depicted in Fig. 2.3, describing the treatment paradigm of migraine sufferers. Patients are categorized as chronic or episodic, and treatment initiation is based on the severity of their condition. The choice of medication is further refined based on the presence or absence of nausea, as shown in Fig. 2.4.

Triptans are crucial for treating acute migraine attacks. However, concerns about medication overuse in headache often led to restrictions on the number of triptan doses per month in prescriptions and reimbursement. This means people with

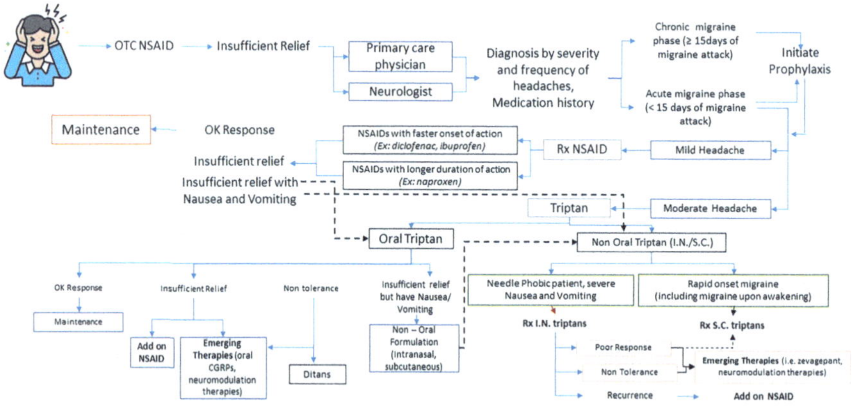

Fig. 2.2 The patient journey of migraine patient. It starts with self management attempts using over-the-counter (OTC) nonsteroidal anti-inflammatory drugs (NSAIDs). Incomplete relief drives the patient to consult primary care physicians or neurologists. They then classify migraine and carefully select medications, NSAIDS or triptans

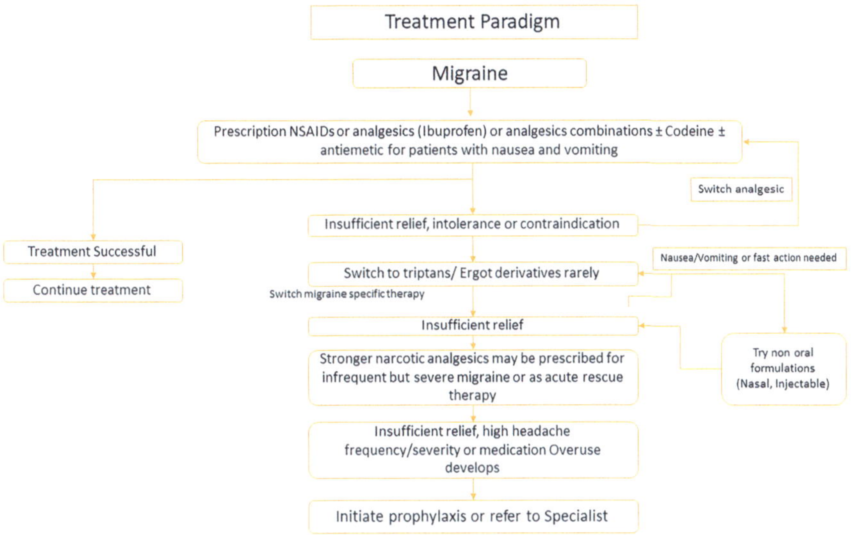

Fig. 2.3 Treatment paradigm of migraine patients

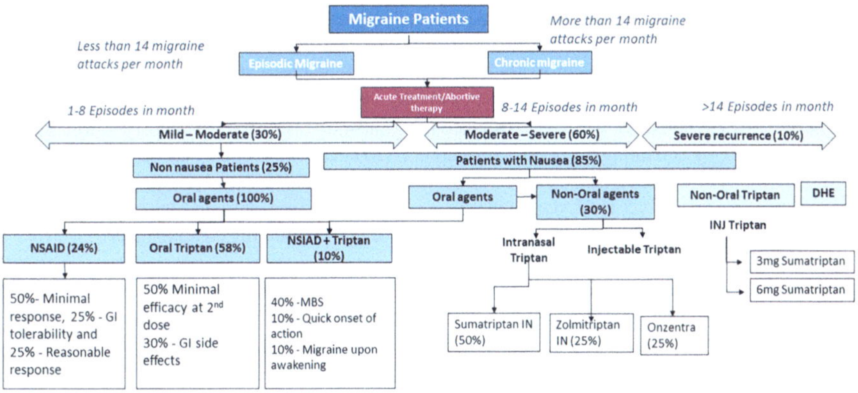

Fig. 2.4 Treatment of migraine, depending upon the patient subtype and type of migraine and its severity

chronic migraines may need additional treatments to handle daily migraines. Reyvow, Ubrelvy, and Nurtec ODT are noteworthy as the first FDA-approved migraine-specific acute therapies in two decades. Comparing them to triptans, they seem to have similar effectiveness in achieving 2-hour pain freedom and pain relief, if not slightly less. Our research suggests these new treatments will likely be reserved for those who don't respond to triptans, can't tolerate them, or for whom triptans are not recommended due to vascular risk factors.

A current trend in developing acute migraine medications involves creating innovative, non-oral delivery systems to enhance efficacy, speed of onset, and minimize

Table 2.2 Therapeutic options for the treatment of migraine

Class	Mechanism of action
5-HT1B/1D receptor agonists	Scientists suggest that serotonin oversees the communication in the trigeminal and spinal pathways, essentially regulating the transmission of pain signals. The 5-HT1B/1D receptor agonists, like triptans, counter migraines by affecting 5-HT1B receptors in the meningeal and coronary vasculature, as well as 5-HT1D receptors located trigeminal nerve terminals. Triptans' action at these 5-HT receptors also hampers the secretion of nociceptive neuropeptides (such as CGRP, neurokinin A, and substance P), thought to be responsible for causing migraine pain
5-HT1F receptor agonists	5-HT1F receptor agonists demonstrate their anti-migraine effects by activating 5-HT1F receptors found on sensory trigeminal neurons, both centrally and peripherally. In contrast to triptans, Lasmiditan shows a greater preference for 5-HT1F receptors compared to vasoconstrictor 5-HT1B receptors. Consequently, this medication does not impact blood vessel constriction, making it a safer option for individuals with migraines who cannot use triptans due to vascular concerns
CGRP receptor antagonists	The α-form of CGRP is instrumental in development of migraines. Antagonists of the CGRP receptor, known as gepants, demonstrate their anti-migraine effects by blocking these receptors. Importantly, CGRP antagonism by gepants does not lead to blood vessel constriction, making these medications safe for individuals with migraines who have concerns about their vascular health. Some oral CGRP antagonists were discontinued due to liver toxicity observed in clinical trials when used preventatively (not during trials for immediate treatment). However, it's noteworthy that this toxicity doesn't seem to be a common issue across the class, as newer therapies like ubrogepant and rimegepant have not been associated with hepatotoxicity
Ergot derivatives	Ergotamine and DHE, produce their anti-migraine effects similarly to triptans by activating 5-HT1B/1D receptors. However, these agents are less specific and bind indiscriminately to various other receptor subtypes, including 5-HT1A, 5-HT1F, 5-HT2, dopaminergic, and adrenergic receptors in blood vessels and smooth muscles. The nonspecific binding of these drugs leads higher occurrence of side effects
NSAIDs	Nonsteroidal anti-inflammatory drugs (NSAIDs) proficiently block the activity of both isoforms of the COX enzyme, COX-1 and COX-2, in a relatively non-specific manner. This action blocks the arachidonic acid pathway, halting the production of prostaglandins and preventing neurogenic inflammation. Prostaglandins cause the initial vasoconstriction of migraines and the onset, as well as heightened sensitivity to, headache pain

side effects. This includes melt-away oral tablets, nasal applications, and injectable forms, catering to specific needs. Pfizer's newly launched intranasal Zavzpret (zavegepant) is an example, being the first non-oral gepant for acute migraine treatment, showing similar effectiveness to other oral gepants (Table 2.2).

2.8 Prophylaxis Overview

Prophylactic treatments for migraines often exhibit partial efficacy over the long term, necessitating the continued need for acute therapy in many patients. This underscores a significant gap in more efficacious preventive treatments. The introduction of monoclonal antibodies targeting calcitonin gene-related peptide (anti-CGRP MAbs) has partially addressed this gap, serving as the first migraine preventives specifically formulated for EM and CM with broad approval. However, most prophylactic agents lack a clear distinction between EM and CM in their approvals, leading physicians to use preventive treatments that overlap for both migraine populations. Botox remains the exception with exclusive approval for CM. Standard-of-care therapies for migraine prevention include the antiepileptic drug topiramate, the beta blocker propranolol, and, though not officially approved for migraine prophylaxis, the tricyclic antidepressant amitriptyline. In select European markets, serotonin antagonists like pizotifen and methysergide find occasional use (Table 2.3).

Table 2.3 Prophylaxis overview

Class	Mechanism of action
Antiepileptic drugs	The precise pathways through which antiepileptic drugs (AEDs) exert their effectiveness in preventing migraines remain incompletely elucidated. It is hypothesized by researchers that these agents influence nociception, or pain perception, by mitigating brain hyperexcitability. This modulation is thought to occur either through the adjustment of GABA- and/or glutamate-mediated neurotransmission or by modulating ion channels (Ca^{2+}, Na^+, and/or K^+) within synaptic terminals
Beta blockers	Beta blockers without intrinsic sympathetic activity restricts extracranial arteries from dilation by increasing their vascular tone (Meiyanti 2016). Furthermore, selective beta 1 blockers may diminish sympathetic tone and hinder platelet activity (platelet aggregation hampers the smooth flow of blood through blood vessels) (Joseph et al. 1988)
Calcium channel blockers	Investigations into the exact mechanism by which calcium channel blockers prevent migraines are still continuing. It is theorized that channelopathies, which are calcium channel dysfunctions, might disrupt serotonin delivery, making patients more susceptible to migraine attacks (Mathew 2001). Drugs like verapamil may block intracellular calcium entry and cell depolarization to inhibit neurogenic inflammation. Additionally, the prophylactic efficacy of flunarizine is believed to stem from the drug's antioxidant properties, preventing nitric oxide (NO)-associated cerebral vasospasm or oxidative reactions during migraines (Ciancarelli et al. 2004; Mathew 2001)

(continued)

Table 2.3 (continued)

Class	Mechanism of action
Antidepressants	Tricyclic antidepressants are proven to be effective in managing migraine. They block the reuptake of serotonin and norepinephrine, which inhibits the nociceptive pathway. Also, muscarinic receptors, alpha-1-adrenergic, cholinergic, and histaminergic receptors are also blocked by these drugs. Their nonselective and nonspecific nature poses challenges in discerning the specific antimigraine effects of these drugs. Similarly, serotonin–norepinephrine reuptake inhibitors and selective serotonin reuptake inhibitors are also effective in managing migraine
Serotonergic antagonists	Serotonin antagonists serve a preventive function by inhibiting the series of events that lead to migraines, rendering them a prophylactic treatment (Raskin 1990). Pizotifen acts on 5-HT2B receptors and additionally possesses antihistamine and calcium-channel-blocking activities (Peroutka et al. 1985). These characteristics mitigate vasodilation and restrict inflammation. Researchers posit that methysergide also acts as an antagonist on 5-HT2 receptors
Neuromuscular blocking agents	The mechanism by which onabotulinumtoxin-A prevents migraine is unclear. One theory suggests that the drug inhibits the release of substance P, CGRP, and glutamate by blocking neurotransmission (Dodick et al. 2010). By facilitating the pain signals' transmission, onabotulinumtoxin-A may inhibit the central sensitization (the CNS's exaggerated response to mild pain) thought to be involved in migraine generation. Another theory purports that paralysis of certain facial muscles deactivates peripheral migraine headache trigger sites (Behmand et al. 2003)
CGRP-targeted therapies	CGRP, a potent vasodilatory neuropeptide, has long captivated attention for its potential in both preventing and treating migraines. The role of CGRP release in migraines remains uncertain—whether it actively triggers migraines or occurs as a consequence during an attack. It has been observed that intravenous CGRP administration induces migraines in people suffering from migraine (Hansen et al. 2010; Lassen et al. 2002), and elevate the baseline levels of peripheral CGRP (Carod-Artal et al. 2011). Additionally, research suggests that the effectiveness of triptans, a common migraine medication, may be partly attributed to their ability to control the release of CGRP and other neuropeptides (Eltorp et al. 2000)

References

Anttila V (2010) Genome-wide association study of migraine implicates a common susceptibility variant on 8q22.1. Nat Genet 42(10):869–873. https://doi.org/10.1038/ng.652

Asghar MS, Hansen AE, Amin FM, van der Geest RJ, van der Koning P, Larsson HBW, Olesen J, Ashina M (2011) Evidence for a vascular factor in migraine. Ann Neurol 69(4):635–645. https://doi.org/10.1002/ana.22292

Aurora SK, Barrodale PM, Tipton RL, Khodavirdi A (2007) Brainstem dysfunction in chronic migraine as evidenced by neurophysiological and positron emission tomography studies. Headache 47(7):996–1003. https://doi.org/10.1111/j.1526-4610.2007.00853.x

Behmand RA, Tucker T, Guyuron B (2003) Single-site botulinum toxin type A injection for elimination of migraine trigger points. Headache 43(10):1085–1089. https://doi.org/10.1046/j.1526-4610.2003.03210.x

Bianchin MM, Londero RG, Lima JE, Bigal ME (2010) Migraine and epilepsy: a focus on overlapping clinical, pathophysiological, molecular, and therapeutic aspects. Curr Pain Headache Rep 14(4):276–283. https://doi.org/10.1007/s11916-010-0121-y

Bigal ME, Lipton RB (2008) Concepts and mechanisms of migraine chronification. Headache 48(1):7–15. https://doi.org/10.1111/j.1526-4610.2007.00969.x

Bigal ME, Lipton RB, Tepper SJ, Rapoport AM, Sheftell FD (2004) Primary chronic daily headache and its subtypes in adolescents and adults. Neurology 63(5):843–847. https://doi.org/10.1212/01.WNL.0000137039.08724.18

Blin O et al (1991) Apomorphine-induced yawning in migraine patients: enhanced responsiveness. Clin Neuropharmacol. 14:91–95

Carod-Artal FJ, Ezpeleta D, Martín-Barriga ML, Guerrero AL (2011) Triggers, symptoms, and treatment in two populations of migraneurs in Brazil and Spain. A cross-cultural study. J Neurol Sci 304(1–2):25–28. https://doi.org/10.1016/j.jns.2011.02.027

Castro M, Nunes B, De Vries B, Lemos C, Vanmolkot K, Van Den Heuvel J, Temudo T, Barros J, Sequeiros J, Frants R, Koenderink J, Pereira-Monteiro J, Van Den Maagdenberg A (2008) Two novel functional mutations in the Na^+, K^+ -ATPase α2-subunit *ATP1A2* gene in patients with familial hemiplegic migraine and associated neurological phenotypes. Clin Genet 73(1):37–43. https://doi.org/10.1111/j.1399-0004.2007.00918.x

Cerbo R et al (1997) Dopamine hypersensitivity in migraine: role of the apomorphine test. Clin Neuropharmacol. 20:36–41

Charbit AR, Akerman S, Goadsby PJ (2010) Dopamine: what's new in migraine? Curr Opin Neurol 23(3):275–281. https://doi.org/10.1097/WCO.0b013e3283378d5c

Chasman DI, Schürks M, Anttila V, de Vries B, Schminke U, Launer LJ, Terwindt GM, van den Maagdenberg AMJM, Fendrich K, Völzke H, Ernst F, Griffiths LR, Buring JE, Kallela M, Freilinger T, Kubisch C, Ridker PM, Palotie A, Ferrari MD et al (2011) Genome-wide association study reveals three susceptibility loci for common migraine in the general population. Nat Genet 43(7):695–698. https://doi.org/10.1038/ng.856

Ciancarelli I, Tozzi-Ciancarelli M, Di Massimo C, Marini C, Carolei A (2004) Flunarizine effects on oxidative stress in migraine patients. Cephalalgia 24(7):528–532. https://doi.org/10.1111/j.1468-2982.2003.00705.x

Connor KM, Shapiro RE, Diener H-C, Lucas S, Kost J, Fan X, Fei K, Assaid C, Lines C, Ho TW (2009) Randomized, controlled trial of telcagepant for the acute treatment of migraine. Neurology 73(12):970–977. https://doi.org/10.1212/WNL.0b013e3181b87942

D'Amico D et al (1991) Genetic transmission of migraine without aura: a study of 68 families. Ital J Neurol Sci. 12:581–584

D'Andrea G, Leon A (2010) Pathogenesis of migraine: from neurotransmitters to neuromodulators and beyond. Neurol Sci 31(S1):1–7. https://doi.org/10.1007/s10072-010-0267-8

De Fusco M, Marconi R, Silvestri L, Atorino L, Rampoldi L, Morgante L, Ballabio A, Aridon P, Casari G (2003) Haploinsufficiency of ATP1A2 encoding the Na+/K+ pump α2 subunit associated with familial hemiplegic migraine type 2. Nat Genet 33(2):192–196. https://doi.org/10.1038/ng1081

de Vries B, Freilinger T, Vanmolkot KRJ, Koenderink JB, Stam AH, Terwindt GM, Babini E, van den Boogerd EH, van den Heuvel JJMW, Frants RR, Haan J, Pusch M, van den Maagdenberg AMJM, Ferrari MD, Dichgans M (2007) Systematic analysis of three FHM genes in 39 sporadic patients with hemiplegic migraine. Neurology 69(23):2170–2176. https://doi.org/10.1212/01.wnl.0000295670.01629.5a

Deen M, Correnti E, Kamm K, Kelderman T, Papetti L, Rubio-Beltrán E, Vigneri S, Edvinsson L, Maassen Van Den Brink A (2017) Blocking CGRP in migraine patients—a review of pros and cons. J Headache Pain 18(1):96. https://doi.org/10.1186/s10194-017-0807-1

Del BE et al (1994) Video assessment of yawning induced by sublingual apomorphine in migraine. Headache. 34:536–538

Di Lorenzo C, Grieco GS, Santorelli FM (2012) Migraine headache: a review of the molecular genetics of a common disorder. J Headache Pain 13(7):571–580. https://doi.org/10.1007/s10194-012-0478-x

Dichgans M, Freilinger T, Eckstein G, Babini E, Lorenz-Depiereux B, Biskup S, Ferrari MD, Herzog J, van den Maagdenberg AM, Pusch M, Strom TM (2005) Mutation in the neuronal voltage-gated sodium channel SCN1A in familial hemiplegic migraine. Lancet 366(9483):371–377. https://doi.org/10.1016/S0140-6736(05)66786-4

Diener H-C, Tassorelli C, Dodick DW, Silberstein SD, Lipton RB, Ashina M, Becker WJ, Ferrari MD, Goadsby PJ, Pozo-Rosich P, Wang S-J, Mandrekar J (2019) Guidelines of the International Headache Society for controlled trials of acute treatment of migraine attacks in adults: fourth edition. Cephalalgia 39(6):687–710. https://doi.org/10.1177/0333102419828967

Dodick DW, Silberstein S (2006) Central sensitization theory of migraine: clinical implications. Headache. 46(Suppl 4):S182–S191

Dodick DW, Turkel CC, DeGryse RE, Aurora SK, Silberstein SD, Lipton RB, Diener H, Brin MF (2010) OnabotulinumtoxinA for treatment of chronic migraine: pooled results from the double-blind, randomized, placebo-controlled phases of the PREEMPT clinical program. Headache 50(6):921–936. https://doi.org/10.1111/j.1526-4610.2010.01678.x

Doods H et al (2007) CGRP antagonists: unravelling the role of CGRP in migraine. Trends Pharmacol Sci. 28:580–587

Drummond P (2006) Tryptophan depletion increases nausea, headache and photophobia in migraine sufferers. Cephalalgia 26(10):1225–1233. https://doi.org/10.1111/j.1468-2982.2006.01212.x

Ducros A, Denier C, Joutel A, Cecillon M, Lescoat C, Vahedi K, Darcel F, Vicaut E, Bousser M-G, Tournier-Lasserve E (2001) The clinical spectrum of familial hemiplegic migraine associated with mutations in a neuronal calcium channel. N Engl J Med 345(1):17–24. https://doi.org/10.1056/NEJM200107053450103

Eigenbrodt AK, Ashina H, Khan S, Diener H-C, Mitsikostas DD, Sinclair AJ, Pozo-Rosich P, Martelletti P, Ducros A, Lantéri-Minet M, Braschinsky M, del Rio MS, Daniel O, Özge A, Mammadbayli A, Arons M, Skorobogatykh K, Romanenko V, Terwindt GM et al (2021) Diagnosis and management of migraine in ten steps. Nat Rev Neurol 17(8):501–514. https://doi.org/10.1038/s41582-021-00509-5

Eikermann-Haerter K, Can A, Ayata C (2012) Pharmacological targeting of spreading depression in migraine. Expert Rev Neurother 12(3):297–306. https://doi.org/10.1586/ern.12.13

Eltorp C, Jansen-Olesen I, Hansen A (2000) Release of calcitonin gene-related peptide (CGRP) from Guinea pig dura mater *in vitro* is inhibited by Sumatriptan but unaffected by nitric oxide. Cephalalgia 20(9):838–844. https://doi.org/10.1046/j.1468-2982.2000.00131.x

Esserlind A-L, Christensen AF, Le H, Kirchmann M, Hauge AW, Toyserkani NM, Hansen T, Grarup N, Werge T, Steinberg S, Bettella F, Stefansson H, Olesen J (2013) Replication and meta-analysis of common variants identifies a genome-wide significant locus in migraine. Eur J Neurol 20(5):765–772. https://doi.org/10.1111/ene.12055

Goadsby PJ, Kullmann DM (2005) Another migraine gene. Lancet 366(9483):345–346. https://doi.org/10.1016/S0140-6736(05)66787-6

Hamberger A, van Gelder NM (1993) Metabolic manipulation of neural tissue to counter the hypersynchronous excitation of migraine and epilepsy. Neurochem Res. 18:503–509

Hansen JM et al (2008) Calcitonin gene-related peptide does not cause the familial hemiplegic migraine phenotype. Neurology. 71:841–847

Hansen JM et al (2011) Calcitonin gene-related peptide does not cause migraine attacks in patients with familial hemiplegic migraine. Headache. 51:544–553

Hansen JM, Thomsen LL, Olesen J, Ashina M (2010) Coexisting typical migraine in familial hemiplegic migraine. Neurology 74(7):594–600. https://doi.org/10.1212/WNL.0b013e3181cff79d

Hauge AW, Asghar MS, Schytz HW, Christensen K, Olesen J (2009) Effects of tonabersat on migraine with aura: a randomised, double-blind, placebo-controlled crossover study. Lancet Neurol 8(8):718–723. https://doi.org/10.1016/S1474-4422(09)70135-8

Honkasalo ML et al (1995) Migraine and concomitant symptoms among 8167 adult twin pairs. Headache. 35:70–78

Ho TW, Ferrari MD, Dodick DW, Galet V, Kost J, Fan X, Leibensperger H, Froman S, Assaid C, Lines C, Koppen H, Winner PK (2008) Efficacy and tolerability of MK-0974 (telcagepant), a new oral antagonist of calcitonin gene-related peptide receptor, compared with zolmitriptan for acute migraine: a randomised, placebo-controlled, parallel-treatment trial. Lancet 372(9656):2115–2123. https://doi.org/10.1016/S0140-6736(08)61626-8

Huang Y, Xiao H, Qin X, Nong Y, Zou D, Wu Y (2017) The genetic relationship between epilepsy and hemiplegic migraine. Neuropsychiatr Dis Treat 13:1175–1179. https://doi.org/10.2147/NDT.S132451

Jansen-Olesen I, Baun M, Amrutkar DV, Ramachandran R, Christophersen DV, Olesen J (2014) PACAP-38 but not VIP induces release of CGRP from trigeminal nucleus caudalis via a receptor distinct from the PAC1 receptor. Neuropeptides 48(2):53–64. https://doi.org/10.1016/j.npep.2014.01.004

Joseph R, Steiner TJ, Schultz LU, Clifford Rose F (1988) Platelet activity and selective beta-blockade in migraine prophylaxis. Stroke 19(6):704–708. https://doi.org/10.1161/01.STR.19.6.704

Joutel A, Bousser M-G, Biousse V, Labauge P, Chabriat H, Nibbio A, Maciazek J, Meyer B, Bach M-A, Weissenbach J, Lathrop GM, Tournier-Lasserve E (1993) A gene for familial hemiplegic migraine maps to chromosome 19. Nat Genet 5(1):40–45. https://doi.org/10.1038/ng0993-40

Kahlig KM, Rhodes TH, Pusch M, Freilinger T, Pereira-Monteiro JM, Ferrari MD, van den Maagdenberg AMJM, Dichgans M, George AL (2008) Divergent sodium channel defects in familial hemiplegic migraine. Proc Natl Acad Sci U S A 105(28):9799–9804. https://doi.org/10.1073/pnas.0711717105

Kaiser EA, Russo AF (2013) CGRP and migraine: could PACAP play a role too? Neuropeptides 47(6):451–461. https://doi.org/10.1016/j.npep.2013.10.010

Kim J, Suh S-I, Seol H, Oh K, Seo W-K, Yu S-W, Park K-W, Koh S-B (2008) Regional grey matter changes in patients with migraine: a voxel-based morphometry study. Cephalalgia 28(6):598–604. https://doi.org/10.1111/j.1468-2982.2008.01550.x

Kruit M, Launer L, Overbosch J, van Buchem M, Ferrari M (2009) Iron accumulation in deep brain nuclei in migraine: a population-based magnetic resonance imaging study. Cephalalgia 29(3):351–359. https://doi.org/10.1111/j.1468-2982.2008.01723.x

Lance JW (1991) 5-Hydroxytryptamine and its role in migraine. Eur Neurol 31(5):279–281. https://doi.org/10.1159/000116754

Lassen L, Haderslev P, Jacobsen V, Iversen H, Sperling B, Olesen J (2002) CGRP may play a causative role in migraine. Cephalalgia 22(1):54–61. https://doi.org/10.1046/j.1468-2982.2002.00310.x

Leao AAP (1944) Spreading depression of activity in the cerebral cortex. J Neurophysiol 7(6):359–390. https://doi.org/10.1152/jn.1944.7.6.359

Lodi R, Iotti S, Cortelli P, Pierangeli G, Cevoli S, Clementi V, Soriani S, Montagna P, Barbiroli B (2001) Deficient energy metabolism is associated with low free magnesium in the brains of patients with migraine and cluster headache. Brain Res Bull 54(4):437–441. https://doi.org/10.1016/S0361-9230(01)00440-3

Marconi R, De Fusco M, Aridon P, Plewnia K, Rossi M, Carapelli S, Ballabio A, Morgante L, Musolino R, Epifanio A, Micieli G, De Michele G, Casari G (2003) Familial hemiplegic migraine type 2 is linked to 0.9Mb region on chromosome 1q23. Ann Neurol 53(3):376–381. https://doi.org/10.1002/ana.10464

Martínez E, Moreno R, López-Mesonero L, Vidriales I, Ruiz M, Guerrero AL, Tellería JJ (2016) Familial hemiplegic migraine with severe attacks: a new report with *ATP1A2* mutation. Case Rep Neurol Med 2016:3464285. https://doi.org/10.1155/2016/3464285

Mathew NT (2001) Pathophysiology, epidemiology, and impact of migraine. Clin Cornerstone 4(3):1–17. https://doi.org/10.1016/S1098-3597(01)90035-3

Mathew NT, Stubits E, Nigam MP (1982) Transformation of episodic migraine into daily headache: analysis of factors. Headache 22(2):66–68. https://doi.org/10.1111/j.1526-4610.1982.hed2202066.x

Mathew R, Andreou AP, Chami L, Bergerot A, van den Maagdenberg AM, Ferrari MD, Goadsby PJ (2011) Immunohistochemical characterization of calcitonin gene-related peptide in the trigeminal system of the familial hemiplegic migraine 1 knock-in mouse. Cephalalgia 31(13):1368–1380. https://doi.org/10.1177/0333102411418847

Mauskop A, Altura BT, Altura BM (2002) Serum ionized magnesium levels and serum ionized calcium/ionized magnesium ratios in women with menstrual migraine. Headache 42(4):242–248. https://doi.org/10.1046/j.1526-4610.2002.02075.x

Meiyanti M (2016) The role of triptans in the management of migraine. Univ Med 28(1):49–58. https://doi.org/10.18051/UnivMed.2009.v28.49-58

Newman LC (2007) Understanding the causes and prevention of menstrual migraine: the role of estrogen. Headache 47(Suppl 2):S86–S94. https://doi.org/10.1111/j.1526-4610.2007.00819.x

Olesen J (2008) The role of nitric oxide (NO) in migraine, tension-type headache and cluster headache. Pharmacol Ther 120(2):157–171. https://doi.org/10.1016/j.pharmthera.2008.08.003

Pardutz A, Schoenen J (2010) NSAIDs in the acute treatment of migraine: a review of clinical and experimental data. Pharmaceuticals (Basel) 3(6):1966–1987. https://doi.org/10.3390/ph3061966

Peroutka SJ, Banghart SB, Allen GS (1985) Calcium channel antagonism by pizotifen. J Neurol Neurosurg Psychiatry 48(4):381–383. https://doi.org/10.1136/jnnp.48.4.381

Prakash S, Belani P, Trivedi A (2014) Headache as a presenting feature in patients with serotonin syndrome: a case series. Cephalalgia 34(2):148–153. https://doi.org/10.1177/0333102413495968

Ramagopalan SV et al (2007) Molecular mechanisms of migraine? J Neurol. 254:1629–1635

Raskin NH (1990) Modern pharmacotherapy of migraine. Neurol Clin. 8:857–865

Rogawski MA (2008) Common pathophysiologic mechanisms in migraine and epilepsy. Arch Neurol 65(6):709–714. https://doi.org/10.1001/archneur.65.6.709

Russell MB (2008) Is migraine a genetic illness? The various forms of migraine share a common genetic cause. Neurol Sci 29(S1):52–54. https://doi.org/10.1007/s10072-008-0887-4

Sánchez del Rio M, Linera JA (2004) Functional neuroimaging of headaches. Lancet Neurol 3(11):645–651. https://doi.org/10.1016/S1474-4422(04)00904-4

Schytz HW, Birk S, Wienecke T, Kruuse C, Olesen J, Ashina M (2009) PACAP38 induces migraine-like attacks in patients with migraine without aura. Brain 132(1):16–25. https://doi.org/10.1093/brain/awn307

Stewart WF et al (1997) Familial risk of migraine: a population-based study. Ann Neurol. 41:166–172

Sundrum T, Walker CS (2018) Pituitary adenylate cyclase-activating polypeptide receptors in the trigeminovascular system: implications for migraine. Br J Pharmacol 175(21):4109–4120. https://doi.org/10.1111/bph.14053

Tuka B, Helyes Z, Markovics A, Bagoly T, Szolcsányi J, Szabó N, Tóth E, Kincses ZT, Vécsei L, Tajti J (2013) Alterations in PACAP-38-like immunoreactivity in the plasma during ictal and interictal periods of migraine patients. Cephalalgia 33(13):1085–1095. https://doi.org/10.1177/0333102413483931

Valfrè W, Rainero I, Bergui M, Pinessi L (2008) Voxel-based morphometry reveals gray matter abnormalities in migraine. Headache 48(1):109–117. https://doi.org/10.1111/j.1526-4610.2007.00723.x

van der Kuy P-HM, Lohman JJHM (2003) The role of nitric oxide in vascular headache. Pharm World Sci 25(4):146–151. https://doi.org/10.1023/A:1024800512790

Vanmolkot KRJ, Kors EE, Turk U, Turkdogan D, Keyser A, Broos LAM, Kia SK, van den Heuvel JJMW, Black DF, Haan J, Frants RR, Barone V, Ferrari MD, Casari G, Koenderink JB, van den Maagdenberg AMJM (2006) Two de novo mutations in the Na, K-ATPase gene ATP1A2 associated with pure familial hemiplegic migraine. Eur J Hum Genet 14(5):555–560. https://doi.org/10.1038/sj.ejhg.5201607

Vetvik KG, MacGregor EA (2021) Menstrual migraine: a distinct disorder needing greater recognition. Lancet Neurol 20(4):304–315. https://doi.org/10.1016/S1474-4422(20)30482-8

Weiller C, May A, Limmroth V, Jüptner M, Kaube H, Schayck RV, Coenen HH, Dlener HC (1995) Brain stem activation in spontaneous human migraine attacks. Nat Med 1(7):658–660. https://doi.org/10.1038/nm0795-658

Welch KMA, Nagesh V, Aurora SK, Gelman N (2001) Periaqueductal gray matter dysfunction in migraine: cause or the burden of illness? Headache 41(7):629–637. https://doi.org/10.1046/j.1526-4610.2001.041007629.x

Woods RP, Iacoboni M, Mazziotta JC (1994) Bilateral spreading cerebral hypoperfusion during spontaneous migraine headache. N Engl J Med 331(25):1689–1692. https://doi.org/10.1056/NEJM199412223312505

Diagnostic Tests for Migraine

3

Rohini Pujari, Shvetank Bhatt, and Urvashi Soni

3.1 Introduction

Migraine derives from the Greek word hemicrania, which means "half of the head" and was previously known as "hypoglycaemic headache" in the early twentieth century. It is the most prevalent interdisciplinary and complex neurologic condition marked by recurrent headache attacks. With 12% of the population affected, migraines can strike up to 17% of women and 6% of men annually (Simmonds et al. 2023). It typically affects girls more often than boys when they are young. North America has the highest adjusted prevalence of migraine, followed by Europe, Asia, Africa, South and Central America, and South America (Amiri et al. 2021). It is the second most common cause of disability globally. A family history of migraines is common. If one or both of the parents have a history of migraines, the reported risk rises to 40% and 75%, respectively (Burch et al. 2019).

It accounts for 3% of all emergency visits annually and is fourth most common cause for visits. Its frequency rises throughout adolescence, but it keeps rising until 35–39 years of age, at which point it starts to decline, especially after menopause. In addition, it is thought to be the second most common cause of impairment in terms of years spent living with a handicap, following back pain (Aguilar-Shea et al. 2022).

Migraine can be episodic or chronic, with or without an aura. Photophobia, phonophobia, nausea, and vomiting are the most prevalent symptoms of the migraine. Prodromal (before the headache phase) symptoms include a yawning, sad mood, food cravings, and fatigue before the beginning of pain (Khan et al. 2021). Postdromal (after headache phase) symptoms can linger up to 48 h after the headache has resolved and commonly include exhaustion, difficulty concentrating, and

R. Pujari (✉) · S. Bhatt · U. Soni
School of Health Sciences and Technology, Dr. Vishwanath Karad MIT World Peace University, Pune, Maharashtra, India

S. Bhatt et al. (eds.), *Management of Migraine Pain*,
https://doi.org/10.1007/978-981-97-4529-6_3

neck stiffness. Prodromal phase, aura, migraine headache, and Postdromal phase are the four phases of migration (Eigenbrodt et al. 2021).

Headache disorders are classified as:

1. long-lasting chronic daily headache,
2. hemicranias continua,
3. chronic migraine,
4. chronic tension headache, and
5. newly developed persistent daily headache (Rizzoli and Mullally 2018) (Fig. 3.1).

Fig. 3.1 Four stages of migraine

3.2 Approaches for Diagnosis of Migraine

3.2.1 Detailed History Taking

The diagnostic process necessitates a thorough and precise understanding of the patient's condition. Patients should be afforded ample time to articulate and elucidate their symptoms, with targeted inquiries used to clarify details. Screening questions are designed to document various aspects of the headache, such as its onset and duration, nature (episodic or continuous), severity, and potential triggers. Details regarding the character, location, and intensity of pain should be documented, along with any associated symptoms like prodrome or aura (e.g., fatigue, yawning, sensory disturbances, vertigo). Symptoms like eye-watering, conjunctival infection, eyelid edema, ptosis, neck stiffness, nasal congestion, fever, sweating, rashes or agitation indicative of primary or secondary types of headaches also must be noted (Weatherall 2015).

Furthermore, a comprehensive history of prior and current treatments, including medications, dosages, and reasons for discontinuation, is essential. Information on complementary or alternative therapies sought by the patient should also be gathered. Medical histories related to depression, sleep disorders, anxiety and allergies as well as details on current medications, social background, family history, lifestyle habits (alcoholism, smoking, caffeine, occupation), and any history of recurring abdominal pain, hangovers, or motion sickness should be documented (Evans 2019). Additionally, a physical examination must be conducted to detect any factors aggravating the susceptibility of an individual to migraines (Sorrell 2006).

3.2.2 Physical Examination

While migraine is primarily diagnosed based on a thorough medical history and the fulfillment of specific diagnostic criteria, a physical examination can also be useful in the diagnostic process (Evans 2019).

The physical examination in suspected cases of migraine typically aims to:

Rule out other conditions: Many medical conditions can mimic migraine symptoms, so a physical examination helps in identifying signs of other disorders. For example, signs of sinusitis, temporomandibular joint dysfunction (TMJ), or cervical spine issues might be evident upon examination (Weatherall 2015).

Neurological assessment: A neurological examination may be conducted to evaluate cranial nerves, reflexes, coordination, and sensory/motor function. This helps in ruling out other neurological conditions and identifying any abnormalities that may suggest a different diagnosis (Sorrell 2006).

Evaluation of trigger points: Some individuals with migraine may have tender points or trigger points in the neck, head, or shoulder region. Palpation of these areas during a physical examination can help identify areas of tenderness that may exacerbate migraine symptoms (Evans 2019).

***Assessment of associated symptoms*:** Migraine is often associated with symptoms like nausea, emesis, sensitivity to sound (phonophobia), and sensitivity to light (photophobia). These symptoms may be evaluated during the physical examination to corroborate the diagnosis (Weatherall 2015).

***Vital signs*:** Measurement of vital signs such as body temperature, pulse rate, heart rate, blood pressure, and blood oxygen levels may be performed to ensure no underlying medical issues are contributing to the symptoms (Weatherall 2015; Evans 2019).

***Evaluation of eye health*:** Eye examination may be included to rule out eye disorders that could be contributing to symptoms such as visual disturbances or eye pain (Sorrell 2006).

***Assessment of posture and muscle tension*:** Posture assessment and evaluation of muscle tension in the neck and shoulders may be helpful, as tension and poor posture can exacerbate migraine symptoms in some individuals.

Therefore, a collaborative approach involving both history-taking and physical examination is crucial in the diagnostic process for migraine (Khan et al. 2021).

3.2.3 Aura Assessment

Nearly 33% of patients with migraine experience aura. It is transient focal neurological symptoms that normally follow or sometimes are associated with the headache phase of the migraine. Usually, aura has visual, sensory, speech, and/or language components, but it can also be motor, brain stem, and retinal. (Eigenbrodt et al. 2021)

Assessing aura in migraine involves understanding the specific symptoms experienced by the individual before or during a migraine attack. Aura symptoms are typically neurological and can manifest as visual disturbances, sensory changes, speech difficulties, motor symptoms, or brainstem symptoms (Tzankova et al. 2023).

Aura in migraine is assessed as follows:

***History-taking*:** A detailed medical history is necessary to assess for the presence of aura symptoms in migraine. The healthcare provider will ask specific questions to determine if the individual experiences aura and, if so, the nature and timing of these symptoms in relation to their headache attacks (Weatherall 2015).

***Aura symptoms*:** Aura symptoms can vary widely among individuals and may include:

- visual disturbances like blind spots or tunnel vision, zigzag lines, and flashes of light;
- sensory changes, such as tingling or numbness in the face, hands, or other parts of the body;
- speech and language difficulties, such as difficulty finding words or slurred speech;
- motor symptoms, such as weakness or paralysis affecting one side of the body; and

- brainstem symptoms, such as vertigo, double vision, or ringing in the ears (Viana et al. 2017).

***Timing of aura*:** Aura symptoms typically precede or occur concurrently with the headache phase of the migraine attack. However, in some cases, aura symptoms may occur without an ensuing headache or may overlap with the headache phase.

***Duration of aura*:** Aura symptoms typically last between 5 and 60 min, although they can vary in duration. Prolonged aura lasting more than 1 h or persistent aura without infarction (PAWOI) lasting more than a week should raise concern for alternative diagnoses.

***Impact on function*:** Assessing how aura symptoms impact daily functioning is important. Severe or prolonged aura symptoms can significantly disrupt activities of daily living and may require specific management strategies.

***Documentation*:** It is essential to document the presence and characteristics of aura symptoms in the medical record to aid in diagnosis and treatment planning. This may include descriptions of the type, frequency, severity, and duration of aura symptoms.

***Differential diagnosis*:** It is crucial to differentiate migraine aura from other neurological conditions that may present with similar symptoms, such as transient ischemic attack (TIA), epilepsy, or vestibular disorders. Overall, a comprehensive assessment of aura symptoms in migraine involves careful history-taking, documentation, and consideration of differential diagnoses to accurately diagnose and manage the condition. Other techniques such as magnetic resonance imaging (MRI) and computerized tomography (CT) scan are also used to evaluate migraine with aura (Holle and Obermann 2013).

3.2.4 Pattern Recognition of Migraine Phases

Recognition of migraine phases and their patterns is important for the diagnosis of migraine. Pattern recognition of migraine phases involves identifying the typical sequence of symptoms that occur during a migraine attack. Migraine attacks generally consist of several distinct phases, including prodrome, aura (not always present), headache (or pain) phase, and postdrome. Different phases of migraine have different intensity of pain, symptoms, and duration. Pattern recognition is an important diagnostic tool for migraine (Messina et al. 2022a). There are different patterns observed in different types of migraine. For example, a new artificial intelligence tool can detect whether surgery will be useful in reducing pain due to nerve compression headaches. Physicians can try to identify the pattern based on history, general physician, and neurological examination. In addition, psychological tools are also useful with various blood tests. Chronic headaches, headaches due to intracranial pressure or lesions and headaches due to other systemic comorbid diseases have different patterns (Nurse 1974).

Each phase is recognized as follows:

Prodrome phase

This phase occurs hours to days prior to the onset of pain associated with headache. Symptoms during the prodrome phase can include mood changes (e.g., depression, irritability), food cravings, increased thirst, fatigue, neck stiffness, yawning, or heightened sensitivity to light, sound, or smell. Recognizing prodromal symptoms can help individuals anticipate the onset of a migraine attack and take preventive measures if possible (Viana et al. 2017).

Aura phase

Not all migraineurs experience aura, but for those who do, it typically occurs before or during the headache phase. Aura symptoms are typically neurological and may include visual disturbances (like blind spots, zigzag lines, flashes of light), sensory changes (like tingling or numbness), speech and language difficulties, motor symptoms, or brainstem symptoms (such as vertigo or double vision). Aura symptoms typically last between 5 and 60 min, although they can vary in duration (Tzankova et al. 2023).

Headache (pain) phase

This phase is manifested by moderate to severe headache pain, which is often throbbing or pulsating in nature. Headache pain is typically unilateral (on one side of the head) but can sometimes be bilateral. Other symptoms associated with the headache phase include nausea, vomiting, photophobia, and phonophobia. The duration of the headache phase can vary from a few hours to 72 h if untreated (Rizzoli and Mullally 2018).

Postdrome phase

Also known as the "migraine hangover," the postdrome phase occurs after the resolution of the headache pain. Symptoms during the postdrome phase can include fatigue, mood changes (such as feeling drained or euphoric), difficulty concentrating, lingering headache, and overall feelings of exhaustion. The postdrome phase can last hours to days after the headache resolves.

Recognizing the typical pattern of migraine phases can aid in diagnosis, treatment planning, and management strategies. Keeping a headache diary to track the timing, duration, and characteristics of each phase can help identify patterns and triggers, as well as guide treatment decisions (Nurse 1974).

3.3 Diagnosis of Migraine Based on Criteria Developed by the International Headache Society in 1988 and Revised by the Society in 2004

The diagnostic criteria for migraine headache, as outlined by the International Headache Society (IHS) in the International Classification of Headache Disorders (ICHD), are as follows (Cano et al. 2000):

1. ***Migraine without aura:***

   This type of migraine is characterized by recurrent headache attacks that last for 4–72 h. The attacks typically have at least two of the following four characteristics, that is, pulsating quality, moderate or severe pain intensity, aggravation, and unilateral location by routine physical activity. Additionally, the attacks are associated with at least one of the following features, that is, nausea and/or vomiting, photophobia, and phonophobia (Khan et al. 2021; Cano et al. 2000).

2. ***Migraine with aura:***

   Migraine with aura is diagnosed when a patient experiences at least two attacks that fulfill specific criteria. These attacks are accompanied by fully reversible aura symptoms, such as visual disturbances, sensory disturbances, language and/or speech disturbances, motor disturbances, brainstem symptoms, or retinal symptoms. The aura is accompanied by a headache within 60 min. Additionally, at least three of six specific characteristics must be present during the aura phase (Khan et al. 2021; Cano et al. 2000).

3. ***Chronic migraine:***

   Chronic migraine is characterised by headache for 15 or more days per month for over 3 months, with at least 8 of these days fulfilling the criteria for migraine. This diagnosis is based on the ICHD-3 criteria and is essential for identifying patients who require specific management strategies due to the frequency and impact of their migraine attacks (Khan et al. 2021; Cano et al. 2000) (Table 3.1).

Table 3.1 Diagnostic criteria for migraine headache as framed by the International Headache Society (Khan et al. 2021; Cano et al. 2000)

Migraine without aura (previously called common migraine)	Migraine with aura (previously called classic migraine)
A. At least five attacks fulfilling criteria B–D B. Headache attacks lasting 4–72 h (untreated or unsuccessfully treated) C. Headache has at least two of the following characteristics: • Unilateral location • Pulsating quality • Moderate to severe intensity • Aggravation by or causing avoidance of routine physical activity (e.g., walking or climbing stairs) D. During the headache, at least one of the following: • Nausea and/or vomiting • Photophobia and phonophobia E. Not attributed to another disorder	A. At least two attacks fulfilling criteria B–D B. One or more of the following fully reversible aura symptoms: • Visual • Sensory • Speech and/or language • Motor • Brainstem • Retinal C. At least three of the following six characteristics: • At least one aura symptom spreads gradually over 5 min or more, and/or two or more symptoms occur in succession • Each aura symptom lasts 5–60 min • At least one aura symptom is unilateral • At least one aura symptom is positive (i.e., flickering lights, spots, or lines) • The aura is accompanied, or followed within 60 min, by a headache • Not attributed to another disorder D. Not better accounted for by another ICHD-3 diagnosis, and transient ischemic attack has been excluded

These criteria are crucial for clinicians to accurately diagnose migraine headache and distinguish it from other types of headaches or medical conditions. It is important to note that these criteria are meant to guide diagnosis and should be applied in the context of a thorough clinical evaluation.

3.4 Ten Step Method for Diagnosis of Migraine by Danish Headache Society (Eigenbrodt et al. 2021)

The Danish Headache Society and its representatives perceived a European Consensus Statement on diagnosis and clinical management which was endorsed by the European Headache Federation (EHF) and the European Academy of Neurology (EAN) in 2021. The approach aimed to support the care and clinical decision-making by primary care practitioners, neurologists, and headache specialists (Eigenbrodt et al. 2021).

The 10 steps of diagnosis are as follows:

1. **When to suspect migraine:** This step focuses on recognizing the clinical characteristics of migraine, including its types and subtypes, to ensure an accurate diagnosis.
2. **Taking a headache history:** A medical history is crucial for diagnosing migraine, and a full history should enable systematic application of the criteria set out in the International Classification of Headache Disorders (ICHD-3).
3. **Making a diagnosis:** This step emphasizes the importance of using the ICHD-3 criteria for diagnosing migraine and its types and subtypes, as well as the use of diagnostic aids such as headache diaries.
4. **Patient education and self-management:** Patient education and involvement in self-management are essential for ensuring treatment adherence and satisfaction with care provision.
5. **Acute treatment of migraine attacks:** This step addresses the best practices for managing acute migraine attacks, including the use of acute medications and non-pharmacological therapies.
6. **Preventive treatment of migraine attacks:** It focuses on the use of preventive medications and non-pharmacological therapies for managing migraine in various patient populations.
7. **Special considerations for specific patient populations:** This step provides recommendations for managing migraine in specific patient populations, including children, adolescents, adults, pregnant and breastfeeding women, and elderly people.
8. **Evaluating treatment response and managing treatment failure:** It addresses the evaluation of treatment response and the management of treatment failure in patients with migraine.
9. **Managing complications and comorbidities:** This step emphasizes the importance of managing complications and comorbidities associated with migraine.

10. **Planning long-term follow-up:** It highlights the importance of planning long-term follow-up for patients with migraine (Eigenbrodt et al. 2021).

3.5 Medical Investigations

3.5.1 Electroencephalography

Electroencephalogram (EEG) investigation is not typically used as a primary diagnostic tool for migraine. EEG may have a role in certain cases of migraine, particularly when there are atypical features or when other neurological conditions need to be ruled out. With recent progress, EEG helps the chances to recognize modified brain physiological patterns and brain network properties of migraines in depth (Zhang et al. 2023).

EEG reports the sudden and periodic EEG activity of the population of the brain nerve cells and has become an important tool to describe the state of activity of the brain before the neuroimaging era. In cases where aura symptoms are prominent or atypical, EEG may be used to evaluate brain function during the aura phase. EEG can help identify abnormalities in brain activity that may be associated with aura symptoms (Sand 2003). At the time of visual aura, either slow waves, depression of activity in background amplitude, or normal EEG have been suggested. The most abnormal EEGs with unilateral /bilateral delta activity have been observed during attacks of hemiplegic migraine and attacks of migraine with disturbed consciousness. The EEG patterns of migraine patients suggest a possible functional link between sleep, hyperventilation, and migraine. The diagnosis of migraine is primarily based on clinical criteria, and EEG findings are considered in conjunction with other clinical information (Sand 1991).

3.5.2 Blood Tests

Blood tests are not typically used as a primary diagnostic tool for migraine. However, blood tests may be ordered in some cases to rule out other medical conditions mimicking migraine symptoms or to identify potential triggers or comorbidities. Certain medical conditions or factors, such as hormonal imbalances, thyroid dysfunction, vitamin deficiencies (e.g., vitamin D deficiency), or metabolic disorders, may trigger or exacerbate migraine attacks. Blood tests may be used to assess these factors and identify potential triggers for migraine. Thyroid function tests, complete blood count (CBC), and erythrocyte sedimentation rate (ESR) are the basic tests done by physicians to identify a headache or migraine (Evans 2019). Migraine has been linked with certain comorbidities, such as cardiovascular disease, stroke, depression, and anxiety. Blood tests may be ordered as part of a comprehensive evaluation to screen for these comorbid conditions and assess overall health such as signs of kidney disease, liver disease, and malnutrition. Infection or inflammation in and around the brain can lead to headaches and neurological symptoms (Eigenbrodt et al. 2021). Tests for the levels of various inflammatory markers such as tumor necrosis factor-α, C-reactive protein, interleukin 1β (IL-1 β), and IL-6 can be useful to detect inflammatory conditions. It is important to note that blood tests are typically used as adjunctive tools in the evaluation of migraine and should be interpreted with respect to the clinical history and symptoms of an individual. The decision to order blood tests should be based on the specific clinical presentation and needs of the patient, as determined by a healthcare professional (Tzankova et al. 2023).

3.6 Lumbar Puncture

Lumbar puncture, also known as a spinal tap, is not a routine diagnostic procedure for migraine. Lumbar puncture may be considered in certain cases of migraine when there are atypical features, unusual neurological symptoms, or when other neurological conditions such as secondary causes such as meningitis, encephalitis, or other infectious or inflammatory conditions affecting the central nervous system need to be ruled out. In cases where migraine symptoms are atypical or accompanied by unusual neurological symptoms, lumbar puncture may be used to assess for signs of increased intracranial pressure, inflammation, or other abnormalities in the cerebrospinal fluid (CSF). Some individuals with chronic migraine may experience cerebrospinal fluid leaks, which can cause positional headaches or other symptoms. Lumbar puncture may be used to detect evidence of CSF leaks and help guide treatment decisions.

Lumbar puncture is typically reserved for cases where there is a clear clinical indication and when the potential benefits outweigh the risks as it is an invasive procedure that carries some risks, including headache, infection, bleeding, and nerve damage.

3.6.1 CT Neuroimaging Studies

CT neuroimaging studies have been utilized in the diagnosis of migraine to assess for abnormalities in the brain. These studies have provided valuable insights into the structural changes and potential pathologies associated with migraine. CT studies have identified various abnormalities in a subset of migraine cases. These abnormalities may include gliomas, cerebral infarcts, focal atrophy, periventricular edema, choroid plexus papilloma, generalized atrophy, low-density areas, ventricular enlargement, acute infarction, metastatic neoplasms, subarachnoid hemorrhage, primary neoplasms, subdural hematoma, and hydrocephalus. Studies have shown that abnormalities such as focal atrophy are more prevalent in older individuals, reflecting the aging process (Holle and Obermann 2013). Ischemia, calcifications in basal ganglia, ventricular enlargement, and atrophy are frequently observed abnormalities in CT neuroimaging studies of migraine patients. White-matter lesions detected through CT scans have been correlated with the demographics and clinical features of migraine patients. Patients with comorbid conditions like cardiovascular diseases or diabetes tend to have a higher incidence of white-matter abnormalities (Evans et al. 2020). In addition to the primary abnormalities associated with migraine, CT neuroimaging studies have also revealed incidental findings such as heterotopy, atrophy of frontoparietal and cortical areas, petrous apex cholesterol cysts, and meningiomas in some migraine patients (Holle and Obermann 2013). While CT scans provide valuable information, MRI studies offer a more detailed assessment of structural and functional alterations in the brain. Combining both CT and MRI scans can provide a comprehensive evaluation of the neurological status of patients with migraine (Evans et al. 2020).

CT neuroimaging studies play a crucial role in identifying structural abnormalities and potential pathologies in migraine patients. These studies contribute to the diagnostic process and help healthcare providers better understand the underlying mechanisms of migraine (Holle and Obermann 2013; Evans et al. 2020).

3.6.2 MRI Neuroimaging Studies

MRI neuroimaging studies are valuable tools in the diagnosis of migraine, providing detailed information about structural and functional changes in the brain. MRI studies have exhibited white-matter lesions that are associated with the demographics and clinical features of migraine patients (Schwedt and Dodick 2009). Patients with migraine with aura (MA) have been found to have significantly more frequent foci in white matter compared to patients with migraine without aura (MO). In addition to white-matter lesions, MRI studies have revealed the presence of alterations such as atrophy and heterotopy of the frontoparietal and cortical areas meningiomas and petrous apex cholesterol cysts in some migraine patients (Karsan et al. 2023). Studies have shown cortical abnormalities in some migraine patients, indicating potential structural changes in specific brain regions. MRI studies have also identified pituitary abnormalities and white-matter hyperintensities (WMHs) in migraine patients, suggesting a broader impact on brain structures beyond the typical migraine symptoms. While CT scans provide valuable information, MRI offers a more detailed assessment of brain structures and abnormalities. Combining both CT and MRI scans can provide a comprehensive evaluation of the neurological status of patients with migraine (Schwedt and Dodick 2009).

MRI neuroimaging studies offer a comprehensive view of anatomical and physiological alterations in the brains of migraine patients. These studies help in understanding the underlying mechanisms of migraine, differentiating between migraine types, and identifying potential comorbidities or incidental findings that may impact patient management and treatment decisions (Karsan et al. 2023).

3.7 Functional Brain Imaging in Migraine

Functional brain imaging techniques are used in migraine to investigate the underlying neurobiology of migraine attacks, including changes in brain activity and connectivity associated with various phases of the migraine cycle. While these techniques are not typically used as routine diagnostic tools in clinical practice, they provide valuable insights into the mechanisms underlying migraine and may inform the development of novel treatment approaches (Messina et al. 2022b).

Some common functional brain imaging techniques used in migraine research include:

***Functional Magnetic Resonance Imaging (fMRI)*:** fMRI measures alterations in blood flow and oxygenation in the brain, providing information about brain activity. During a migraine attack, fMRI studies have revealed alterations in activity in areas of the brain involved in pain processing, such as the trigeminal system, thalamus, and cortical regions associated with sensory perception and emotional processing (Schramm et al. 2023; Li et al. 2020).

***Positron Emission Tomography (PET)*:** PET imaging can be used to measure brain metabolism, neurotransmitter levels, and receptor binding. PET studies in migraine have identified abnormalities in neurotransmitter systems, such as serotonin, dopamine, and GABA, as well as changes in cortical spreading depression (CSD), a phenomenon believed to underlie migraine aura (Marino et al. 2023).

Single-Photon Emission Computed Tomography (SPECT): SPECT imaging can provide information about cerebral blood flow and perfusion. SPECT studies in migraine have demonstrated alterations in blood flow in specific brain regions during migraine attacks and interictal periods, highlighting the dynamic changes in brain function associated with the migraine cycle (Cho et al. 2020).

Electroencephalography (EEG): EEG measures electrical activity in the brain and can be used to study cortical excitability, event-related potentials, and connectivity patterns. EEG studies in migraine have revealed abnormalities in brain rhythms, sensory processing, and cortical excitability, particularly during the aura phase of migraine attacks (de Tommaso 2019).

Magnetoencephalography (MEG): MEG measures magnetic fields generated by neuronal activity in the brain and provides high temporal and spatial resolution. MEG studies in migraine have identified alterations in cortical excitability, sensory processing, and functional connectivity, shedding light on the pathophysiology of migraine attacks (Liu et al. 2015).

These functional brain imaging techniques have advanced our understanding of the complex neurobiology of migraine and have the potential to guide the development of targeted therapies for migraine prevention and treatment. While these techniques are primarily used in research settings, ongoing advancements in neuroimaging technology may lead to their eventual integration into clinical practice for personalized management of migraine (Schramm et al. 2023).

3.8 Computerized Migraine Diagnostic Tools

Digital tools play an important role in overcoming the gap between correct diagnosis and the growing burden of headaches. Digital technologies aid in reducing diagnostic mistakes caused by recollection bias by allowing for daily tracking of headache frequency and related parameters via e-diaries. Data-driven machine-learning approaches also help to automate the prediction of headache episodes and drug abuse. Wearables can identify real-time digital indicators for the early detection of headache episodes, such as sleep patterns, galvanic skin reaction, and heart rate variability. These wearables can also help with treatment adherence by offering treatments such as smartphone-based biofeedback, behavioral therapy, and smartphone-controlled electrical pulses (Woldeamanuel and Cowan 2022).

The substantial link between mental comorbidities, traumatic brain injury, and headache disorders such as migraines emphasises the significance of a thorough examination and treatment. Digital health solutions provide a solution to the complicated comorbidities associated with migraines by allowing for early and accurate real-time detection of clinical symptoms and indications (Cowan et al. 2022). They also help with symptom management by using technologies such transcranial Doppler, photoplethysmography to analyze temporal artery pulse waves and quantitative EEG wearing headsets. Ongoing research is focusing on the development and validation of digital tools for traumatic brain injury and headaches, using physiological biomarkers such as vestibular-ocular function, nystagmus, intracranial pressure, saccades, pupillary reflex and reaction time and optic nerve sheath diameter (Peng and Wang 2012).

3.9 Conclusion

Diagnosing migraine involves a wide-ranging approach that includes imaging studies, clinical evaluation and adherence to established diagnostic criteria. Diagnosis of migraine primarily relies on a detailed clinical history, including the duration, frequency, and characteristics of headaches. Symptoms such as pulsation, unilateral location, intense pain, and sensitivity to light or sound are commonly reported by migraine patients. The International Headache Society has formulated diagnostic criteria for migraine, distinguishing between MA and MO. These criteria include specific features such as the number of headaches, associated symptoms like nausea and photophobia, and the presence of aura symptoms. CT and MRI neuroimaging studies are used to assess structural and functional alterations in the brain of migraine patients. These studies can help identify abnormalities such as white-matter lesions, cortical abnormalities, and incidental findings like atrophy or cysts. It is essential to differentiate between primary headache diseases like migraine and secondary headaches caused by underlying medical conditions. Imaging studies can aid in ruling out other possible causes of headaches and confirming a diagnosis of migraine. Recognizing patterns of headache evolution, such as transformed migraine or new daily persistent headaches, can provide insights into the underlying cause of chronic or severe headaches. Understanding these patterns is crucial for accurate diagnosis and treatment planning. In conclusion, the diagnosis of migraine involves a multidimensional approach that combines clinical assessment, adherence to diagnostic criteria, and the use of imaging studies to confirm and differentiate migraine from other headache disorders. A thorough understanding of the patient's history, symptoms, and imaging findings is essential for accurate diagnosis and effective management of migraine.

References

Aguilar-Shea AL, Membrilla Md JA, Diaz-de-Teran J (2022) Migraine review for general practice. Aten Primaria 54(2):102208

Amiri P, Kazeminasab S, Nejadghaderi SA, Mohammadinasab R, Pourfathi H, Araj-Khodaei M, Sullman MJM, Kolahi AA, Safiri S (2021) Migraine: a review on its history, global epidemiology, risk factors, and comorbidities. Front Neurol 12:800605

Burch RC, Buse DC, Lipton RB (2019) Migraine: epidemiology, burden, and comorbidity. Neurol Clin 37(4):631–649

Cano A, Palomeras E, Alfonso S, Ortega D, Sanz P, Fossas P (2000) Migraine without aura and migrainous disorder in children; International Headache Society (IHS) and revised IHS criteria. Cephalalgia 20(7):617–620

Cho PG, Kim TW, Ji GY, Park SH, Yun MJ, Shin DA (2020) Diagnostic value of single-photon emission computed tomography/computed tomography scans with Tc-99m HDP in cervicogenic headache. J Clin Med 9(2):399

Cowan RP, Rapoport AM, Blythe J, Rothrock J, Knievel K, Peretz AM, Ekpo E, Sanjanwala BM, Woldeamanuel YW (2022) Diagnostic accuracy of an artificial intelligence online engine in migraine: a multi-center study. Headache 62(7):870–882

de Tommaso M (2019) An update on EEG in migraine. Expert Rev Neurother 19(8):729–737

Eigenbrodt AK, Ashina H, Khan S, Diener H-C, Mitsikostas DD, Sinclair AJ, Pozo-Rosich P, Martelletti P, Ducros A, Lantéri-Minet M, Braschinsky M, del Rio MS, Daniel O, Özge A, Mammadbayli A, Arons M, Skorobogatykh K, Romanenko V, Terwindt GM, Paemeleire K, Sacco S, Reuter U, Lampl C, Schytz HW, Katsarava Z, Steiner TJ, Ashina M (2021) Diagnosis and management of migraine in ten steps. Nat Rev Neurol 17(8):501–514

Evans RW (2019) Diagnostic testing for migraine and other primary headaches. Neurol Clin 37(4):707–725

Evans RW, Burch RC, Frishberg BM, Marmura MJ, Mechtler LL, Silberstein SD, Turner DP (2020) Neuroimaging for migraine: the American Headache Society systematic review and evidence-based guideline. Headache 60(2):318–336

Holle D, Obermann M (2013) The role of neuroimaging in the diagnosis of headache disorders. Ther Adv Neurol Disord 6(6):369–374

Karsan N, Silva E, Goadsby PJ (2023) Evaluating migraine with typical aura with neuroimaging. Front Hum Neurosci 17:1112790

Khan J, Asoom LIA, Sunni AA, Rafique N, Latif R, Saif SA, Almandil NB, Almohazey D, AbdulAzeez S, Borgio JF (2021) Genetics, pathophysiology, diagnosis, treatment, management, and prevention of migraine. Biomed Pharmacother 139:111557

Li Z, Zhou J, Lan L, Cheng S, Sun R, Gong Q, Wintermark M, Zeng F, Liang F (2020) Concurrent brain structural and functional alterations in patients with migraine without aura: an fMRI study. J Headache Pain 21(1):141

Liu H, Ge H, Xiang J, Miao A, Tang L, Wu T, Chen Q, Yang L, Wang X (2015) Resting state brain activity in patients with migraine: a magnetoencephalography study. J Headache Pain 16:525

Marino S, Jassar H, Kim DJ, Lim M, Nascimento TD, Dinov ID, Koeppe RA, DaSilva AF (2023) Classifying migraine using PET compressive big data analytics of brain's μ-opioid and D2/D3 dopamine neurotransmission. Front Pharmacol 14:1173596
Messina R, Cetta I, Colombo B, Filippi M (2022a) Tracking the evolution of non-headache symptoms through the migraine attack. J Headache Pain 23(1):149
Messina R, Gollion C, Christensen RH, Amin FM (2022b) Functional MRI in migraine. Curr Opin Neurol 35(3):328–335
Nurse EG (1974) Pattern recognition as a guide in diagnosing headache. Can Fam Physician 20(12):55–61
Peng K-P, Wang S-J (2012) Migraine diagnosis: screening items, instruments, and scales. Acta Anaesthesiol Taiwanica 50(2):69–73
Rizzoli P, Mullally WJ (2018) Headache. Am J Med 131(1):17–24
Sand T (1991) EEG in migraine: a review of the literature. Funct Neurol 6(1):7–22
Sand T (2003) Electroencephalography in migraine: a review with focus on quantitative electroencephalography and the migraine vs. epilepsy relationship. Cephalalgia 23(Suppl 1):5–11
Schramm S, Börner C, Reichert M, Baum T, Zimmer C, Heinen F, Bonfert MV, Sollmann N (2023) Functional magnetic resonance imaging in migraine: a systematic review. Cephalalgia 43(2):03331024221128278
Schwedt TJ, Dodick DW (2009) Advanced neuroimaging of migraine. Lancet Neurol 8(6):560–568
Simmonds L, Mehta D, Cheema S, Matharu M (2023) Chapter 3—Epidemiology of migraine. In: Swanson JW, Matharu M (eds) Handbook of clinical neurology, vol. 198. Elsevier, pp 31–38
Sorrell MR (2006) The physical examination of migraine. Curr Pain Headache Rep 10(5):350–354
Tzankova V, Becker WJ, Chan TLH (2023) Diagnosis and acute management of migraine. CMAJ 195(4):E153–E158
Viana M, Sances G, Linde M, Ghiotto N, Guaschino E, Allena M, Terrazzino S, Nappi G, Goadsby PJ, Tassorelli C (2017) Clinical features of migraine aura: results from a prospective diary-aided study. Cephalalgia 37(10):979–989
Weatherall MW (2015) The diagnosis and treatment of chronic migraine. Ther Adv Chronic Dis 6(3):115–123
Woldeamanuel YW, Cowan RP (2022) Computerized migraine diagnostic tools: a systematic review. Ther Adv Chronic Dis 13:20406223211065235
Zhang N, Pan Y, Chen Q, Zhai Q, Liu N, Huang Y, Sun T, Lin Y, He L, Hou Y, Yu Q, Li H, Chen S (2023) Application of EEG in migraine. Front Hum Neurosci 17:1082317

Lifestyle Modifications for the Management of Migraine Pain

4

Aryaa Nigade, Gauri Pathak, Moushumi Baidya, and Shvetank Bhatt

4.1 Introduction

A migraine is a type of headache that often affects one side of the head and can cause excruciating throbbing pain or a pulsating sensation. Severe light and sound sensitivity, nausea, and vomiting are frequently present. The pain from migraine attacks can be excruciating, interfering with everyday tasks for hours or even days at a time (Dodick 2018).

Some patients experience an aura, which is a warning symptom, either before or together with their headache. A person may have visual disturbances like blind spots or light flashes, as well as nonvisual disturbances like tingling in one arm or leg, difficulty speaking, or tingling on one side of the face. Certain migraine attacks can be less unpleasant and prevented with the use of medications. The appropriate medications, along with lifestyle modifications and self-help techniques, may be helpful. Recurrent episodes of impairing headaches, frequently coupled with sensory and motor abnormalities, are the hallmark of migraine disease (Agbetou and Adoukonou 2022). Dietary practices and components have an impact on migraine clinical symptoms. Numerous dietary factors for migraine have been found, which has led to the development of tactics including ketogenic diets, exclusion diets, and comprehensive diets, which are primarily intended to help avoid migraine (Gazerani 2020).

A. Nigade · G. Pathak · S. Bhatt (✉)
School of Health Sciences and Technology, Dr. Vishwanath Karad MIT World Peace University, Pune, Maharashtra, India

M. Baidya
Bharat Pharmaceutical Technology, Agartala, West Tripura, India

S. Bhatt et al. (eds.), *Management of Migraine Pain*,
https://doi.org/10.1007/978-981-97-4529-6_4

Maintaining proper hydration is crucial for managing migraine, since dehydration has been identified as a frequent trigger. As a stimulant or an analgesic, depending on personal sensitivity and consumption habits, caffeine, which is present in both drinks and pharmaceuticals, demonstrated a double ability. A higher risk of migraine has been linked to some alcoholic beverages, especially red wine, in those who are vulnerable (Haghdoost and Togha 2022).

Although being well hydrated is always mentioned as being important, different drinks have different effects on different people. It is crucial to comprehend these subtleties in order to customize migraine care techniques to meet the demands of each patient.

4.2 Stages and Symptoms

Four stages of migraine progression can occur in children and teenagers as well as adults: prodrome, aura, migraine attack, and post-drome (Dodick 2018). Every migraine sufferer may not experience each stage.

4.2.1 Prodrome

In this phase, we might observe minor alterations 1 or 2 days prior to a migraine that indicate an impending headache, such as constipation; mood changes, from depression to euphoria; food cravings; neck stiffness; increased urination; fluid retention; and frequent yawning (Luciani et al. 2000).

4.2.2 Aura

Some individuals may have an aura either prior to or during a migraine. Reversible neurological symptoms include auras. Though, they might also involve other disturbances, they are typically visual. Every symptom typically starts out mildly, intensifies over a few minutes, and lasts for up to an hour. Auras caused by migraine include perceptual phenomena, such perceiving different forms, bright spots or flashes of light, visual impairment, sensations of pins and needles in the leg or arm, numbness or weakness on one side of the body or in the face, and having trouble in speaking (Hansen et al. 2012).

4.2.3 Migraine Attack

An untreated migraine often lasts 4–72 h. Each patient has migraine at different frequencies. A migraine may happen infrequently or multiple times per month (Walter 2022).

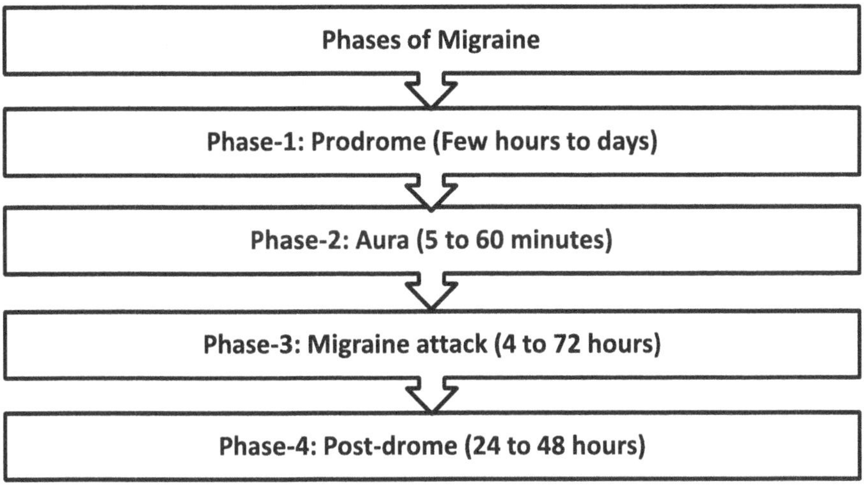

Fig. 4.1 Various phases of migraine attack with their duration

4.2.4 Post-drome

It's possible to feel exhausted, disoriented, and washed out for up to a day following a migraine attack. A few claims to feel ecstatic. A sudden movement of the head could momentarily reactivate the pain (Giffin et al. 2016) (Fig. 4.1).

4.2.5 Causes

Exact causes of migraine are not known. However, heredity and environmental factors seem to be involved. Possible causes include alterations in the brainstem and how it interacts with the trigeminal nerve, a significant pain channel. Inequalities in brain chemistry, such as serotonin, which aids in controlling pain in the nervous system, may also be a factor for migraine. Scientists are investigating the function of serotonin in migraine. Other factor, such as calcitonin gene-related peptide (CGRP), is involved in migraine pain (Dussor 2014).

4.2.6 Triggers of Migraine (Kesserwani 2021)

There are a number of migraine triggers, including:

- **Women's hormonal fluctuations**: Many women seem to experience headaches before or during menstruation, during pregnancy, and throughout menopause, which are all associated with fluctuations in estrogen. Oral contraceptives are one example of a hormonal medicine that can exacerbate migraine. However, some women report that using these drugs reduces the frequency of their migraines (Sacco et al. 2012).

- **Drinks**: They include excessive caffeine intake, such as from coffee, and alcohol, particularly from wine (Nowaczewska et al. 2020a).
- **Stress**: Migraine can result from stress at work or at home (Maleki et al. 2012).
- **Sensory stimuli**: Both loud noises and bright, flashing lights can cause migraine. Some people have migraine when they are around strong scents, such paint thinner, perfume, secondhand smoke, and others (Schwedt 2013).
- **Sleep changes**: For certain people, migraine can be brought on by either sleeping too little or not enough (Ahn and Goadsby 2013).
- **Physical strain**: Sexual activity and other forms of intense physical effort may cause migraine (Ściślicki et al. 2021).

In addition to above triggers sudden change in weather, certain medications like nitroglycerin, food containing high cheese and salts and certain food additives like aspartame and monosodium glutamate is also associated with initiation of migraine attacks.

4.3 Dietary Habits and Diet on Migraine

While triggers can differ from person to person, diet can help some people manage their migraine. Note that any major modifications to one's diet should be done cautiously and that any dietary adjustments should be made after consulting a healthcare expert. The following dietary factors are frequently linked to migraine. In such instances, recognition of food triggers is important. Certain foods such as cheese, chocolate, coffee, alcohol, processed meats, and artificial sweeteners are common triggers for migraine attacks (Nowaczewska et al. 2020a, b). Food items like caffeine can help some people with their migraine, while withdrawal symptoms can cause migraines in others. Moderate consumption may show beneficial effects while sudden withdrawal may worsen the condition of patients (Nowaczewska et al. 2020a). In addition, dehydration of our cells leads to trigger the migraine attack even in winter seasons and proper hydration of our body is required (Arca and Halker Singh 2021). We need to hydrate our body during hot weather or when playing sports, and food containing high amount of fluids should be preferred to avoid migraine. Irregular food habits also lead to migraine attack. One should take regular and balanced meal to avoid migraine. For example, tyramine-containing foods include aged cheeses, cured meats, and several fermented foods; some people find relief from migraine by consuming less of these foods. As mentioned earlier, some people may have sensitivity to gluten or additives like monosodium glutamate. In such cases, one should consume gluten-free diet, and use of monosodium glutamate should be avoided (Obayashi and Nagamura 2016). One should consume a balanced diet with vitamins, Coenzyme Q10, and minerals like magnesium, which may help in avoiding migraine (Nematgorgani et al. 2022). A recent literature review reported that ketogenic, high-folate, low-fat, modified Atkins, and high omega-3/low omega-6 diets demonstrated beneficial effects (Razeghi Jahromi et al. 2019). In addition, ketogenic and modified Atkins diet have been involved in protection of neurons; improve mitochondrial function; and compensate for serotonergic dysfunction, reduction in CGRP levels, and suppression of neuroinflammation (Nematgorgani et al. 2022). Moreover, levels of sodium in our diet also have influence on migraine. A diet low in sodium may be appropriate for patients with vascular risk factors such as hypertension, whereas a high-sodium diet may be suitable for patients with comorbidities such as postural tachycardia syndrome or in those with low blood pressure or low BMI.

4.4 Drinking Behavior on Migraine

Drinking habits and migraine attack have complicated association. Many patients with migraine may not have a problem with moderate alcohol consumption but excessive or specific forms of alcohol intake have been found to be possible triggers for some migraine sufferers (Vives-Mestres et al. 2022). Alcohol may induce migraine by dehydrating our body. Alcohol acts as diuretic and increase the loss of water. Consumption of water in between alcohol drink may reduce the severity of migraine attacks. Moreover, red wine and some beers include histamine, tyramine, and sulfites, which have been linked to migraine attacks. Alcohol produces dose-dependent effect on migraine, and excessive alcohol drink is also associated with increased risk of migraine attacks. Sudden withdrawal of alcohol or hangover after excessive alcohol uses may also be linked with migraine (Zlotnik et al. 2014). Some people discover that the combination of caffeine and alcohol, which is frequently included in beverages like energy drinks and cocktails, can make migraine worse. It's critical to understand the sensitivity of individual person towards these combinations (Heinz et al. 2013).

4.5 Sleeping Habits on Migraine

Migraine and sleep have a comparatively complex relationship that is not fully understood. Many studies are still being conducted worldwide to determine the exact connection between the two. Most of the patients, who suffer from migraine complain of poor sleeping habits and repeated insomnia. This raises the question of whether or not sleep is an important factor in migraine and how it can help control symptoms. Patients suffering from migraine are often prescribed medication with sedative effects to prevent migraine attacks (Giffin et al. 2016). They are reported to suffer from insomnia with poor sleep quality, disturbed sleep patterns, and excessive daytime sleepiness and drowsiness including lack of refreshment after sleep. Clinical studies conducted on both children and adults demonstrated that migraine occurring in the morning is associated with sleep apnea, breathing problems, and snoring. Moreover, the association between bruxism and temporomandibular joint dysfunction (TMD), which is a comorbidity to migraine, has been brought to light by studies. One patient sought treatment for TMD, and it was found that bruxism during sleep was associated with a diagnosis of chronic migraine, particularly episodic migraine, but not tension-type headaches (Vgontzas and Pavlović 2018). There are possibilities that show that bruxism and TMD trigger migraine attacks due to increased peripheral activation of the trigeminal nerve.

Serotonin, or 5-HT, plays an important role in appetite, mood, pain, sexual characteristics, and, most importantly, sleep, in which it inhibits REM sleep and promotes wakefulness (Schuh-Hofer et al. 2007). There is evidence that serotonin levels are low in migraine, as patients with sleep-related migraine are more likely to have disturbed sleep with headaches, which may suggest that serotonin could be a marker for this condition. The hypothalamus is thought to influence the pathophysiology of migraine, as it has been observed that positron emission tomography (PET) scans 24 h before a migraine attack showed a change in hypothalamic activity (Schulte and May 2016).

A new system known as the glymphatic system has been introduced and is said to be responsible for the elimination of central nervous system (CNS) waste (Reddy and van der Werf 2020). Since it is active during sleep, it can be assumed that sleep disturbances lead to changes in CNS waste removal, causing waste to accumulate, which in turn can lead to migraine. If the patient suffers from chronic migraine, the accumulation of CNS waste products will be even greater (Iliff et al. 2012). From this, it can be concluded that sleep and migraine are related, but more experiments need to be conducted to get an accurate data of the relationship between the two.

In a cross-sectional study of 357 patients from a headache clinic in Taiwan, sleep disturbance was examined in patients with mild, moderate, and high-frequency migraine as well as chronic migraine with or without aura. They were divided into four groups according to the frequency of migraine attacks per month: 1–4, 5–8, 9–14, and more than 14 days. Sleep quality was assessed using the Pittsburgh Sleep Quality Index (PSQI). It was found that the index score was highest in patients with frequent migraine attacks and lowest in the group with the lowest frequency of attacks per month (Lin et al. 2016). This study found the relationship between sleep and migraine frequency and concluded that sleep quality is indeed related to the frequency of migraine episodes, regardless of whether they occur with or without aura. Many more studies are being conducted to gain more insight into the relationship between sleep and migraine and how sleep habits affect migraine sufferers.

4.6 Stress on Migraine

According to statistics, stress is a trigger for almost 70% of migraine attacks. It is a very common factor in patients suffering from migraine, and compared to all other triggers, stress is the most frequent trigger of migraine attacks (Dodick 2009). There are various interactions between stress and migraine that lead to biochemical changes that may be part of the stress response, for example, changes due to the psychological response to stressors and so forth. Stress is thought to trigger the onset of migraine, which is clinically symptomatic, and may also increase the duration and intensity of migraine attacks that occur. This shows that stress can be considered a risk factor that increases the frequency of migraine attacks by triggering them, which can also lead to chronic migraine (Sauro and Becker 2009).

It is said that stress and migraine are linked in many ways and are related on several aspects and levels. One of these is the physiological stress response, which involves the sympathetic nervous system and the hypothalamic-pituitary-adrenal axis. When these two systems are activated, behavioral and physiological changes are observed in the patient, which are considered a response to stress. These changes can also lead to the onset of migraine or possibly trigger migraine attacks (Maleki et al. 2012). The link between chronic stress and headaches may be due to changes in the immune system. Chronic stress activates the immune system in a way that facilitates the transmission of pain across neurons. Pain mediators such as IL-6, IL-1β, TNFα, and nitrous oxide can sensitize the pain matrix as they are inflammatory mediators. There are prophylactic migraine treatments that reduce the activity of such inflammatory mediators to strengthen the link between stress, the immune system, and migraine (Maier 2003).

Patients suffering from chronic headaches have more stressful events in the year prior to the onset of headaches. In a study conducted by the Pain Research and Treatment Unit involving about 63 headache patients, it was observed that patients suffering from migraine headaches had experienced comparatively more life-altering events in the year prior to the onset of headaches than the headache-free controls (De Benedittis et al. 1990). This difference made it clear that the year prior to the stressful event is a factor influencing the occurrence of headaches with or without migraine attacks, as no difference was found for 2 or 5 years prior to the onset of headaches, similar to the frequency of events 1 year earlier. Migraine sufferers are said to have more stressors compared to healthy people because they perceive them as more stressful and threatening than healthy people who do not perceive them as stressful and threatening, making it difficult for patients to cope with these stressors. Studies have also shown that migraine attacks triggered by stress have a higher pain intensity than attacks caused by other triggers (Hassinger et al. 1999). Changes in the female hormone balance can also affect the occurrence of migraine in women (Sacco et al. 2012). For example, menarche has been associated with an increased likelihood of migraine occurrence while most women have been observed to experience a decrease in migraine attacks and episodes after the menopause and the drop in estrogen levels (Granella et al. 1993).

The influence of stress on migraine can be explained by another mechanism involving the effects of the migraine attack on the mast cells in the dura. There is ample evidence that mast cells are involved in many nonallergic inflammatory diseases, especially those that worsen under stress. Stress leads to the release of corticotropin-releasing hormone (CRH), which activates the degranulation of mast cells located near the trigeminal nerve endings. They release inflammatory mediators that activate the nociceptors (receptors that recognize signals from damaged tissue or respond to chemicals released by damaged tissue) at the nerve endings, which perpetuates the development of pain in the migraine attack (Theoharides and Cochrane 2004). Reportedly, about 6% of patients with chronic pain have experienced abuse in the form of physical and/or sexual abuse. There is said to be a similar relationship in migraine sufferers. Patients who suffer from post-traumatic stress disorder (PTSD) have been exposed to a traumatic stressful situation and relive the same event or stressor due to some trigger in their daily lives. One study found that chronic migraine is more common in patients with PTSD than episodic migraine (Bailey et al. 2003).

A case-control study was conducted to observe the stress perception of migraine patients. A total of 227 patients were included; 103 of whom suffered from chronic migraine. Patients were asked to complete questionnaires such as the Perceived Stress Scale (PSS), the Migraine Disability Assessment Scale (MIDAS), and many others; after which, their perception of stress and its impact on their quality of life was analyzed, which was also determined. The results showed that the mean PSS score was higher in chronic migraine sufferers compared to the control group. The study concluded that perceived stress affects the quality of life of migraine patients (Moon et al. 2017). The level of perceived stress had a strong impact on migraine patients, and chronic stress plays a major role as a trigger for migraine attacks and episodes. More research and experiments need to be conducted to clarify the exact relationship between stress and migraine, but there is strong evidence of the relationship between the two and that stress is a very important aspect in the field of migraine.

4.7 Smoking Habits on Migraine

It has been observed that people who smoke suffer from worse migraine attacks compared to nonsmoking migraine sufferers. There is also evidence that smoking can worsen the effects of migraine such as stroke and so forth. A Swedish study found a higher frequency of recurrent headaches or migraine attacks in people who smoke, regardless of their gender, compared to nonsmokers who have a lower intensity and frequency of headaches or migraine attacks (Molarius et al. 2008). A study was also conducted in Denmark to investigate the link between smoking and migraine in twins who were smokers and nonsmokers, with the smokers having more migraine episodes and headaches (Le et al. 2011). Nowadays, electronic nicotine delivery systems (ENDS), such as e-cigarettes, are heavily used by both adolescents and adults, and it has been reported that such individuals who used ENDS throughout their lives suffered from headaches. Approximately 22–25% reported headaches or migraine with ENDS use, and 44% of adolescents showed the same results with increased frequency and intensity (Park-Lee et al. 2021).

Migraine sufferers are expected to experience similar side effects of smoking as people who do not suffer from migraine, but there are some additional levels of severity experienced by those who smoke and suffer from migraine. For example, if you compare a migraine sufferer and a non-migraine sufferer, they both have a risk of stroke because of their smoking habits, but the person who suffers from migraine have an increased risk of stroke according to the observations of the study. It was found that migraine sufferers who smoke have a nine-fold increased risk of stroke compared to other migraine sufferers (Bigal et al. 2009). Another example is that 1413 migraine patients who smoke were diagnosed with cutaneous allodynia (pain caused by stimulation of the skin that should not normally cause a sensation of pain); 117 migraine patients from the USA who have smoked all their lives were found to have signs of various autonomic disorders such as eyelid or orbital swelling, nasal congestion, rhinorrhea, and so forth. These symptoms occurred in 70% of lifelong smokers and in about 42% of nonsmoking migraine patients (Tietjen et al. 2009). Out of 100 patients at a migraine clinic in the UK, around 20% smoked and believed that smoking was a trigger for the migraine attack, as it increased the intensity of the pain. Another study conducted in Spain with 58 medical students suffering from migraine, the majority of whom smoked, indicated that smoking was a very strong trigger for the attacks and also increased the frequency of migraine attacks. These studies provided results indicating that cigarette smoking triggers migraine attacks and is also associated with more severe pain during attacks (López-Mesonero et al. 2009). In Saudi Arabia, a study of 247 individuals was conducted and found that migraine sufferers who smoked had a negative psychological state and various psychosocial consequences such as stress and depression. An Italian study with 80 participants also showed that migraine patients who smoked suffered from various psychological problems such as lack of sleep, lack of energy, lack of appetite, poor memory, and difficulties in performing other activities such as walking, sexual behavior, and pain endurance (Al-Hayani et al. 2023).

Since smoking increases the risk of various cardiovascular diseases through inflammation, it may trigger migraine through the swelling of blood vessels that leads to the onset of migraines. One theory states that the level of carboxyhemoglobin (COHb) plays a role in the development of migraine. An increase in COHb levels was only found in patients who smoked, whereas no such effect was observed in nonsmokers. Carbon monoxide is associated with migraines because smoke inhalation increases carbon monoxide levels, which leads to the onset of headaches as it also increases COHb levels. The increase in COHb levels leads to anorexia due to the displacement of oxygen from hemoglobin. Anorexia leads to vasodilation, which raises the possibility that smoking is associated with migraine (Volans and Castleden 1976).

It has been observed that people who suffer from migraine find it harder to stop smoking than other individuals. The mechanisms linking smoking and migraine are not yet confirmed, but one of them could be the effect of smoking on the cardiovascular system. Smoking increases the risk to the cardiovascular system through inflammatory processes that can lead to the onset of migraine (Ambrose and Barua 2004). The brain of migraine sufferers has an increased sensitivity to smells, which means that strong odors act as triggers for attacks. Smoking reduces the perception of smell and has long-term effects on people's perception of smell. Therefore, it is possible that migraine sufferers tend to choose smoking as a coping mechanism to avoid triggering attacks and that headaches can be avoided in this way (Weinberger and Seng 2023). Since smoking increases the risk of various cardiovascular diseases through inflammation, it may trigger migraine through the swelling of blood vessels

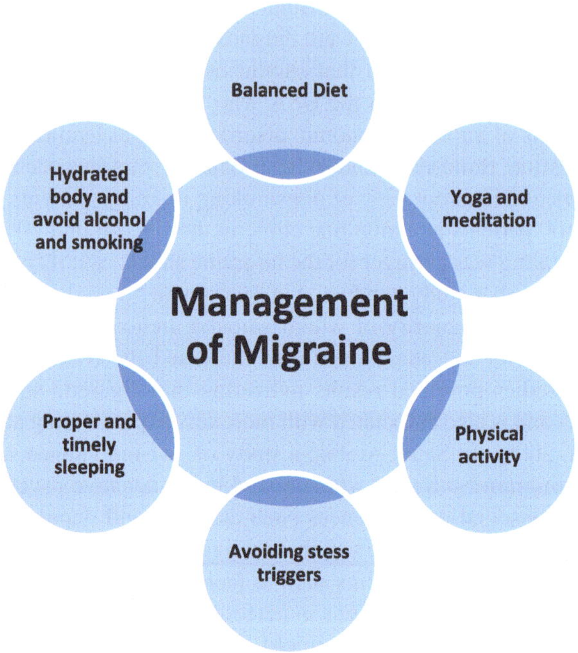

Fig. 4.2 Important factors for the effective management of migraine

that leads to the onset of migraines. Many more studies are still underway to determine the exact mechanism of the effects of smoking on migraine (Fig. 4.2).

4.8 Effect of Physical Activity on Migraine

Physical Activity is any type of voluntary body movements produced by contraction of skeletal muscles (Caspersen et al. 1985), in order to promote a healthy lifestyle, prevent obesity, prevent any cardiovascular disease, and also to freshen up the mood. It exists in several forms as that of running, yoga, cycling, dancing, and trekking. Although all of these are already prevalent in the whole world, these days, yoga or even power yoga have been immensely popular in other countries besides India. The words migraine and exercise have not been used frequently together, even though exercise can help tremendously in the management of migraine pain or its frequency. Nowadays, people have been more and more dependent on the use of medication, which results in an instant relief from the pain and throbbing. It may not have a short-term effect; but it definitely does have a long-term one.

Various studies were organized to review the association of migraine with exercise. The average conclusion of all those studies was that daily physical activity or exercise can result in it having a prophylactic effect on the occurrence of migraine headaches (Amin et al. 2018).

A clinical trial conducted to check the effect of aerobic exercise in migraine and tension type of headaches which sometimes coexist. Here, a control group and test group were made, and the people in test group were asked to carry out aerobic exercises in terms of cross training or cycling for 45 min, 3 times a week for 3 months (once with a trainer, and twice at home or in the gymnasium), whereas the control group was asked to carry out their daily normal routine and activities. Within the test group, there was a significant decrease in the frequency and intensity of migraine headaches and reduced the coexisting tension type headache (Krøll et al. 2018).

There are certain beliefs that being physically active can trigger migraine headaches, which results in intentionally avoiding any kind of physical activity to be spared from the pain. A study evaluated the deliberated evasion of exercise in patients suffering from migraine headaches at different intensity levels. All the volunteers and participants were women. This study concluded stating that almost 78% participants avoided physical activity due to the beliefs that it may lead to severe migraine pain, so it can be said that avoiding exercise can be considered as a non-pharmacological method of managing migraine aches. However, taking into account a case of no physical activity can also lead to weight gain and related disorders. Therefore, an alternative strategy to overcome this is slow exposure to the triggers, where a person can start exercising for a few minutes and they gradually increase the activity (Farris et al. 2018).

The mechanisms through which triggering of migraine headaches and physical activity are linked are still not studied properly, but some of them include increased cardiac output and systolic blood pressure (Sağlı Diren et al. 2023), increased release of neuropeptides, and lactate metabolism, whereas the mechanism involves in prevention of pain in migraineurs is associated with high levels of beta-endorphins, endocannabinoid in plasma post workout (Amin et al. 2018).

Yoga is the most commonly used method to reduce stress. From a study, where people were divided in two groups, one which underwent yoga practice five times a week and the other did not, it was concluded that the group who practiced yoga showcased an improved quality of life along with reduction in frequency and severity of migraine pain. This determines that yoga can be efficacious as a mode of therapy in migraine patients (Kisan et al. 2014).

Yoga Nidra, an act of sleeping consciously is often practiced by migraineurs as it had been proved effective in controlling stress level and neuropsychiatric symptoms in the patients by asserting parasympathetic system dominance hence blocking the sympathetic system activation and thus induces calmness and tranquility by the enhanced alpha waves; which are responsible for improving memory, concentration and boosts serotonin release (Shashikiran et al. 2022).

Exercise triggers the release of endorphins. Endorphins ensure an improved mood, reduction in stress. This may benefit those sufferers whose migraine are generally triggered by stress. A study suggests that doing aerobic exercise for 30 min, 3 times a week for at least 8 weeks improved the physical strength as well as the fitness of patients, and hence, aerobics can be useful as an alternative therapeutic treatment (Agbetou and Adoukonou 2022). A Sweden-based randomized study comprised of a certain number of men and women in the age range on 18–79, covered a broad extensive area of 58 municipalities all over the country. The aim of this experiment was to study the association of socioeconomic factors, lifestyle habits, and recurrent headaches and migraine in the population. The results stated that physical inactiveness led to higher occurrence of such throbbing headaches (Molarius et al. 2008).

Along with the trials and studies mentioned above, it was also proven that the prevalence of migraine headaches in men and women who play basketball was lesser than that of the general population (Kinart et al. 2002).

It has already been determined that exercise is treated as an important method in managing chronic pain related to arthritis, back pain, and fibromyalgia, so its significance in long-term management of headaches in migraineurs can also be considered to be one of the most effective treatment plans (Daenen et al. 2015).

A controlled clinical trial was conducted in 2003 to assess the changes in the nitric oxide (NO) levels in blood due to changes in physical activity, where two groups, one control and the other who went through 1 h of exercise thrice a week, which resulted in patients who were having complaints of pain related to migraine, have lesser complaints in regards to the intensity, duration, and occurrence of the headaches along with an increase in the blood NO levels post therapy (Narin et al. 2003).

Based on the current evidence that has been gathered from various articles and research papers, it is safe to establish that physical activity and migraine are related to each other; but it is unreliable to assume that lack of exercise is the main etiological factor of migraine headaches (Lippi et al. 2018). Therefore, it is advised to be physically active except for the cases where exercise triggers the onset of headache due to the release of NO, which causes vasodilation and increases the blood flow to the brain and hence increases the intensity of pain, inhibiting the long-term effects of exercise in pain management (Narin et al. 2003).

4.9 Effect of Meditation on Migraine

Meditation, now popularized in many western countries, is a practice of mindfulness that originated in India. A large lot of patients suffering from migraine (migraineurs) are said to have frequent migraine attacks that are characterized by throbbing headache, photophobia, and nausea. Severe migraine result in chronic pain and feeling of restlessness, which can be regulated and improved by meditation. Spiritual meditation may also be of help. It has been suggested that prayer and spiritual meditation has resulted in decreased frequency of severe migraine (Wachholtz et al. 2017).

Research studies have proved that people who tend to be religious have an improved quality of live whether it may be physiologically or psychologically. According to a meta-analytical study conducted, it was certified that individuals who are religious are protected against the depressive symptoms that are associated with migraines, and those who are atheists resulted in having more depressive symptoms (Smith et al. 2003).

Meditation leads to relaxation of mind and body; which in turn reduces the frequency and severity of migraine (Wachholtz and Pargament 2008). The prospect of mindfulness reduces stress and pain by engrossing the certain parts of brain that are often associated with cognitive and pain modulation (Wells et al. 2021).

A study also took place where 92 individuals were segregated in 4 groups and were asked to maintain their respective headache diaries. Thirty days into the study, it was observed that the occurrence of headaches had significantly reduced in the group consisting of people who practiced spiritual meditation. The study chiefly concluded that only the occurrence and not the severity of such migraine headaches was diminished (Wachholtz et al. 2017).

A reanalysis of a trial regarding the effect of active meditation on migraine pain was conducted where three vital meditative forms were compared with distractive meditation to check the meditation's effect on pain when the same is practiced for a time period of 30 days and the session lasting at least 20 min every day. The study resulted in a profound decrease in throbbing during migraine and decreased irritation over a certain period of time, although it is said to be helpful only if mindful meditation is practiced for minimum 20 days consistently (Wachholtz et al. 2019).

4.10 Conclusion

Migraine is critical neurological condition that requires crucial examination and management measures. These migraine attacks may sometime comorbid with other disorders such as hypertension, obesity, depression, diabetes mellitus, and so forth. One should take proper and balanced diet to manage frequent attacks of migraine. In addition, our body should be properly hydrated to avoid migraine. Chronic alcoholism, and excessive intake of caffeine also worsens the condition of the migraine sufferer. Proper sleeping habits, regular meditation, and yoga are useful to avoid migraine. Smoking also leads to worsening the condition of a person suffering with migraine. In addition, regular physical activity and avoiding stress is also considered as useful strategy to avoid migraine that is specially associated with many comorbid conditions. Based on the above information, we can conclude that modification of lifestyle is very useful strategy for the effective management of migraine.

References

Agbetou M, Adoukonou T (2022) Lifestyle modifications for migraine management. Front Neurol 13:719467. https://doi.org/10.3389/fneur.2022.719467

Ahn AH, Goadsby PJ (2013) Migraine and sleep: new connections. Cerebrum 2013:15

Al-Hayani M, AboTaleb H, Bazi A, Alghamdi B (2023) Depression, anxiety and stress in Saudi migraine patients using DASS-21: local population-based cross-sectional survey. Int J Neurosci 133(3):248–256. https://doi.org/10.1080/00207454.2021.1909011

Ambrose JA, Barua RS (2004) The pathophysiology of cigarette smoking and cardiovascular disease: an update. J Am Coll Cardiol 43(10):1731–1737. https://doi.org/10.1016/j.jacc.2003.12.047

Amin FM, Aristeidou S, Baraldi C, Czapinska-Ciepiela EK, Ariadni DD, Di Lenola D, Fenech C, Kampouris K, Karagiorgis G, Braschinsky M, Linde M (2018) The association between migraine and physical exercise. J Headache Pain 19(1):83. https://doi.org/10.1186/s10194-018-0902-y

Arca KN, Halker Singh RB (2021) Dehydration and headache. Curr Pain Headache Rep 25(8):56. https://doi.org/10.1007/s11916-021-00966-z

Bailey BE, Freedenfeld RN, Kiser RS, Gatchel RJ (2003) Lifetime physical and sexual abuse in chronic pain patients: psychosocial correlates and treatment outcomes. Disabil Rehabil 25(7):331–342. Taylor & Francis. https://doi.org/10.1080/0963828021000056866

Bigal ME, Kurth T, Hu H, Santanello N, Lipton RB (2009) Migraine and cardiovascular disease: possible mechanisms of interaction. Neurology 72(21):1864–1871. https://doi.org/10.1212/WNL.0b013e3181a71220

Caspersen CJ, Powell KE, Christenson GM (1985) Physical activity, exercise, and physical fitness: definitions and distinctions for health-related research. Public Health Rep 100(2):126–131

Daenen L, Varkey E, Kellmann M, Nijs J (2015) Exercise, not to exercise, or how to exercise in patients with chronic pain? Applying science to practice. Clin J Pain 31(2):108–114. https://doi.org/10.1097/AJP.0000000000000099

De Benedittis G, Lorenzetti A, Fieri A (1990) The role of stressful life events in the onset of chronic primary headache. Pain 40(1):65–75. https://doi.org/10.1016/0304-3959(90)91052-K

Dodick DW (2009) Review of comorbidities and risk factors for the development of migraine complications (infarct and chronic migraine). Cephalalgia 29(Suppl 3):7–14. https://doi.org/10.1177/03331024090290S303

Dodick DW (2018) A phase-by-phase review of migraine pathophysiology. Headache 58(Suppl 1):4–16. https://doi.org/10.1111/head.13300

Dussor G (2014) Serotonin, 5HT1 agonists, and migraine: new data, but old questions still not answered. Curr Opin Support Palliat Care 8(2):137–142. https://doi.org/10.1097/SPC.0000000000000044

Farris SG, Thomas JG, Abrantes AM, Godley FA, Roth JL, Lipton RB, Pavlovic J, Bond DS (2018) Intentional avoidance of physical activity in women with migraine. Cephalalgia Rep 1:2515816318788284. https://doi.org/10.1177/2515816318788284

Gazerani P (2020) Migraine and diet. Nutrients 12(6):1658. https://doi.org/10.3390/nu12061658

Giffin NJ, Lipton RB, Silberstein SD, Olesen J, Goadsby PJ (2016) The migraine postdrome: an electronic diary study. Neurology 87(3):309–313. https://doi.org/10.1212/WNL.0000000000002789

Granella F, Sances G, Zanferrari C, Costa A, Martignoni E, Manzoni GC (1993) Migraine without aura and reproductive life events: a clinical epidemiological study in 1300 women. Headache 33(7):385–389. https://doi.org/10.1111/j.1526-4610.1993.hed3307385.x

Haghdoost F, Togha M (2022) Migraine management: non-pharmacological points for patients and health care professionals. Open Med (Wars) 17(1):1869–1882. https://doi.org/10.1515/med-2022-0598

Hansen JM, Lipton RB, Dodick DW, Silberstein SD, Saper JR, Aurora SK, Goadsby PJ, Charles A (2012) Migraine headache is present in the aura phase: a prospective study. Neurology 79(20):2044–2049. https://doi.org/10.1212/WNL.0b013e3182749eed

Hassinger HJ, Semenchuk EM, O'Brien WH (1999) Appraisal and coping responses to pain and stress in migraine headache sufferers. J Behav Med 22(4):327–340. https://doi.org/10.1023/a:1018722002393

Heinz AJ, de Wit H, Lilje TC, Kassel JD (2013) The combined effects of alcohol, caffeine, and expectancies on subjective experience, impulsivity, and risk-taking. Exp Clin Psychopharmacol 21(3):222–234. https://doi.org/10.1037/a0032337

Iliff JJ, Wang M, Liao Y, Plogg BA, Peng W, Gundersen GA, Benveniste H, Vates GE, Deane R, Goldman SA, Nagelhus EA, Nedergaard M (2012) A paravascular pathway facilitates CSF flow through the brain parenchyma and the clearance of interstitial solutes, including amyloid β. Sci Transl Med 4(147):147ra111. https://doi.org/10.1126/scitranslmed.3003748

Kesserwani H (2021) Migraine triggers: an overview of the pharmacology, biochemistry, atmospherics, and their effects on neural networks. Cureus 13(4):e14243. https://doi.org/10.7759/cureus.14243

Kinart CM, Cuppett MM, Berg K (2002) Prevalence of migraines in NCAA division I male and female basketball players. National Collegiate Athletic Association. Headache 42(7):620–629. https://doi.org/10.1046/j.1526-4610.2002.02149.x

Kisan R, Sujan M, Adoor M, Rao R, Nalini A, Kutty BM, Chindanda Murthy B, Raju T, Sathyaprabha T (2014) Effect of yoga on migraine: a comprehensive study using clinical profile and cardiac autonomic functions. Int J Yoga 7(2):126–132. https://doi.org/10.4103/0973-6131.133891

Krøll LS, Hammarlund CS, Linde M, Gard G, Jensen RH (2018) The effects of aerobic exercise for persons with migraine and co-existing tension-type headache and neck pain. A randomized, controlled, clinical trial. Cephalalgia 38(12):1805–1816. https://doi.org/10.1177/0333102417752119

Le H, Tfelt-Hansen P, Skytthe A, Kyvik KO, Olesen J (2011) Association between migraine, lifestyle and socioeconomic factors: a population-based cross-sectional study. J Headache Pain 12(2):157–172. https://doi.org/10.1007/s10194-011-0321-9

Lin Y-K, Lin G-Y, Lee J-T, Lee M-S, Tsai C-K, Hsu Y-W, Lin Y-Z, Tsai Y-C, Yang F-C (2016) Associations between sleep quality and migraine frequency: a cross-sectional case-control study. Medicine 95(17):e3554. https://doi.org/10.1097/MD.0000000000003554

Lippi G, Mattiuzzi C, Sanchis-Gomar F (2018) Physical exercise and migraine: for or against? Ann Transl Med 6(10):181. https://doi.org/10.21037/atm.2018.04.15

López-Mesonero L, Márquez S, Parra P, Gámez-Leyva G, Muñoz P, Pascual J (2009) Smoking as a precipitating factor for migraine: a survey in medical students. J Headache Pain 10(2):101–103. https://doi.org/10.1007/s10194-009-0098-2

Luciani R, Carter D, Mannix L, Hemphill M, Diamond M, Cady R (2000) Prevention of migraine during prodrome with naratriptan. Cephalalgia 20(2):122–126. https://doi.org/10.1046/j.1468-2982.2000.00030.x

Maier SF (2003) Bi-directional immune-brain communication: implications for understanding stress, pain, and cognition. Brain Behav Immun 17(2):69–85. https://doi.org/10.1016/s0889-1591(03)00032-1

Maleki N, Becerra L, Borsook D (2012) Migraine: maladaptive brain responses to stress. Headache 52 Suppl 2(Suppl 2):102–106. https://doi.org/10.1111/j.1526-4610.2012.02241.x

Molarius A, Tegelberg A, Ohrvik J (2008) Socio-economic factors, lifestyle, and headache disorders—a population-based study in Sweden. Headache 48(10):1426–1437. https://doi.org/10.1111/j.1526-4610.2008.01178.x

Moon H-J, Seo J-G, Park S-P (2017) Perceived stress in patients with migraine: a case-control study. J Headache Pain 18(1):73. https://doi.org/10.1186/s10194-017-0780-8

Narin SO, Pinar L, Erbas D, Oztürk V, Idiman F (2003) The effects of exercise and exercise-related changes in blood nitric oxide level on migraine headache. Clin Rehabil 17(6):624–630. https://doi.org/10.1191/0269215503cr657oa

Nematgorgani S, Razeghi-Jahromi S, Jafari E, Togha M, Rafiee P, Ghorbani Z, Ahmadi ZS, Baigi V (2022) B vitamins and their combination could reduce migraine headaches: a randomized double-blind controlled trial. Curr J Neurol 21(2):105–118. https://doi.org/10.18502/cjn.v21i2.10494

Nowaczewska M, Wiciński M, Kaźmierczak W (2020a) The ambiguous role of caffeine in migraine headache: from trigger to treatment. Nutrients 12(8):2259. https://doi.org/10.3390/nu12082259

Nowaczewska M, Wiciński M, Kaźmierczak W, Kaźmierczak H (2020b) To eat or not to eat: a review of the relationship between chocolate and migraines. Nutrients 12(3):608. https://doi.org/10.3390/nu12030608

Obayashi Y, Nagamura Y (2016) Does monosodium glutamate really cause headache?: a systematic review of human studies. J Headache Pain 17:54. https://doi.org/10.1186/s10194-016-0639-4

Park-Lee E, Ren C, Sawdey MD, Gentzke AS, Cornelius M, Jamal A, Cullen KA (2021) Notes from the field: E-cigarette use among middle and high school students—National Youth Tobacco Survey, United States, 2021. MMWR Morb Mortal Wkly Rep 70(39):1387–1389. https://doi.org/10.15585/mmwr.mm7039a4

Razeghi Jahromi S, Ghorbani Z, Martelletti P, Lampl C, Togha M (2019) Association of diet and headache. J Headache Pain 20(1):106. https://doi.org/10.1186/s10194-019-1057-1

Reddy OC, van der Werf YD (2020) The sleeping brain: harnessing the power of the glymphatic system through lifestyle choices. Brain Sci 10(11):868. https://doi.org/10.3390/brainsci10110868

Sacco S, Ricci S, Degan D, Carolei A (2012) Migraine in women: the role of hormones and their impact on vascular diseases. J Headache Pain 13(3):177–189. https://doi.org/10.1007/s10194-012-0424-y

Sağlı Diren G, Kaya Ciddi P, Ergezen G, Şahin M (2023) Effect of physical activity level on pain, functionality, and quality of life in migraine patients. Agri 35(4):212–219. https://doi.org/10.14744/agri.2022.26504

Sauro KM, Becker WJ (2009) The stress and migraine interaction. Headache 49(9):1378–1386. https://doi.org/10.1111/j.1526-4610.2009.01486.x

Schuh-Hofer S, Richter M, Geworski L, Villringer A, Israel H, Wenzel R, Munz DL, Arnold G (2007) Increased serotonin transporter availability in the brainstem of migraineurs. J Neurol 254(6):789–796. https://doi.org/10.1007/s00415-006-0444-0

Schulte LH, May A (2016) The migraine generator revisited: continuous scanning of the migraine cycle over 30 days and three spontaneous attacks. Brain 139(Pt 7):1987–1993. https://doi.org/10.1093/brain/aww097

Schwedt TJ (2013) Multisensory integration in migraine. Curr Opin Neurol 26(3):248–253. https://doi.org/10.1097/WCO.0b013e328360edb1

Ściślicki P, Sztuba K, Klimkowicz-Mrowiec A, Gorzkowska A (2021) Headache associated with sexual activity—a narrative review of literature. Medicina (Kaunas) 57(8):735. https://doi.org/10.3390/medicina57080735

Shashikiran HC, Shetty P, Akshay R, Venugopal A, Shetty S (2022) Effect of yoga nidra on the brain activity in individuals with migraine. Yoga Mimamsa 54(1). https://journals.lww.com/yomi/fulltext/2022/54010/effect_of_yoga_nidra_on_the_brain_activity_in.4.aspx:18

Smith TB, McCullough ME, Poll J (2003) Religiousness and depression: evidence for a main effect and the moderating influence of stressful life events. Psychol Bull 129(4):614–636. https://doi.org/10.1037/0033-2909.129.4.614

Theoharides TC, Cochrane DE (2004) Critical role of mast cells in inflammatory diseases and the effect of acute stress. J Neuroimmunol 146(1–2):1–12. https://doi.org/10.1016/j.jneuroim.2003.10.041

Tietjen GE, Brandes JL, Peterlin BL, Eloff A, Dafer RM, Stein MR, Drexler E, Martin VT, Hutchinson S, Aurora SK, Recober A, Herial NA, Utley C, White L, Khuder SA (2009) Allodynia in migraine: association with comorbid pain conditions. Headache 49(9):1333–1344. https://doi.org/10.1111/j.1526-4610.2009.01521.x

Vgontzas A, Pavlović JM (2018) Sleep disorders and migraine: review of literature and potential pathophysiology mechanisms. Headache 58(7):1030–1039. https://doi.org/10.1111/head.13358

Vives-Mestres M, Casanova A, Puig X, Ginebra J, Rosen N (2022) Alcohol as a trigger of migraine attacks in people with migraine. Results from a large prospective cohort study in English-speaking countries. Headache 62(10):1329–1338. https://doi.org/10.1111/head.14428

Volans GN, Castleden CM (1976) The relationship between smoking and migraine. Postgrad Med J 52(604):80–82. https://doi.org/10.1136/pgmj.52.604.80

Wachholtz AB, Pargament KI (2008) Migraines and meditation: does spirituality matter? J Behav Med 31(4):351–366. https://doi.org/10.1007/s10865-008-9159-2

Wachholtz AB, Malone CD, Pargament KI (2017) Effect of different meditation types on migraine headache medication use. Behav Med 43(1):1–8. https://doi.org/10.1080/08964289.2015.1024601

Wachholtz A, Vohra R, Metzger A (2019) A reanalysis of a randomized trial on meditation for migraine headaches: distraction is not enough but meditation takes time. Complement Ther Med 46:136–143. https://doi.org/10.1016/j.ctim.2019.08.011

Walter K (2022) What is migraine? JAMA 327(1):93. https://doi.org/10.1001/jama.2021.21857

Weinberger AH, Seng EK (2023) The relationship of tobacco use and migraine: a narrative review. Curr Pain Headache Rep 27(4):39–47. https://doi.org/10.1007/s11916-023-01103-8

Wells RE, O'Connell N, Pierce CR, Estave P, Penzien DB, Loder E, Zeidan F, Houle TT (2021) Effectiveness of mindfulness meditation vs headache education for adults with migraine: a randomized clinical trial. JAMA Intern Med 181(3):317–328. https://doi.org/10.1001/jamainternmed.2020.7090

Zlotnik Y, Plakht Y, Aven A, Engel Y, Am NB, Ifergane G (2014) Alcohol consumption and hangover patterns among migraine sufferers. J Neurosci Rural Pract 5(2):128–134. https://doi.org/10.4103/0976-3147.131652

Importance of Diet and Other Nutrients for the Management of Migraine Pain

5

Kalpesh Ramdas Patil and Shiv Kumar Yadav

5.1 Introduction

Migraine is a common neurological condition that causes unbearable headaches along with a wide range of somatosensory and momentary motor abnormalities (Charles 2018). Migraine appears in episodic or chronic forms, with and without aura (Olesen 2018). The factors or mechanisms behind migraine pathology have been studied, and it shows the activation of a trigeminovascular system involving the release of pro-inflammatory mediators and oxidative stress (Andreou and Edvinsson 2019; Burstein et al. 2015). Migraine is a multifaceted and complex disorder that is affected by both hereditary and environmental factors (Buse et al. 2019; Nyholt et al. 2017; Burstein et al. 2015). Investigations have been conducted for effective and safe approaches to the prevention and management of migraine (Bertels and Pradhan 2019). Management of migraine is challenging as the available medications or non-pharmacological approaches are only partially beneficial (Vikelis et al. 2018; Puledda and Goadsby 2017). Researchers have attempted to develop methods for making lifestyle modifications to avoid or manage headaches (Robblee and Starling 2019). In this direction, the role of dietary triggers in migraine has been spotted, leading to the development of various dietary strategies for the management of headaches, including migraine (Robblee and Starling 2019; Buse et al. 2019) The area of precision nutrition is emerging as a means of developing an individualized dietary plan for several diseases (Nyholt et al. 2017; Bertels and Pradhan 2019). It is noteworthy that migraine has been linked to metabolism or a

K. R. Patil (✉)
Department of Pharmacology, R. C. Patel Institute of Pharmaceutical Education and Research, Dhule, Maharashtra, India

S. K. Yadav
Department of Pharmacy Practice, R. C. Patel Institute of Pharmaceutical Education and Research, Dhule, Maharashtra, India

S. Bhatt et al. (eds.), *Management of Migraine Pain*,
https://doi.org/10.1007/978-981-97-4529-6_5

metabolic endocrine disorder (Puledda and Goadsby 2017). An elimination diet approach has been introduced in the field, and it became apparent that dietary substances play a role in the development of headaches (Bertels and Pradhan 2019). It is generally acknowledged that certain food ingredients can cause migraine attacks and that migraines are sensitive to diet.

5.2 Migraine Triggers

Food and nutrition status are significant pain triggers, especially among children and teenagers. A wide range of variables may trigger a migraine, including stress, menstruation, inadequate sleep, climate variations, skipping meals, alcohol, cigarette smoke, odor, direct sunlight, high altitude, coughing, and certain foods like aspartame, chocolate, cheese, and alcohol. Common migraine triggers comprise cheese, chocolate, alcohol, and fasting; fasting is the most prevalent, followed by chocolate and alcohol. Some variables like premenstrual syndrome, pregnancy, hypertension, obesity, and personal issues may cause pain. Adolescents who experience frequent headaches may find themselves experiencing these triggers regularly. Certain food components cause vasoconstriction by facilitating the release of neurotransmitters like serotonin and norepinephrine or directly through the cortical and brain-stem nervous systems. Fasting and missing meals can cause a relapse of pain in migraine patients. Food colorant, sodium nitrite induces vasodilatation via the release of nitric oxide (NO); hence, sensitive individuals should stay away from such substances. Alcohol intake could aggravate headaches in some individuals which suggests its avoidance in such populations (Özturan et al. 2016; Sun-Edelstein

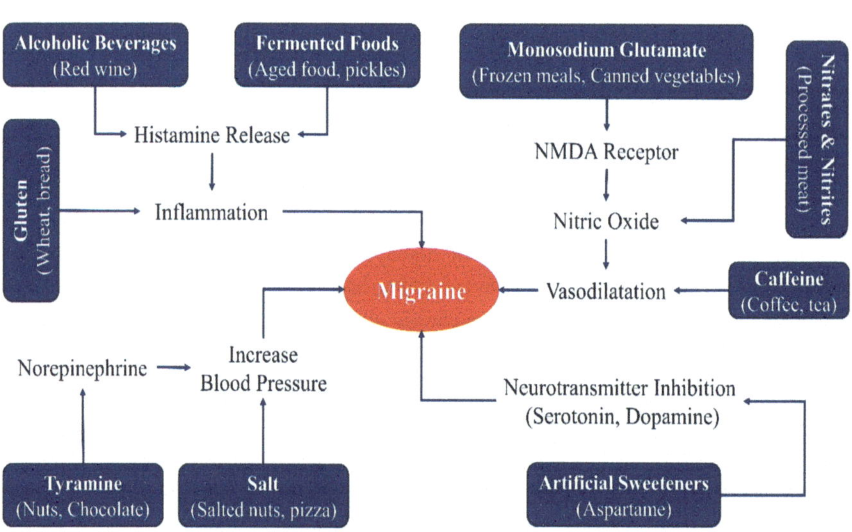

Fig. 5.1 Food components as migraine triggers

and Mauskop 2009). Various food components which are identified as migraine triggers, and their mechanisms are presented in Fig. 5.1.

5.3 Types of Migraines

The headaches or migraine are manifested in various intensities and frequencies. Apart from migraine pathology, erroneous glasses prescriptions, stress, loud noises, lack of sleep, and tight headwear can all trigger and exacerbate migraine episodes. Menstrual migraine, chronic migraine, hemiplegic migraine, migraine with aura, and migraine without aura are some of the types of headaches. According to the World Health Organization, about 50% of the population has experienced at least one headache in the past 12 months. The characteristics of various types of migraines are represented in Table 5.1.

Table 5.1 Characteristics of migraine types

S. no.	Type	Characteristic	Symptoms
1	Migraine with aura (Classic migraine)	Recurring headaches with sensory disturbances (aura)	Sensory disturbances like light flashes, blind spots, and other vision changes or tingling in the hand or face
2	Migraine without aura	Absence of aura, that is, warning sign of a migraine before the onset of migraine episode. Usually, headache on one side of head, worsen by movements, interfere with daily activities	Throbbing pain. nausea or vomiting, photophobia, and phonophobia
3	Chronic migraine	Headache on at least 15 days per month, with 8 of these having migraine symptoms, for at least 3 months	Frequent headache, increased sensitivity to light, sound or smells, nausea, and vomiting
4	Hemiplegic migraine	Motor weakness affecting only one side of the body, supplemented by aura. Impaired vision, speech, or sensation	Visual disturbances including colored spots. Slurring words, inability to speak clearly. Difficulties with reading, listening, speaking, and writing
5	Menstrual migraine (Hormone headache)	Menstrual migraines can occur every month before or during a woman's period. Feeling sick and vomiting repeatedly. Usually at the same time of day, week, or month, these attacks last over a week and last from 1 h to 10 days. Normal feeling in between the attacks	Dull throbbing or severe pulsing headache, sensitivity to light, nausea, fatigue, and dizziness
6	Medication overuse headache	Chronic headache triggered by regular excessive use of acute headache medicines for over 3 months	Nausea, restlessness, trouble concentrating, memory problems, and irritability

5.4 Mechanisms/Pathophysiology of Migraine

Trigeminovascular activation, including peripheral trigeminal nerve terminals, is thought to trigger migraines. Nitric oxide (NO), substance P, and calcitonin gene-related peptides are released by this system, mediating pain perception. Chemicals such as potassium and arachidonic acid may induce anhidrotic neurogenic inflammation in the dura mater, which transmits signals to pain-processing regions. The immune system has a role in migraines, with cytokines that include adiponectin, interleukin-1, and tumor necrosis factor associated with inflammation, pain threshold alterations, and migraine precipitation. Compared to healthy controls, migraine patients demonstrate increased levels of TNF-α, IL-6, and IL-10 (Goadsby et al. 2017). Additionally, various dietary components including certain micronutrients have been identified to play a significant role in the modification of migraine episodes. Serotonin and dopamine are also involved in the pathology of migraine. Serotonin levels have also been found to be lower in migraine patients. An effective therapy for migraines involves the use of serotonin agonists, which have been identified in the trigeminal nerve and cranial vessels. Dopamine relates to the pathophysiology of migraine over the three decades. The trigeminovascular system, which is thought to be implicated in headache pain, contains dopamine receptors. Dopamine agonists limit neuronal firing in this region of the brain (Aggarwal et al. 2012; Akerman and Goadsby 2007).

5.5 Nutrients for the Management of Migraine Pain

Several factors influence migraines, including excessive cholesterol, mitochondrial dysfunction, minimal serotonin, improved calcitonin, MMP-9, vitamin D, and inflammation. A migraine introduces peptides like substance P and calcitonin gene-related peptide (CGRP), which trigger neurogenic inflammation and vasodilation. For acute migraines, triptans are essential for preventing or reducing the severity of migraine episodes. Antidepressants, beta-blockers, calcium channel blockers, and NSAIDs are also common medications for migraine treatment. Inconsistencies in magnesium, Coenzyme Q10 (CoQ10), a-lipoid acid, vitamins, and carnitine are common findings in migraineurs. Relative to those who suffer chronic migraines, those who suffer periodic headaches at random times are less prone to lacking these nutrients (Nattagh-Eshtivani et al. 2018).

5.5.1 Magnesium

The prevalent intracellular cation that exists in all tissues is magnesium. It helps in the preservation and expense of intracellular energy, acts as a cofactor for various enzymes, is essential for nucleic acid synthesis, maintains cell membrane balance, and contributes a role in cell division and growth, additionally controlling ion channels, receptors, and cellular transport (Nattagh-Eshtivani et al. 2018). Magnesium is crucial for migraine therapy since it impacts both peripheral and central nervous system excitability (Domitrz and Cegielska 2022). The role of magnesium in migraine pathogenesis is well recognized, with deficiencies identified to induce cortical spreading depression, modify nociceptive signals processing, and release of neurotransmitters, and promote platelet hyper-aggregation; all of which constitute key components of migraine development. Researchers noted that magnesium may be a safe, effective, and economical migraine treatment option. Magnesium, for specific patient subsets, may successfully manage migraines, tension headaches, and cluster headaches (Yablon and Mauskop 2011). Multiple theories have been suggested to explain the connection between magnesium deficiency and migraines. Many guidelines suggest magnesium as a potentially beneficial therapy for migraine. Most of the studies acknowledge the efficacy of magnesium as a preventive and symptomatic remedy following its oral and intravenous administration, respectively. However, this finding needs to be supported by further extensive studies (Savi 2023).

Magnesium deficiency is linked with multiple facets of migraine pathophysiology, notably platelet aggregation, cortical spreading depression, substance-P release, release of neurotransmitters, and vasoconstriction (Yablon and Mauskop 2011). The NMDA receptors have a role in nociception and neuroplastic modifications in the trigeminal nociceptive neurons, as well as cerebral blood flow regulation. Magnesium ions inhibit the NMDA receptor, which prohibits calcium ions from entering the cell thereby reducing calcium's impact on neurons and cerebral vessels. Therefore, magnesium deficiency favors calcium entry into neurons and cerebral vascular muscles, hastening NMDA activation (Nattagh-Eshtivani et al. 2018). NO is a synaptic modulator, influencing the nociceptive processing and vasodilatation. Serotonin, released from platelets during migraines, promotes cerebral vasoconstriction and triggers nausea and vomiting. The CGRP, a neuropeptide, is crucial in migraine development, is involved in vasodilation, and could enhance nociceptive transmission. Thus, magnesium could help migraine patients by reducing vasospasm, hindering platelet accumulation, sustaining cell membranes, and lowering the production of inflammatory mediators, thereby enhancing mitochondrial oxidative phosphorylation, 5-HT neurotransmission, and the NO system (Nattagh-Eshtivani et al. 2018).

5.5.2 Coenzyme Q10

Despite the considerable financial and health implications related to migraine headaches, only a small number of medications are currently approved for their protective use, and nearly all of these substances have a likelihood of causing severe side effects. A dietary supplement called CoQ10 has demonstrated potential in the prevention of migraines (Shoeibi et al. 2017). CoQ10 is a natural vitamin like lipophilic substance that exhibit antioxidant potential and acts as mitochondrial electron transporter through the electron transfer from complex I and complex II to cytochrome C. Earlier studies demonstrated a notable prophylactic impact of CoQ10, which reduces migraineurs' frequency, duration of episode, decreases the severity, and makes their attacks easier to tolerate. Additionally, it was shown to lessen the patients' phonophobia, photophobia, and nausea symptoms (Dahri et al. 2017). Migraine physiopathology is unclear, but mitochondrial dysfunction may cause lessened oxygen metabolism. CoQ10, has a crucial role in mitochondrial energy stores and may be a natural prophylactic cure with negligible side effects in migraine patients. Due to its antioxidant properties, CoQ10 is a beneficial therapy for patients with mitochondrial dysfunction. Raised matrix metalloproteinase-9 (MMP-9) is associated with dysfunction of the blood-brain barrier and neuroinflammation aggravates migraine episodes. MMP-9 and blood-brain barrier dysfunctions are also linked with migraine attacks. Interleukin-6 (IL-6), tumor necrosis factor-alpha (TNF- α), and H_2O_2, which is a vital reactive oxygen species, maintain MMP expression and its regulation. One of the most essential antioxidants, CoQ10 decreases H_2O_2 and suppresses the expression of MMPs and cytokines. CoQ10 is used as an over-the-counter dietary supplement in the United States. The Canadian Headache Society's guidelines strongly recommend CoQ10 as a migraine preventative agent, and the American Academy of Neurology rates it to be useful in migraine prevention (grade C quality evidence). It is advised to take 1–3 mg/kg of CoQ10 everyday, regardless if the precise beneficial quantity is uncertain (Nattagh-Eshtivani et al. 2018).

5.5.3 Vitamins (B2, B6, B12, and D)

Riboflavin also referred to as vitamin B2, is necessary for the metabolism of lipids, carbohydrates, and protein. Riboflavin has a role in the electron transport chain, and it is flavin mononucleotide and flavin adenine dinucleotide precursor. As riboflavin is involved in mitochondrial energy metabolism and cellular energy production, it has demonstrated its potential in the prevention and management of migraines. Mitochondrial dysfunction, inadequate oxygen metabolism, and altered mitochondrial energy metabolism are causative mechanisms of migraine pathology. High-dose riboflavin therapy is recommended as a promising treatment for migraine prophylaxis. About 400 mg daily dose of riboflavin is advised in adults with migraine. The riboflavin administration for up to 3–6 months in adult migraineurs demonstrated efficacy comparable to propranolol. Owing to its tolerability and safety, the Canadian Headache Society and AAN suggest riboflavin consumption in adults having migraine. It was found to be ineffective in pediatric migraine, and hence, its use is not endorsed in this patient population. Extensive clinical trials are still needed to confirm this migraine-preventive potential of riboflavin in adults (Nattagh-Eshtivani et al. 2018; Orr 2016).

The trio of mitochondrial dysfunction, brain energy deficiency, and oxidative stress is involved in the development of migraine. Niacin is a nutritive precursor of cofactors essential for mitochondrial energy metabolism. Niacin deficiency caused decreased oxidative phosphorylation and altered mitochondrial respiration. Clinical studies support the oral and parenteral administration of niacin for the reduction of several migraine episodes (Liu et al. 2022). Niacin is a vital nutrient that has a role in the dilatation and widening of blood vessels. Niacin, an essential nutrient, plays a crucial role in blood vessel dilatation and widening. Deficiency of serotonin is recognized in the development of migraine. Niacin is a negative feedback regulator in the kynurenine pathway and causes a rise in serotonin levels in the plasma. Although niacin has been studied for its effectiveness against migraine, further trials are needed to sustain this claim. Trigeminovascular complex involved in acute migraine causes intracranial vasoconstriction and aura. Oral administration of niacin shows intracranial vasodilation and extracranial vasoconstriction which could prevent the symptoms of migraine. The cutaneous blushing and lateral dilation caused by niacin due to its vasodilator properties could benefit tension-type headaches in its acute stage. Niacin may be a migraine-preventative agent as it promotes mitochondrial energy metabolism and improves blood flow and oxygen delivery to the skeletal muscles. Reduced incidence of muscular tension and soreness following niacin therapy is attributed to a declining level of lactic acid (Nattagh-Eshtivani et al. 2018).

Along with conventional migraine treatments such as ergotamine, triptans, and monoclonal antibodies addressing calcitonin gene-related peptide receptors, cobalamin commonly known as vitamin B12 has been investigated as a possible preventative medication. The reported effects of vitamin B12 on NO and homocysteine urge further research into its underlying processes related to migraine pathogenesis. Several studies endorse the efficacy of cobalamin in migraine patients (Urits et al. 2020). Vitamin B12 or cobalamin has an important role in the prevention of migraines. Increased NO levels can alter the respiratory chain and produce NO radicals. NO, an important mediator of chronic pain, pain transmission, hyperalgesia, and inflammation is scavenged by cobalamin, which leads to reduced episodes of migraine in patients. In addition to cobalamin, vitamins like B3, B6, and coenzymes are advocated for migraine prevention. Vitamin B12 synthesis need riboflavin, which showed anti-inflammatory and analgesic properties in preclinical studies. Dysfunctional vitamin B12 pathway is implicated in pathophysiology of migraine. Vitamin B6, B12, and folate level decline is associated with increased levels of homocysteine. Elevated homocysteine triggers inflammation, drive electrophysiological hyperactivity negatively, and aggravates migraine episodes (Nattagh-Eshtivani et al. 2018). Hence, selected vitamins including vitamin B6, B12, and folate targeting raised homocysteine are effective options for migraine prevention. Further clinical trials are recommended to sustain the efficacy of these anti-homocysteine vitamins in adult and pediatric migraine subjects (Urits et al. 2020; Liampas et al. 2020).

Vitamin D has an important role in the development of the adult brain and embryogenesis. Additionally, it possesses antioxidant and anti-inflammatory actions, which implies the possibility of vitamin D as a treatment of migraine. Earlier studies suggest that vitamin D could decrease the frequency of migraine episodes; however, there is a scarcity of reports on vitamin D prescriptions for the management of migraines. Calcium supplements and vitamin D are discovered to significantly decrease migraine episodes following 2 months of treatment. However, the precise link between headaches and deficiency of vitamin D remains uncertain. The intestinal magnesium absorption is dependent on vitamin D. Hence, vitamin D deficits may cause decreased magnesium absorption and subsequent tension headaches. Furthermore, the existence of vitamin D receptors and binding protein within the brain is linked with tension headaches (Ghorbani et al. 2019; Dell'Isola et al. 2021; Nattagh-Eshtivani et al. 2018).

5.5.4 Alpha Lipoic Acid

Alpha-lipoic acid (ALA) is involved in protein, carbohydrate, and lipid metabolism in addition to being a cofactor for complex multienzymes. ALA is an amphipathic antioxidant with anti-inflammatory properties that has the capacity to regenerate endogenous antioxidants, purify reactive oxygen and nitrogen species as well as nitrogen, and contribute to modulation of NF-κB signaling. Moreover, it safeguards endothelial function and is appropriate for supplementation in many adult and pediatric medical conditions. Abnormal levels of ALA are reported in migraine patients. According to recent study, ALA is suggested as additive therapy for migraine due to its favorable actions on endothelial and mitochondrial functions (Kelishadi et al. 2022). Insulin resistance is prevalent in migraine subjects that have been reduced by ALA. In an exploratory study, ALA exhibited a decreased frequency of migraine episodes in patients of migraine with insulin resistance. However, further clinical studies are recommended to endorse this possibility (Cavestro et al. 2018).

5.5.5 L-Carnitine

L-carnitine assists in transporting fatty acids through the membrane and facilitates their metabolism in mitochondria. L-carnitine deficiency modifies the oxidation of fatty acid and accumulates toxins following nociceptive stimuli. Additionally, L-carnitine diminishes beta-oxidation, which is connected to migraines. Previous studies support the utility of L-carnitine in the management of migraines. Moreover, L-carnitine transports free fatty acids to the mitochondria, controls enzyme activity, and offers protection against oxidative damage. Some studies demonstrated that supplementation with CoQ10 and L-carnitine improved migraine headache symptoms and serum mitochondrial metabolic disorders markers. To endorse these results, crossover trials are desirable (Hajihashemi et al. 2019; Charleston et al. 2021).

5.5.6 Nutraceuticals

Tanacetum parthenium L., commonly referred to as feverfew has been known for its use in several human diseases for centuries. A sesquiterpene lactone, parthenolide, has a lactone structure, and it is a chief active ingredient of feverfew having interesting effects on the central nervous system (Özturan 2016). Multiple biological actions of feverfew and parthenolide imply a promising mechanism of action against migraine prevention. These mechanisms include inhibition of Fos-induced stimulation of the nucleus trigeminalis caudalis, which is central to migraine pathophysiology, and partial agonism of TRPA1 channels. It is well-recognized that parthenolide inhibits the production of platelets in capillaries, which suppresses the prostaglandins that represent pain. Adult migraine prophylaxis has been thoroughly studied with feverfew, a nutraceutical. Five controlled trials with different levels of efficacy were identified in a 2004 Cochrane review (Orr 2016; Özturan 2016). Feverfew was studied along with vitamins, minerals, acupuncture, and other nutraceuticals for migraine prevention and treatment. In accordance with a thoroughly executed randomized controlled trial, feverfew is more efficient than a placebo in avoiding migraines in adults. According to level B evidence, the American Academy of Neurology stated that feverfew is likely effective in preventing episodes of migraines (Orr 2016).

Petasites hybridus or butterbur belongs to the Asteraceae family has been used traditionally for a variety of ailments. The conventional use of butterbur includes the treatment of migraine and tension headache pain, stomach discomfort, asthma, allergic rhinitis, and gastrointestinal spasms. The pharmacologically active elements of butterbur are referred as petasins, which are categorized as sesquiterpenes, or petasol and angelic acid esters. The anti-inflammatory properties of butterbur are credited, at least partly, to the capacity of petasin to inhibit the production of prostaglandins, leukotrienes, and cyclooxygenase-2 (Orr 2016, Özturan 2016). Petasins display multiple features that render it an attractive choice for the prevention of migraines, such as its capacity to inhibit L-type voltage-gated calcium channels, which have vasodilatory effects. The application of butterbur in the management of migraine pain was made logical because of the role that inflammation performs in pain associated with migraines and the anti-inflammatory characteristics of petasins. Among the different extracts of butterbur, Petadolex® is a widely recognized medicinal product for controlling migraine episodes. Butterbur (Petadolex®), a migraine prophylaxis drug, has shown hopeful results in adults and pediatric subjects, but concerns about its safety are evolving due to its hepatotoxic pyrrolizidine alkaloids. Butterbur extracts offer pain relief in migraine via desensitization of TRPA1 and TRPV1 receptors and reduction of CGRP release from trigeminal afferents. Future research will conclude suitable petasin for blocking CGRP-dependent transmission; discussion will also include its safety and tolerability (Borlak et al. 2022; Orr 2016) (Table 5.2).

5.6 Migraine/Headache and Diet

Diet has a significant role in migraine; therefore, dietary modifications might be advantageous in the prevention and management of migraine [8–10]. Specific foods can trigger migraine attacks. Dietary interventions can influence migraine traits through multiple mechanisms which include neuroinflammation, serotonergic

Table 5.2 Effects of various nutrients on migraine

Sr. no.	Nutrient	Actions/Mechanisms	Advantages/Disadvantage
1	Magnesium	Blockage of NMDA receptor, and calcium channel, improves 5-hydroxytryptamine activity	An affordable option for migraine prevention. The most common side effects are abdominal cramping, vomiting, and diarrhea
2	Coenzyme Q10	Antioxidant, suppression of matrix metalloproteinase-9, and cytokines	Recommended by the Canadian Headache Society and American Academy of Neurology as a migraine preventative agent. Side effects include a burning sensation in the mouth, diarrhea, loss of appetite, and nausea
3	Riboflavin (Vitamin B2)	Improve mitochondrial energy productivity	Limit cardiovascular risk in migraine patients. Good safety and tolerability. Side effects at higher doses include diarrhea, polyuria, and yellow-orange urine
4	Vitamin B3 (Niacin)	Improve serotonin levels and promote mitochondrial energy metabolism	Oral and parenteral administration is possible. Side effects may include diarrhea, headache, and stomach discomfort. Most common side effect is niacin flush
5	Vitamin B12 (Cobalamin)	Nitric oxide scavenger and decrease homocysteine	Reduce the severity and frequency of migraine attacks. Side effects may include headache, nausea, and vomiting
6	Vitamin D	Antioxidant and anti-inflammatory actions. Promote intestinal magnesium absorption	Reduce headache frequency, especially in those with vitamin D deficiency. Side effects include headache, musculoskeletal pain, and osteomalacia
7	Alpha lipoic acid (ALA)	Improves mitochondrial and endothelial functions and clinical symptoms	Decrease migraine severity and frequency. It is usually well-tolerated. The common side effects are headache, heartburn, nausea, and vomiting
8	L-carnitine	Transport of free fatty acids to the mitochondria, control enzyme activity and offer protection against oxidative damage	The side effects include stomach upset, nausea, vomiting, dry mouth, headache, and restlessness It can also cause a fishy odor of the urine, breath, and sweat
9	Feverfew (*Tanacetum parthenium*)	Parthenolide lactone is active ingredient. Parthenolide inhibit prostaglandins, and decreases platelet production	American Academy of Neurology stated feverfew as likely effective in preventing episodes of migraines

(continued)

Table 5.2 (continued)

Sr. no.	Nutrient	Actions/Mechanisms	Advantages/Disadvantage
10	Butterbur (*Petasites hybridus*)	Analgesic and anti-inflammatory effects by inhibition of leukotriene biosynthesis, inhibition of L-type calcium channels, TRPA1 and TRPV1 receptor desensitization	Production of hepatotoxic pyrrolizidine alkaloids

dysfunction, brain mitochondrial function, hypothalamic function, neuronal excitation, leptin, adiponectin, calcitonin gene-related peptide, and NO levels (Soodeh et al. 2019; D'Onofrio et al. 2017). Several diets are proposed to reduce migraine and headache frequency; however, only a few diets have been extensively studied. Many diets known as elimination diets claim to prevent migraine/headaches through the exclusion of certain foods that trigger migraine attacks. However comprehensive diets are believed to prevent headaches or migraine through the inclusion of certain nutrients in the diet (Martin and Vij 2016a). Comprehensive diets involve the maintenance of adequate quantities of certain macro-nutrients (carbohydrates, fats, and proteins) and micronutrients (vitamins and ions). The comprehensive diets include a ketogenic diet (KD), low-fat diet, high-folate diet, modified Atkins, and omega-3 (high)/omega-6 (low) diets (Martin and Vij 2016b). Patients experiencing headaches or migraine reported many food items as dietary triggers. Some foods initiate migraine immediately within an hour after its consumption, whereas others cause headaches after 12 h of food consumption. Dietary triggers include, but are not limited to, chocolate, caffeine, citrus fruits, coffee, tomatoes, onions, milk products, some preservatives, artificial sweeteners, and alcoholic beverages (markedly red wine and beer). Several factors determine the effect of dietary triggers on initiation of migraine such as quantity and timing of triggering food item consumption. Although the identification of these triggers is challenging, an accurate identification of dietary triggers will help in the prevention of migraine attacks in susceptible patients (Zaeem et al. 2016; Martin and Vij 2016a, b; Gazerani 2020). Various diets for migraine prevention and management are proposed based on the connection between dietary migraine triggers and the beneficial effects of specific food components (Table 5.3).

5.6.1 Elimination Diet

The concept of an elimination diet is based upon the exclusion of food components in the diet which are identified as a trigger for migraine or headaches. Identification of such dietary triggers is challenging; however, the personal approach involving monitoring of food items whose consumption is followed by migraine attacks as well as maintenance of food diaries are promising approaches. Long-term maintenance of patients on the elimination diet is subjected to the risk of undernutrition because of inadequate supply of protein, micronutrients, and energy (Gazerani

Table 5.3 Elimination and comprehensive migraine diets

Diet type	Restricted food	Allowed food
Elimination diet	Wheat, soybeans, rye, barley, corn, soybeans, chickpeas, peanuts, citrus fruits, bananas, apples, tomatoes, eggplant, peppers, potatoes, onion, yams, celery, nuts, chocolate, sugar, coffee, tea, alcohol	Oats, rice, buckwheat, amaranth, sorghum, millet, lentils, pears, apricots, blueberries, plums, asparagus, broccoli, cauliflower, brussels sprouts, cabbage, carrots, chard, kale, collards, mustard, spinach, lettuce, zucchini, olive oil, vanilla extract, maple syrup
Ketogenic diet	Dairy products (except butter and some cheeses), meats, fruit, grains, chips, crackers, cookies, ice cream, soda, potatoes, sweet potatoes, beets, sugar	Broccoli, asparagus, mushrooms, avocado, butter, cheese, coconut, eggs, fatty fish, lard, meats, nuts and seeds, and olives
Dietary approaches to stop hypertension (DASH) diet	Tyramine-containing foods (cheeses, processed meats, and fermented foods), caffeine, alcohols, MSG, aspartame, and certain food colorants	Lean proteins (chicken, Turkey, fish, eggs), brown rice, quinoa, whole wheat bread, skim milk, low-fat yogurt
Mediterranean diet	Cheeses, chocolate, oranges, lemon, onions and garlic, peanuts	Olive oil, nuts, and fish like salmon, green leafy vegetables, legumes, nuts, seeds, and whole grains, blueberries
Modified Atkins	High-sugar foods and beverages	Olive oil, avocado slices, cherry tomatoes, cucumber, vinegar, peanut butter, salmon, fatty fish, strawberries, blueberries, almond milk, coconut milk. Poultry, red meat, eggs, cheese, cream, mayonnaise, oils. Almonds, and nuts
Weight-loss diets	Processed meats, caffeine, alcohol, tomatoes, citrus fruits, dairy, salted, fish	Fruits, vegetables, whole grains, lean proteins, omega-3 fatty acids (fish, seeds, and legumes)
Low-sodium diets	Tomatoes, mushrooms, caffeine, chocolate, oranges, lemons, limes, grapefruits, aged cheeses, cured meats, smoked fish, yeast, sausages, nitrates and nitrites monosodium glutamate, nuts and seeds, alcohol	Dark leafy greens, nuts, seeds, whole grain, bananas, avocados, potatoes
Low-fat diets	Processed meat, MSG-containing food, butter, palm oil, chocolate, nuts, seeds, avocado, and caffeine	Seafood, fresh fruits, vegetables, whole grains, lean proteins, ketogenic diet, omega 3 fatty acid, Atkins diet
Gluten-free diet	Barley, rye, oats, bread and baked goods, pasta, bread and baked goods, beer, soy sauce	Chicken, fish, eggs, grains, rice, buckwheat, dairy alternatives almond milk, coconut milk, olive oil, avocado, and nuts
High-folate diet	Oranges, lemons, grapefruits, broccoli	Folic acid, vitamin B6 and B12 spinach, lettuce, avocado

2020). Several food items are considered as migraine triggers. Alcoholic beverages especially red wine is an important and common trigger in many migraine patients. There is a complex liaison between various headaches and caffeine. Recent hypotheses suggest that red wine is rich in phenolics. The main component of red wine, quercetin-3-glucuronide, inhibits mitochondrial aldehyde dehydrogenase, causing a rise in acetaldehyde levels and headaches in susceptible individuals (Devi et al. 2023). Caffeine has been historically known to be able to trigger or cure migraines. Caffeine is a valuable adjuvant remedy for acute headaches, like migraine. Moreover, studies have revealed that caffeine overuse relates to medication overuse on headaches and chronic migraines. In a crossover study, sudden cessation of caffeine triggered migraines (Moskatel and Zhang 2022; Scher et al. 2004).

5.6.2 Epigenetic Diet

The epigenetic diet emphasizes that specific components of diet through mechanisms have the potential to alter disease pathology. The targets for epigenetic diet are either molecules like DNA or certain cellular structures like mitochondria. In the case of migraine, epigenetic diet intervention is proposed to inhibit the migraine pathogenesis and/or promote migraine inhibition (Gazerani 2020). Recent epigenetic studies on migraine linked an unusual methylation of DNA with the migraine. Methylation of several genes could be associated with the development of migraine. These genes include, but are not limited to, zinc finger protein 234, SH2 domain containing 5, diacylglycerol kinase gamma, suppressor of cytokine signaling 1, and so forth (Mehta et al. 2023). The dietary component folate is known for its favorable effects on migraine. Folate is also implicated in DNA methylation which has set an impetus for the role of epigenetic diet in migraine. It is hypothesized that a diet targeting DNA methylation like a folate-rich diet could be beneficial for the management of migraine. However, further research is warranted to sustain the role of epigenetic diets in the prevention and management of migraine in susceptible patients. Epigenetic profiling of migraine patients will be beneficial to define the appropriate epigenetic diet for migraine. Alterations of cellular proteins and RNA products via histone modification or noncoding RNA actions could be the additional mechanisms of action of certain components of an epigenetic diet (Gazerani 2020). The use of an epigenetic diet in the prevention and treatment of diseases like migraine is linked to epigenetic profile alterations is rather unclear as it is challenging to forecast the final effect of diet due to its little specificity to the epigenome and an abundance of interactions among the active ingredients of diet (Fila et al. 2019).

5.6.3 Ketogenic Diet

Ketogenic diet (KD) have been increasingly shown to be beneficial for preventing migraines (Tereshko et al. 2023). Many preclinical and clinical studies reinforce the significance of the KD as a preventive treatment for the migraine (Valente et al.

2022). In various neurological diseases, ketogenic diets have therapeutic implications, affecting neuroplasticity and exercising neuroprotective effects. KD enhances brain excitability and metabolism, modifying neuronal inflammation via the formation of ketone bodies. A KD demonstrated its usefulness against cluster headaches and drug resistant migraine (Tereshko et al. 2023). A KD combines a robust restriction of carbohydrates with boosted protein and lipid intake (Moskatel and Zhang 2022). There are many KDs suggested for migraine treatment, which limit carbohydrate intake but offer adequate protein consumption, with deviations based on the lipid-to-carbohydrate-protein ratio. These diets include classical KD, low glycemic index diet, low-calorie KD, very low-calorie KD and Modified Atkins Diet (Tereshko et al. 2023). It is reasoned that the efficacy of the KD on migraine might extend above its potential applications for weight loss, as migraine is linked with weight gain (Valente et al. 2022). The KD is a fasting-like protocol that promotes the generation of ketone bodies. Carbohydrate restriction lowers insulin secretion and improves glucagon secretion. It also mobilizes fatty acids from adipose tissue to the liver for β-oxidation and formation of ketone bodies like acetone, acetoacetate, and β-hydroxybutyric acid (Valente et al. 2022). These ketone bodies are then circulated to tissues like the brain, heart, and skeletal muscle (Tereshko et al. 2023). Despite the need for wider investigations, the KD is a promising and accepted intervention for migraine prevention and management (Moskatel and Zhang 2022).

5.6.4 Gluten-Free Diet

When wheat flour is washed with water, gluten is left behind as a sticky mass (Ensor 2019). The link between headaches, particularly migraine, and gluten is now established in celiac disease patients (Beuthin et al. 2020). This diet prevents items with gluten like barley, rye, wheat, and malt. Rather, it emphasizes gluten-free alternatives including rice, quinoa, corn, and potatoes, alongside naturally gluten-free items such as vegetables, fruits, seafood, nuts, meat, legumes, and milk products. Within a few days to weeks of following a gluten-free diet, gastrointestinal symptoms in patients with celiac disease may significantly subside (Moskatel and Zhang 2022).

5.6.5 Fatty Acids Diet

A diet high in omega-3 fatty acids could decrease the incidence and intensity of migraines, according to a randomized clinical study (Burch 2021). Current industrialized diets usually have inadequate amounts of omega-3 fatty acids and high levels of omega-6 fatty acids; both of which are precursors to oxylipins, which are lipid mediators that control pain and are linked to migraine pathology. Omega-3 oxylipins decrease pain, whereas omega-6 oxylipins enhance pain and may trigger migraines (Slomski 2021). The Mediterranean diet includes nutrients that are vital to human health. Mediterranean diet consists of antioxidants, monounsaturated fats, and a balanced ratio of omega-6 to omega-3 fatty acids (Marchetti et al. 2022). Recent studies have investigated whether modifying fatty acid intake may benefit migraine patients.

5.6.6 DASH/Mediterranean Diet/MIND

The Dietary Approaches to Stop Hypertension (DASH) diet, which initially emerged to treat hypertension, builds a focus on vegetables, fruits, and whole grains while lowering sugar, sodium, and foods with elevated saturated fat (Moskatel and Zhang 2022). Higher quantities of fiber and minerals, like calcium, magnesium, and potassium, are found in the DASH diet. The anti-inflammatory benefits of DASH may be supported by these food and dietary properties. Inflammatory cytokines are suppressed by high-fiber diets since they delay glucose absorption and modify gut microflora. The 1, 25-dihydroxycholecalciferol and the angiotensin-converting enzyme are inhibited by calcium present in dairy items, which could influence inflammation (Mirzababaei et al. 2020). Additionally, previous studies indicate an association between some comorbidities and migraine, which notably increased blood pressure. As a result, a blood pressure-lowering diet could enhance health outcomes in migraine patients. DASH diets were initially developed to alleviate blood pressure in patients with hypertension. Emerging research shows that DASH diets could alleviate symptoms of migraines (Arab et al. 2022). The Mediterranean diet represents one of the most beneficial dietary patterns for preventing and reducing neuroinflammation (Marchetti et al. 2022). Identical to the DASH diet, the Mediterranean diet prioritizes whole grains, vegetables, beans, fruits, nuts, olive oil, and low protein from animals. Mediterranean diet in combination with DASH referred as the Mediterranean-DASH Intervention for Neurodegenerative Delay (MIND) could benefit patients from migraine attacks (Moskatel and Zhang 2022).

References

Aggarwal M, Puri V, Puri S (2012) Serotonin and CGRP in migraine. Ann Neurosci 19:88–94

Akerman S, Goadsby PJ (2007) Dopamine and migraine: biology and clinical implications. Cephalalgia 27:1308–1314

Andreou AP, Edvinsson L (2019) Mechanisms of migraine as a chronic evolutive condition. J Headache Pain 20:117

Arab A, Khorvash F, Kazemi M, Heidari Z, Askari G (2022) Effects of the dietary approaches to stop hypertension (DASH) diet on clinical, quality of life and mental health outcomes in women with migraine: a randomised controlled trial. Br J Nutr 128:1535–1544

Bertels Z, Pradhan AAA (2019) Emerging treatment targets for migraine and other headaches. Headache 59(Suppl 2):50–65

Beuthin J, Veronesi M, Grosberg B, Evans RW (2020) Gluten-free diet and migraine. Headache 60:2526–2529

Borlak J, Diener HC, Kleeberg-Hartmann J, Messlinger K, Silberstein S (2022) Petasites for migraine prevention: new data on mode of action, pharmacology and safety. A narrative review. Front Neurol 13:864689

Burch RJB (2021) Dietary omega 3 fatty acids for migraine. British Medical Journal Publishing Group. 374:n1535
Burstein R, Noseda R, Borsook D (2015) Migraine: multiple processes, complex pathophysiology. J Neurosci 35:6619–6629
Buse DC, Greisman JD, Baigi K, Lipton RB (2019) Migraine progression: a systematic review. Headache 59:306–338
Cavestro C, Bedogni G, Molinari F, Mandrino S, Rota E, Frigeri MC (2018) Alpha-lipoic acid shows promise to improve migraine in patients with insulin resistance: a 6-month exploratory study. J Med Food 21:269–273
Charles A (2018) The pathophysiology of migraine: implications for clinical management. Lancet Neurol 17:174–182
Charleston LT, Khalil S, Young WB (2021) Carnitine responsive migraine headache syndrome: case report and review of the literature. Curr Pain Headache Rep 25:26
D'Onofrio F, Raimo S, Spitaleri D, Casucci G, Bussone G (2017) Usefulness of nutraceuticals in migraine prophylaxis. Neurol Sci 38:117–120
Dahri M, Hashemilar M, Asghari-Jafarabadi M, Tarighat-Esfanjani A (2017) Efficacy of coenzyme Q10 for the prevention of migraine in women: a randomized, double-blind, placebo-controlled study. Eur J Integr Med 16:8–14
Dell'Isola GB, Tulli E, Sica R, Vinti V, Mencaroni E, Di Cara G, Striano P, Verrotti A (2021) The vitamin D role in preventing primary headache in adult and pediatric population. J Clin Med 10:5983
Devi A, Levin M, Waterhouse AL (2023) Inhibition of ALDH2 by quercetin glucuronide suggests a new hypothesis to explain red wine headaches. Sci Rep 13:19503
Domitrz I, Cegielska J (2022) Magnesium as an important factor in the pathogenesis and treatment of migraine—from theory to practice. Nutrients 14:1089
Ensor KR (2019) Efficacy of a gluten free diet to help relieve migraines in chronic migraine patients: a case series. Mississippi State University
Fila M, Chojnacki C, Chojnacki J, Blasiak J (2019) Is an "epigenetic diet" for migraines justified? the case of folate and DNA methylation. Nutrients 11:2763
Gazerani P (2020) Migraine and diet. Nutrients 12:1658
Ghorbani Z, Togha M, Rafiee P, Ahmadi ZS, Rasekh Magham R, Haghighi S, Razeghi Jahromi S, Mahmoudi M (2019) Vitamin D in migraine headache: a comprehensive review on literature. Neurol Sci 40:2459–2477
Goadsby PJ, Holland PR, Martins-Oliveira M, Hoffmann J, Schankin C, Akerman S (2017) Pathophysiology of migraine: a disorder of sensory processing. Physiol Rev 97:553–622
Hajihashemi P, Askari G, Khorvash F, Reza Maracy M, Nourian M (2019) The effects of concurrent coenzyme Q10, L-carnitine supplementation in migraine prophylaxis: a randomized, placebo-controlled, double-blind trial. Cephalalgia 39:648–654
Kelishadi MR, Naeini AA, Khorvash F, Askari G, Heidari Z (2022) The beneficial effect of alpha-lipoic acid supplementation as a potential adjunct treatment in episodic migraines. Sci Rep 12:271
Liampas IN, Siokas V, Aloizou AM, Tsouris Z, Dastamani M, Aslanidou P, Brotis A, Dardiotis E (2020) Pyridoxine, folate and cobalamin for migraine: a systematic review. Acta Neurol Scand 142:108–120
Liu H, Wang L, Chen C, Dong Z, Yu S (2022) Association between dietary niacin intake and migraine among American adults: National Health and Nutrition Examination Survey. Nutrients 14:3052
Marchetti M, Gualtieri P, De Lorenzo A, Trombetta D, Smeriglio A, Ingegneri M, Cianci R, Frank G, Schifano G, Bigioni G, Di Renzo L (2022) Dietary omega-3 intake for the treatment of morning headache: a randomized controlled trial. Front Neurol 13:987958
Martin VT, Vij B (2016a) Diet and headache: part 1. Headache 56:1543–1552
Martin VT, Vij B (2016b) Diet and headache: part 2. Headache 56:1553–1562
Mehta D, De Boer I, Sutherland HG, Pijpers JA, Bron C, Bainomugisa C, Haupt LM, Van Den Maagdenberg A, Griffiths LR, Nyholt DR, Terwindt GM (2023) Alterations in DNA methyla-

tion associate with reduced migraine and headache days after medication withdrawal treatment in chronic migraine patients: a longitudinal study. Clin Epigenetics 15:190

Mirzababaei A, Khorsha F, Togha M, Yekaninejad MS, Okhovat AA, Mirzaei K (2020) Associations between adherence to dietary approaches to stop hypertension (DASH) diet and migraine headache severity and duration among women. Nutr Neurosci 23:335–342

Moskatel LS, Zhang N (2022) Migraine and diet: updates in understanding. Curr Neurol Neurosci Rep 22:327–334

Nattagh-Eshtivani E, Sani MA, Dahri M, Ghalichi F, Ghavami A, Arjang P, Tarighat-Esfanjani A (2018) The role of nutrients in the pathogenesis and treatment of migraine headaches: review. Biomed Pharmacother 102:317–325

Nyholt DR, Borsook D, Griffiths LR (2017) Migrainomics—identifying brain and genetic markers of migraine. Nat Rev Neurol 13:725–741

Olesen J (2018) Headache classification committee of the international headache society (IHS) the international classification of headache disorders. Cephalalgia 38:1–211

Orr SL (2016) Diet and nutraceutical interventions for headache management: a review of the evidence. Cephalalgia 36:1112–1133

Özturan A, Şanlier N, Coşkun Ö (2016) The relationship between migraine and nutrition. Turk J Neurol 22:044–050

Puledda F, Goadsby PJ (2017) An update on non-pharmacological neuromodulation for the acute and preventive treatment of migraine. Headache 57:685–691

Robblee J, Starling AJ (2019) SEEDS for success: lifestyle management in migraine. Cleve Clin J Med 86:741–749

Savi L (2023) Magnesium and migraine. In Treatments, nutraceuticals, supplements, and herbal medicine in neurological disorders. (pp. 421–439). Academic Press.

Scher AI, Stewart WF, Lipton RB (2004) Caffeine as a risk factor for chronic daily headache: a population-based study. Neurology 63:2022–2027

Shoeibi A, Olfati N, Soltani Sabi M, Salehi M, Mali S, Akbari Oryani M (2017) Effectiveness of coenzyme Q10 in prophylactic treatment of migraine headache: an open-label, add-on, controlled trial. Acta Neurol Belg 117:103–109

Slomski AJJ (2021) Diets high in omega-3 fatty acids might ease migraines. JAMA 326:691–691

Soodeh RJ, Zeinab G, Paolo M, Lampl C, Toga M (2019) Association of diet and headache. J Headache Pain 20:106

Sun-Edelstein C, Mauskop A (2009) Foods and supplements in the management of migraine headaches. Clin J Pain 25:446–452

Tereshko Y, Dal Bello S, Di Lorenzo C, Pez S, Pittino A, Sartor R, Filippi F, Lettieri C, Belgrado E, Garbo R, Merlino G, Gigli GL, Valente M (2023) 2:1 ketogenic diet and low-glycemic-index diet for the treatment of chronic and episodic migraine: a single-center real-life retrospective study. J Headache Pain 24:95

Urits I, Yilmaz M, Bahrun E, Merley C, Scoon L, Lassiter G, An D, Orhurhu V, Kaye AD, Viswanath O (2020) Utilization of B12 for the treatment of chronic migraine. Best Pract Res Clin Anaesthesiol 34:479–491

Valente M, Garbo R, Filippi F, Antonutti A, Ceccarini V, Tereshko Y, Di Lorenzo C, Gigli GL (2022) Migraine prevention through ketogenic diet: more than body mass composition changes. J Clin Med 11:4946

Vikelis M, Spingos KC, Rapoport AM (2018) A new era in headache treatment. Neurol Sci 39:47–58

Yablon LA, Mauskop A (2011) Magnesium in headache. In: Magnesium in the Central Nervous System. University of Adelaide Press, Adelaide (AU); 2011. Available from https://www.ncbi.nlm.nih.gov/books/NBK507271. PMID: 29920023.

Zaeem Z, Zhou L, Dilli E (2016) Headaches: a review of the role of dietary factors. Curr Neurol Neurosci Rep 16:101

Microbial Players: Unveiling the Role of Gut Microbiome in Migraine

6

Priyanka Pandey and Vinod Tiwari

Questions to Be Answered

1. What are the key pathways of migraine pain?
2. How has been the connection between gut and migraine evolved through the times?
3. How do the metabolites factors, gut microbes, and gut dysbiosis served in precipitation of migraine pain?
4. What are the potential neurotransmitters involved in the migraine pathophysiology?
5. What are the key mechanisms involved in histamine, serotonin, sphingomyelin pathways, and vasoactive neuropeptides involved in migraine pain?
6. What are the treatment options available in migraine management? What are their serving properties in the gut homeostasis and migraine treatment?
7. Is diet really effective in gut homeostasis and related migraine management?

6.1 Migraine Pain: What We Know

Migraine is a prevailing neurological disorder that causes palpitating pain at the partial side of the brain. The pathomechanism of trigeminal system plays a crucial role in triggering the nausea, vomiting, and extreme sensitivity to the light. Migraine symptoms can last up to a few hours to multiple days (Spekker and Nagy-Grócz 2023). This headache disorder is more prevalent in women (17%) in comparison to men (5–8%) with global burden of third most common neurological condition

P. Pandey · V. Tiwari (✉)
Department of Pharmaceutical Engineering and Technology, Indian Institute of Technology (B.H.U), Varanasi, Uttar Pradesh, India
e-mail: vtiwari.phe@itbhu.ac.in

S. Bhatt et al. (eds.), *Management of Migraine Pain*,
https://doi.org/10.1007/978-981-97-4529-6_6

(Cámara-Lemarroy et al. 2016). Although there are multiple factors involved in the migraine including pathological (hormones, neurotransmitters, immunogenic, and metabolite factors), epigenetic (genetic alterations), and environmental factors, the exact mechanism behind this is not clear. Recently, it has been discovered that the gastrointestinal (GI) tract is a major component for the migraine pain, which alterations can lead to Helicobacter pylori infection, celiac disease, gastroparesis, irritable bowel syndrome, hepatobiliary disorders, and gut-microbiota alterations (Cámara-Lemarroy et al. 2016). Gut-microbiota alterations among them have been prominently linked to both episodic and chronic migraine condition (Kim et al. 2023; Yong et al. 2023). Pathological mechanisms involving microbiota modulations lead to immunological milieu alterations, which may further incorporate dysfunction of the autonomic and enteric nervous system (Cámara-Lemarroy et al. 2016). One of the mechanisms behind the gut-related migraine is earlier antibiotic therapy, which may lead to the dysbiosis condition. It alters the intestinal components, which can lead to metabolic alterations that may be responsible for migraine (Kappéter et al. 2023). According to the report of mSystems in 2016, people with the symptoms of migraine were found to have more bacteria in their mouth, which produce the nitric oxide gas, a leading cause for migraine pain.

6.2 Understanding the Gut–Microbiota–Migraine Interaction

Gut–brain axis dysfunction indicates towards the bidirectional pathways and connects the GI tract with central nervous systems. Gut microflora microbes in the GI tract can affect the central physiological mechanisms which may alter the behavioral and cognitive functions including migraine (Yong et al. 2023). The gut–brain axis constitutes for the bidirectional communication through the key interaction of vagus nerves. Vagus nerve facilitates the swift response to gastrointestinal homeostasis with various physiological role such as sensory and motor functions (Chandra et al. 2023; Akagi et al. 2024). One of the mechanisms behind this may be vagus nerve carrying the change of brain physiology due to elimination of the certain type of bacteria from the intestinal gut flora (Suzuki et al. 2024; Qu et al. 2023; Steidel et al. 2021). The potential implications between the gut microflora and vagus nerve may further affect the neurological functions such as behavioral and cognition complexity such as occurrence of migraine (Crawford et al. 2022; Pekary and Sattin 2022). Apart from this sensory and motor responsiveness, vagal nerve goes in for multiple associations such as neurotransmitters pathway and humoral pathways involved in migraine pathogenesis. The bidirectional complexion of the vagal nerve permits for the signal transmission of gut-microbiota to the brain and further back to the brain to the intestinal microflora (Hassan et al. 2023; Stasiłowicz et al. 2021). The interconnectivity signal systems have been less traveled yet; however, some community of microorganism including bacteria, virus, and fungi precipitation came into the picture till date. The microbes through bidirectional communication also influence at the cardiovascular systems, neurological systems, endocrine systems, immunological systems, and metabolic systems. Most of the ongoing research studies have been aimed for the development of the strategies which could provide the desired therapeutic purposes.

6.3 Risk Factors in Migraine Pain

1. *Metabolite-related risk factor*: Gut–brain axis influences the gastroenteric micro-environment through intestinal movement and mucosal immunity. Vagus nerve plays a crucial role in the intestinal movement through brain functionality. Severe comorbidities such as anxiety and depression express positive correlation with the intestinal alterations and migraine pain with the 15% of the population (Kang et al. 2022; Georgescu et al. 2017; Dai et al. 2017). Such risk factors facilitate the sodium butyrate and sodium propionate-based intestinal permeability with the alterations in the immune system. Formation of the sodium butyrate and sodium propionate is a result of the fiber fermentation, which may further lead to the gut barrier integrity, glucose and lipid metabolism, and fatty acid oxidation (Nogal et al. 2021; Lanza et al. 2022). Similarly, metabolite modules yield enrichment in the g-aminobutyric acid (GABA) synthesis and kynurenine degradation, and they further create differences in the gut microbiota composition than healthy group (Chen et al. 2020). Similarly, inflammatory soup-induced migraine decreases the lactobacillus with elevation in the gut microbiota metabolite including propanoate, butanoate, tryptophan, and related metabolites, indole-3-acetamide, and tryptophol (Miao et al. 2022). Likewise, tryptophan metabolites and microbiome products such as short chain fatty acids or peptidoglycan also communicate in this regard. Further, the downstream consequences of the gut alterations are the downregulation in the glucose transporters (GLUT1) at the brain region, which consequently results in the cerebral hypometabolism (Diether et al. 2023; Zhou et al. 2022; Eastwood et al. 2023). The glucose-related hypometabolism is linked with the energy deficit due to mitochondrial related oxidative stress. Melatonin is known for its active transportation into the mitochondria through organic anion transporter-3 (OAT-3) and oligopeptide transporter (PEPT)1/2 (Anderson 2019). The proinflammatory TNF-a elevates the phosphorylation of α-amino-3-hydroxy5-methyl-4-isoxazolepropionic acid (AMPA) subtypes of glutamate receptor in the trigeminal nucleus caudalis in the intraspinal region. GLUT1-related melatonin are responsible for the migraine precipitation (Zhang et al. 2022; Tang et al. 2018; Wu et al. 2017; Nakaya et al. 2016). Due to metabolic decrement in the GLUT1, and related reduction in the mitochondrial functioning, melatonin level also gets reduced. These mechanism-related energy crises in the cortex further underpin the migraine attacks (particularly migraine with aura). Moreover, GLUT1 is evidently expressed in the satellite glial and nociceptive ganglia cells; thus, in response to migraine, it may further spread the pain through different brain centers (Guo et al. 2023; Tham et al. 2020). The energy-based nociception is particularly related to the dura mater and trigeminal ganglia of the brain cells; thus, the primary widespread of

nociception are found in this area only. In response to the widespread pain, it can regulate the satellite-related calcium gene-related peptide (CGRP) (Figs. 6.1 and 6.2) (Anderson 2019).

2. *Gastrointestinal microbiomes-related risk factor*: Gastrointestinal area influences majorly in gut dysbiosis. Some gastrointestinal diseases such as irritable bowel syndrome, Helicobacter pylori infection, gastroparesis, celiac disease, and hepatobiliary disorder are prominently involved in the vasoactive circulatory system as well as chronic response. They modulate the intestinal microbiota of the enteric immunologic system (Arzani et al. 2020; Cámara-Lemarroy et al. 2016). Gastroparesis and functional dyspepsia are linked with gastric stasis and associated nausea and vomiting, which are prime factors for migraine attacks. GI comorbidities are directly associated with epilepsy, and migraine attacks occur in between the interictal period (Aurora et al. 2021). Interestingly, single nucleotide polymorphism data also suggest some microbes' role in migraine pain. These microbes include *Coprococcus*, *Anaerotruncus*, Bifidobacteriaceae, and Bifidobacteriales (He et al. 2023). 16S ribosomal RNA-based epigenesis is considered an effective tool for taxonomical classification. It contains nine variable regions (V1–V9), separated by highly conserved sequences (Matsuo et al. 2021). In case of migraine, V4–V5 regions provide relevant coverage upon the alterations of gut-archaeal and gut-bacterium domain including proteobacteria, firmicutes, and Bacteroidetes (Bai et al. 2023). Similarly, V3–V4 regions target the gut-microbiome alterations including PAC000195 (Yong et al. 2023). The oral microbiota is directly correlated with migraine with the heart failure and angina pectoris. Oral bacteria breakdown nitrates converting it to the nitric oxide precipitating migraine pain in 80% of the patient (Jiang et al. 2021). Moreover, patient with ulcerative colitis and inflammatory bowel disease is significantly associated with intestinal diamine oxidase (DAO). Mechanism behind deficiency in DAO is associated with lactose malabsorption. Similarly, Celiac disease and inflammatory bowel diseases (IBD) are also correlated with extraintestinal-induced migraine; among them, IBD are the major risk factors (Gabrielli et al. 2021). Patient with consumptions of TCA, SSRI, NSAIDS, 5-ASA, and azathioprine have higher risk of IBD-related migraine (Chehel Cheraghi et al. 2016).
3. *Gut dysbiosis and migraine pain*: It is noteworthy that gut dysbiosis enhance the gut permeability, through pro-inflammatory-mediated cytokines such as high-sensitivity C-reactive protein (CRP), IL-6, IL-8 IL-1b, and TNF-a, which can further lead the activation of hypothalamic–pituitary–adrenal axis (Vajdi et al. 2023; Sasso et al. 2023). Gut permeability and gut dysbiosis trigger systemic inflammation through mitochondrial-induced oxidative stress due to nonselective gut-permeability. The mechanism may be intestinal elevation in the high mobility group box 1 (HMGB1) along with the same effect in the cortex, the central sensitization further precipitates in the migraine with aura. The migraine nociception in this mechanism, follow the microglial related hypertrophy pathways. It further downstream the toll like receptor (subtypes 2/4) and related lipopolysaccharide at gut barrier lining. The early development etiology suggests

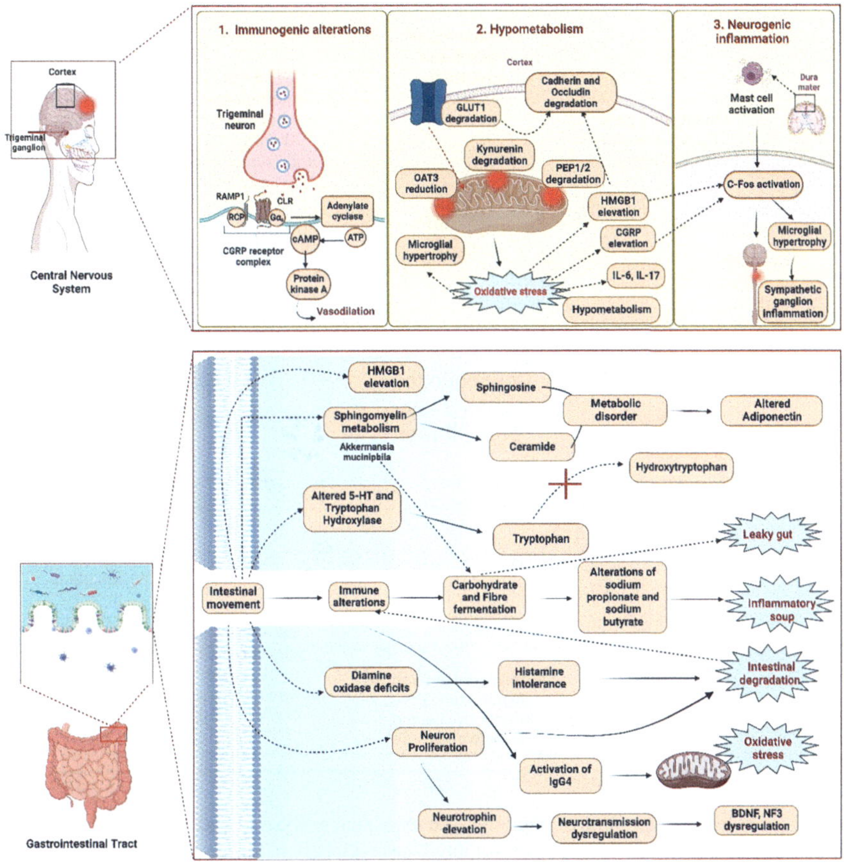

Fig. 6.1 The relationship between gut and brain alterations in migraine: The interplay in between gut and brain can lead to heighten the metabolic and immunogenic modifications. Gut dysbiosis can cause the neurotransmission dysregulations, which subsequently can alter the other homeostatic factors. For example, it can elevate the neutrophins through neuron proliferations, which further can influence the expression of brain derived neurotropic factors (BDNF) and nuclear factor kappa-light-chain enhancer of activated B cells (NF-κβ). Furthermore, diamine oxidase deficiency can lead to the histamine intolerance, further discomposing the intestinal microbiome. The fermentation process of carbohydrates and fibres produce the resultant sodium propionate and sodium butyrate and subsequently inflammatory milieu generations take place. Interestingly, 5-HT (serotonin) and sphingomyelin metabolism disrupt the tryptophan metabolism and adiponectin precipitation respectively. These alterations enhance the excitability and precipitation the inflammasome, for example, HMGB1 (high mobility group box1) in the gastrointestinal tract. These process underscores the complexity of gut–brain axis and influence the spinal and supraspinal pathways. In the brain, they trigger the vasodilation and neurogenic inflammation through calcitonin gene-related peptide (CGRP) alterations and mast cell activations respectively. These intricacies also impede the energy regulation through mitochondrial interruptions. Some of the hypometabolism markers are organic anion transporters 3 (OAT3) reductions, phosphoenolpyruvate (PEP1/2), and kynurenine degradations and consequential inflammasome markers such as altered HMGB1, CGRP, and proinflammatory cytokines (interleukin-6 and interleukin-17), and microglial hypertrophy

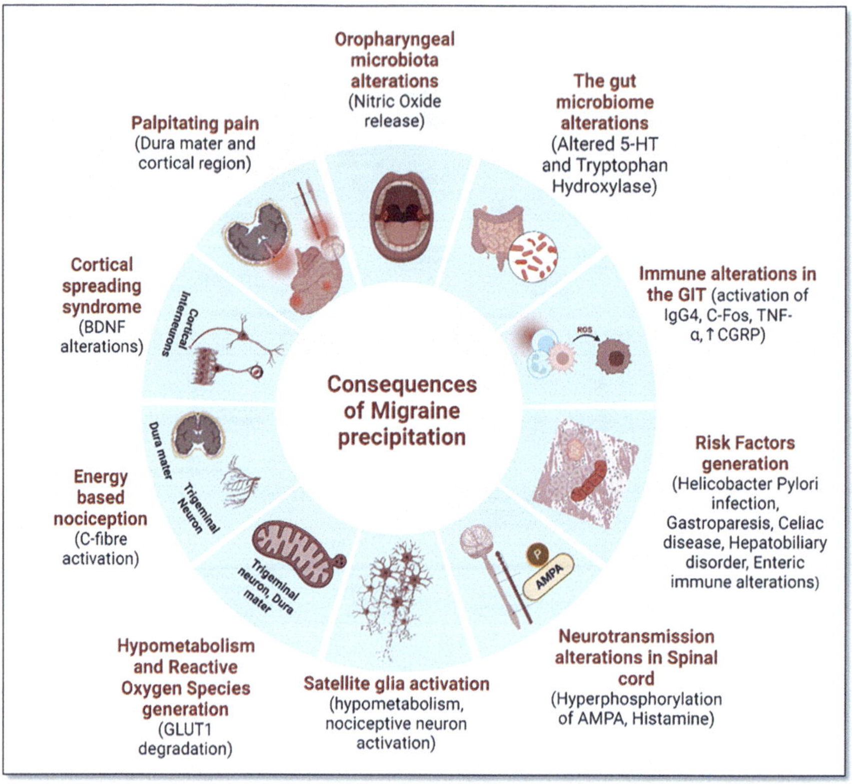

Fig. 6.2 Step by step algorithm of migraine: The sequential analysis of gastrointestinal symptoms and comorbidities explains the pathways for migraine precipitation. It is created based on different clinical and preclinical experimentation, such as 5-HT (5-hydroxytryptamine), IgG4 (immunoglobulin G4), TNF-a (tumor necrosis factor-a), CGRP (calcitonin gene related peptide), AMPA (α-amino-3-hydroxy-5-methyl-4-isoxazolepropionic acid), GLUT1 (glucose transporter 1, BDNF (brain-derived neurotropic factor)

their linkage through the gut–amygdala–cortex region, which also suggests towards the inflammatory bowel disease and endometriosis risk factor (follow section B) (Anderson 2019). Patient with consumptions of TCA, SSRI, NSAIDS, 5-ASA, and azathioprine have higher risk of IBD-related migraine (Chehel Cheraghi et al. 2016). Particularly, NSAIDS, for example, Piroxicam, regulate for the leaky gut through negatively regulation of the lipopolysaccharide binding protein levels (LBP), which further alter the adherence junction protein-cadherin, tight junction protein- occludin, CGRP, high mobility group box-1 (HMGB1), and IL-6 & IL-17 levels (Dağıdır et al. 2023). Enhancement in the TNF-a is responsible for the microbiota dysbiosis-induced migraine like pain (Figs. 6.1 and 6.2).

6.4 Biochemical Pathways

1. *Sphingomyelin pathway*: The sphingolipid pathways are the structural component of gut epithelial cells, where they maintain the cellular integrity at the gastrointestinal tract. It balances signaling at cellular membrane which is important for the gut lumen protection, which alterations led to the leaky gut (Johnson et al. 2020; Rohrhofer et al. 2021; Brown et al. 2019; Hasegawa et al. 2017). Sphingomyelin metabolism is majorly responsible for the neurovascular-mediated inflammatory processes (Nixon 2009; Maceyka and Spiegel 2014). Some gut microflora component such as Bacteroides, Prevotella, Clostridium, Bifidobacterium (Probiotics beneficial for gut epithelial), Akkermansia muciniphila act as sphingolipid metabolizing enzymes and regulate the breakdown the complex dietary fibers and carbohydrates degradation, which further results into the leaky gut (Fujisaka et al. 2023; Shim et al. 2023). This intracellular calcium alterations led to neurotransmitter-mediated neuronal excitability. Antibiotic-induced community dysbiosis also leads to disruption in the 5-HT (serotonin), tryptophan hydroxylase, and metabolite enzymes such as bile acids (Nan et al. 2023; Sun et al. 2018; MacEachern et al. 2018). This hypersensitivity further activates the inflammatory oxidative stress mediated through mitochondrial dysfunctioning. All these pathways are cumulatively responsible for migraine precipitation (Kilinc et al. 2023). Metabolism of sphingomyelin is responsible for the gut-microbiota alterations. The enzymatic cleavage of the Sphingomyelin into sphingosine and ceramide and produce the short chain fatty acids, including acetate, butyrate, and propionate (Parada Venegas et al. 2019), Hannun and Obeid 2018). These short chain fatty acids influence the immunity of the gut and negatively affect the mucosal barrier into the gut. The immunomodulation further is responsible for the inflammation, lipid metabolism, insulin sensitivity, and migraine precipitation. Similarly, the phosphorylation of the Sphingolipid lead to the metabolic disorder through the alterations in the adiponectin level. Both episodic and chronic migraine are getting affected by these alterations (Figs. 6.1 and 6.2) (Peterlin et al. 2015).
2. *Neurotransmitters pathways*: In the migraine pathophysiology, neurotransmitters play a crucial role in the gut–brain bidirectional communication. Neurotransmission alterations are also one of the major cues, which have been implicated in the gut–brain induced migraine processes. Gastrointestinal symptoms are directly associated with the serotonin, noradrenaline, dopamine, calcitonin, glutamate, histamine, acetylcholine, substance P and neurotensin alterations, which later on precipitates with migraine like consequences in the brain.
   (a) *Serotonin*: The major proportions of the serotonin (95%) are found in the intestinal lining particularly in the enterochromaffin cells. Along with the transmission of hormonal, endocrine, autocrine, and paracrine system, it is also recognized for its intestinal motility and inflammation (Terry and

Margolis 2017). These intricacies further play with neurotransmitters precursors, such as tryptophan (precursor for serotonin) alterations in gut-microflora is directly associated with migraine precipitation. Alterations in the enterochromaffin serotonin are associated with irritable bowel syndrome, a common coexistence factor for migraine. Multiple microbiota-related dysbiosis and intestinal alterations are responsible for the altered serotonin pathway (Kappéter et al. 2023). Gut serotonin is influenced by the different routine factors such as diet and stress, anxiety, nausea, and vomiting which potentially impact on the gut and associated central factors, such as migraine. Interestingly, lower serotonin exerts vasodilation through its receptor 5-HT1B & 5-HT1D in the blood, which is a positive factor for the migraine progression (Liu et al. 2024; Prakash et al. 2024). It influences at the cortical spreading depression, which involves the neuronal depolarization at the brain cortex region. Cortex depolarization is responsible for the initiation of the migraine aura. Decrement in the 5-HT particularly 5-HT3 and 5-HT4 signaling pathway enhance the GI motility and associated alterations in the mucosal permeability and related immune activation. N-acetyl serotonin One of the mechanisms may be alteration of Aryl hydrocarbon receptor activity, which can disrupt the balance between N-acetyl serotonin (NAS) and melatonin synthesis. The imbalance further may impair the mitochondrial GLUT1 level and potentially contribute in the neurological pathway disruption, possibly results into the migraine (Anderson 2019). The consequences are related to the visceral hypersensitivity and inflammations in the intestinal linings and presynaptic positions in the sensory and mesenteric neurons. 5-HT enhances the peristaltic reflexes in the GI tract, which further regulates the ascending and descending contraction in the limbic and segmental motor portions in the small intestinal regions (Lucarini et al. 2020; Ceuleers et al. 2016; Nezami and Srinivasan 2010). Tryptophan hydroxylase is important for conversion of the tryptophan to hydroxytryptophan, which imbalance is also an important trigger factor for migraine. Additionally, tryptophan deficiency is responsible for the immature dopamine and related altered gastrointestinal moment (Figs. 6.1 and 6.2) (Roth W et al. 2021).

(b) *Histamine*: Histamine regulates the intestinal homeostasis through enzyme diamine oxidase (DAO). DAO deficits lead to the histamine intolerance and associated intestinal degradations. Further, these deficits are directly linked with onset of migraine (Izquierdo-Casas et al. 2018). However, more study needs to clarify this particular mechanism.

(c) *Neurotensin*: The correlation in between neurotensin and related migraine are still an emerging area of research. Neurotensin plays crucial role in the gut motility, gut secretion, gut absorptions, as well as other neurotransmitters secretions. Gut bacteria may participate in the regulation of the neurotensin, which may further precipitate in the neuropsychiatric disorders such as schizophrenia, Parkinson's, and migraine. Intestinal neurotrophins are responsible for regulating the gut homeostasis, which alterations may lead to the nerve proliferations. Nerve proliferations further through intestinal dysregulation leads to the abruptly elevate the neurotrophins, which may dysregulate the further neurotransmission process. These further reduce the BDNF as well as NT-3 expressions (Figs. 6.1 and 6.2) (Lanza et al. 2021).

3. *Vasoactive neuropeptides*: Vasoactive neuropeptide influences the vascular tone through different neuropeptide markers such as CGRP, substance P, vasoactive intestinal peptide, CRP, matrix metallopeptidase (MMP-9), nuclear factor kappa-light-chain-enhancer of activated B cells (NF-kb), inducible nitric oxide synthase (iNOS), pituitary adenylate cyclase-activating polypeptide-38 (PACAP-38), cytokines, and adhesion molecules (Chehel Cheraghi et al. 2015; Amin et al. 2014). Particularly, vasoactive intestinal peptides are involved in the different phases of migraine, including interictal migraine. It modulates the activity through the mast cell behavior. Mast cells through dura mater and c-Fos through trigeminal nucleus are the mechanism behind the migraine in sympathectomy in sympathetic ganglion (Kilinc et al. 2015). Co-enhancement in the PACAP-38 and intestinal peptide are involved in the pediatric patients of migraine (Hanci et al. 2021; Waschek et al. 2018). Neurogenic inflammation is also important in the pathophysiology for migraine through trigeminal nociception through directly activation of nociceptive fibers in the dura mater region. Likewise, oropharyngeal microbiota imbalance triggers the immunogen-induced migraine in 29% of patients, with the involvement of bacteria like viruses Herpes spp., Epstein-Barr, and microscopic fungi (Erdener et al. 2021; Spekker et al. 2021; Malhotra 2016). These patients suffer with food allergens like wheat flour 5%, cow's milk 13%, egg white 47% or yolk 26%, sweet pepper 6%, and quail egg 15%; among those, white egg food allergen is the common complication patient may feel. Among these patients, 48.5% was evidenced with activation in the IgG4 (Zhao Z et al. 2021). Similarly, nitroglycerine is responsible for mechanical pain sensations. Migraine-related hyperalgesia is responsible for the increment in the c-Fos and TNF-a and reduction in the CGRP at trigeminal nucleus caudalis (Kang et al. 2022). Similarly, sterile meningeal inflammation is the important mechanism of trigeminocervical C-fibers related migraine. Ganglionic and central pathophysiology suggest its role in the meningeal nociception over the occipital cortex region. In this mechanism, synaptic mismatch initiates the parenchymal inflammation, propagating it towards the long-lasting headache with or without aura of migraine (Figs. 6.1 and 6.2) (Erdener et al. 2021).

Treatment: The interrelation between the migraine, gastric motility, and vomiting shares important clinical implications as empty stomach and vomiting may affect the absorptions and therapeutic efficacy of the migraine treatments. For this reason, many patients do not receive adequate pain relief due to oral treatment. This reason further necessitates to switch upon the multiple treatment (Aurora et al. 2021).

1. *Antibiotic treatment*: Antibiotic treatment is an effective approach for the treatment of the migraine-like pain. The trigeminal nociceptive system increases with prolonged migraine-pain with the genetic deletion of the TNF-α or its antagonist in the trigeminal nucleus caudalis (Sp5C) with the migraine-inducing factor

nitroglycerin (NTG) (Tang et al. 2020). Chronic non-atropic gastritis is a common metabolic gastrointestinal disorder. Wuzhuyu (Chinese-based decoction) has been proven as an effective treatment in this regard (Hu et al. 2023). It facilitates the stool retention with the dysbiosis clearance. The same is responsible for the antimigraine effect with the reduction in the CGRP and c-FOS inflammatory factors (Kang et al. 2022).

2. *Probiotic and migraine management*: Probiotics play an important physiological role in maintaining gut microbiota, which modulation may alter the systematic effects, including chronic low-grade inflammations and neurological conditions such as anxiety, depression, and migraine. Probiotics in gut-related-migraines influence diverse community of microbes from digestive tracts including immunological precipitations (Ghavami et al. 2021; Tang et al. 2020; Georgescu et al. 2019). Probiotics inhibit the growth of harmful bacteria through beneficial bacterial growth. They also treat low-grade chronic inflammations by modulating the inflammatory responses and homeostasis of neurotransmitters, particularly for serotonin, which imbalances are crucial in migraine precipitation (Mazziotta et al. 2023; Cristofori et al. 2021). Through these processes, they maintain the integrity of intestinal barrier and treat the leaky gut. Interestingly, prebiotics, for example, fermented fibers, reduce the level of Bifidobacteria and butyrate-producing bacteria (Pham et al. 2018; Clapp et al. 2017). Prebiotics also elevate the short chain fatty acids in the gut, which are important for the gut homeostasis and strengthen the intestinal epithelia's integrity through suppression of the nuclear factor kappa (NF-kb) pathways and related leaky gut markers such as Hs-CRP and Zonulin. Although they potentially alleviated the migraine through different pathways, further investigations are still an early stage of research. Therefore, more conscientious studies are needed to detailed explanations. According to the phase I/II studies, probiotic treatment is a prominent alternative in providing the relieve with migraine-like symptoms (Sasso et al. 2023; Dai et al. 2017). Metabolite treatment is also an important approach in gut microbiota-induced migraine. Metabolite such as topiramate is responsible for treating the inflammation-induced migraine-like behaviors (Miao et al. 2022). Probiotics account for 45% alleviation in the chronic migraine and 40% in episodic migraine. Similarly, DAO supplementation has been significantly impacted in the treatment of migraine attack through alleviation of frequency of pain (Izquierdo-Casas et al. 2019). Natural Chinese medicine such as *Uncaria rhynchophylla* (Miq.) Miq. Ex Havil. (GU) and *Gastrodia elata* blume have also shown to have potential antimigraine therapies through gut microbial homeostasis (Yang et al. 2020). It elevates the relative abundance of *Coprococcus* and *Bacteroides* and reduces the *Escherichia-Shigella* and *Prevotella* species. GU also modulate the metabolomics level by reducing glutamic acid, L-tyrosine, 3-indoxyl sulfate, L-ariginine levels and through elevation of L-tryptophan, 5-HIAA, and Linoleic acid (Yang et al. 2020).

3. *Fecal microbiota transplantation*: Fecal microbiota transplantation (FMT) has been proven as effective approach in treating the metabolite alterations and subsequently migraine pain alleviation. It modulates the alteration in the microbio-

logical bacteriophage levels, composition of microbiota, and short chain fatty acids such as valerate, propionate, butyrate, and acetate (Halaweish et al. 2022). These short chain fatty acids result in regulations of immune reactions including T-cell regulations and further results into the inflammations (Vinolo et al. 2011). Post-FMT enhances the potential for the immunomodulatory changes through IL-10 and IL-17 producing cells which further improve the mucosal immunity. Some of the key short chain fatty acids for migraine is propionate and butyrate. Propionate shields the blood-brain barrier through oxidative stress inhibition process. Species responsible for propionate production are *B. uniformis*, *B. vulgatus*, *Prevotella copri*, *Roseburia Veilonella* spp., *Alistipes putredinis*, *Akkermansia inulinivorans*, and *A. muciniphila*. While short chain fatty acids result with the neural inflammation through the immunity alterations, certain bacteria help in the essential metabolite (neuroactive) productions in the central nervous systems (Qin et al. 2010; Zhernakova et al. 2016; Hoyles et al. 2018; Park et al. 2019; Cryan and Dinan 2012). Likewise, butyrate impedes with the serotonin release through the intestinal enterochromaffin cells. Low level of serotonin is well-documented with the vasodilation related migraine precipitation. Firmicutes family are well known for the serotonin release. Some of the examples are *Faecalibacterium prausnitzii*, *Eubacterium hallii*, *Coprococus* spp., *Lachnosiraceae* spp., *Roseburia* spp., Clostridial Clusters IV and XIVa (Kaur et al. 2019; Vecsey et al. 2007; Stefanko et al. 2009; Reigstad et al. 2015; Louis and Flint 2017; Qin et al. 2010).

4. *Biochemical molecules*
   (a) *Short Chain neuropeptides*: Small neuropeptides such as short length fatty acids (Sodium propionate and sodium butyrate) and CGRP has become the prominent target for the treatment of migraines (Vinolo et al. 2011). CGRP antagonist such as Gepants, Rimegepant, Ubrogepant and, CGRP antibodies such as Fremanezumab, Erenumab, and Galcanezumab decrease the frequency of migraine attacks (Cohen et al. 2022). In the gut, CGRP are expressed by extrinsic as well as intrinsic nervous system through regulation of primary afferent fibres as well as enteric nervous systems respectively. Majorily, it is responsible for the regulation of ascending (excitatory) and descending (inhibitory) enteric sensory systems (Browning and Travagli 2014). Further, it accelerates the intestinal transit through its motor stimulations, which are responsible for prosecretory secretions (resultant diarrhea). Therefore, the inhibitory activities of CGRP control peristaltic activity through these pathways, which results into diarrhea management (Holzer and Holzer-Petsche 2022). Likewise, short chain fatty acids including sodium propionate and sodium butyrate are associated with the episodic pain attacks in the trigeminal nucleus. They counteract at the damaged trigeminal nucleus and alleviate the mucosal alterations through reducing the inflammatory markers such as TNF-a and IL-1b. These consequences are responsible for pain-like symptoms (Lanza et al. 2021). Short chain fatty acids formed through fermentation are further responsible for the inflammatory-mediated migraine pain (Browning and Travagli 2014). Dietary nutrition

potentially impacts at the gut microbiota and related immune aspects, expanding it to the central nervous systems. Diet, such as ketogenic plan, is helpful in the management of migraine attack, through mitochondrial and short chain fatty acid (butyrate) regulations. It improves the central metabolic energy with betahydroxybutyrate regulations. Inulin supplementation is also a potential diet plan in regulating the intestinal microbes and related immune parameters particularly in migraine condition. It reverses the oxidative stress and inflammation in the gut, which is one of the prime responsive factors for gut induced migraine. Furthermore, synthetic antioxidants, including 5-substituted-1,3,4-thiadiazole-2-thiols and 5-(5-substituted-1,3,4-oxadiazole-2-yl) benzene, have also been oxidative stress modulators in gut dysbiosis-induced migraine attack (Vajdi et al. 2023).

6.5 Conclusion

The evidence presented in this book chapter, strongly supports the notion that the gut microbiota plays a significant role in influencing migraine susceptibility, frequency, and severity. The comprehensive exploration of microbial communities, their metabolites, and the intricate communication within the gut–brain axis has provided valuable insights into the mechanisms underlying migraine pathophysiology.

Furthermore, the discussion of potential therapeutic interventions targeting the gut microbiome offers a glimpse into the future of migraine management. Manipulating the gut microbiota through dietary interventions, probiotics, or other innovative strategies holds promise for alleviating migraine symptoms and improving overall patient well-being.

As we conclude, it is evident that the gut microbiome is not merely a passive bystander but an active participant in the complex puzzle of migraine disorders. This chapter serves as a stepping stone for researchers, clinicians, and healthcare professionals, urging them to explore and embrace the potential of microbiome-based approaches in understanding and managing migraine. The journey into the gut–brain axis continues, offering exciting prospects for advancing our understanding and treatment of this prevalent neurological condition (Figs. 6.1 and 6.2).

References

Akagi Y, Takayama Y, Nihashi Y, Yamashita A, Yoshida R, Miyamoto Y, Kida YS (2024) Functional engineering of human iPSC-derived parasympathetic neurons enhances responsiveness to gastrointestinal hormones. FEBS Open Bio 14(1):63–78. https://doi.org/10.1002/2211-5463.13741

Amin FM, Hougaard A, Schytz HW, Asghar MS, Lundholm E, Parvaiz AI, de Koning PJH, Andersen MR, Larsson HBW, Fahrenkrug J, Olesen J, Ashina M (2014) Investigation of the pathophysiological mechanisms of migraine attacks induced by pituitary adenylate cyclase-activating polypeptide-38. Brain 137(3):779–794. https://doi.org/10.1093/brain/awt369

Anderson G (2019) Integrating pathophysiology in migraine: role of the gut microbiome and melatonin. Curr Pharm Des 25(33):3550–3562. https://doi.org/10.2174/1381612825666190920114611

Arzani M, Jahromi SR, Ghorbani Z, Vahabizad F, Martelletti P, Ghaemi A, Sacco S, Togha M (2020) School of Advanced Studies of the European Headache Federation (EHF-SAS). Gut-brain Axis and migraine headache: a comprehensive review. J Headache Pain 13;21(1):15. https://doi.org/10.1186/s10194-020-1078-9. PMID: 32054443; PMCID: PMC7020496.

Aurora SK, Shrewsbury SB, Ray S, Hindiyeh N, Nguyen L (2021) A link between gastrointestinal disorders and migraine: insights into the gut-brain connection. Headache 61(4):576–589. https://doi.org/10.1111/head.14099

Bai J, Shen N, Liu Y (2023) Associations between the gut microbiome and migraines in children aged 7–18 years: an analysis of the American gut project cohort. Pain Manag Nurs 24(1):35–43. https://doi.org/10.1016/j.pmn.2022.06.002

Brown EM, Ke X, Hitchcock D, Jeanfavre S, Avila-Pacheco J, Nakata T, Arthur TD, Fornelos N, Heim C, Franzosa EA, Watson N, Huttenhower C, Haiser HJ, Dillow G, Graham DB, Finlay BB, Kostic AD, Porter JA, Vlamakis H, Clish CB et al (2019) Bacteroides-derived sphingolipids are critical for maintaining intestinal homeostasis and Symbiosis. Cell Host Microbe 25(5):668–680.e7. https://doi.org/10.1016/j.chom.2019.04.002

Browning KN, Travagli RA (2014) Central nervous system control of gastrointestinal motility and secretion and modulation of gastrointestinal functions. Compr Physiol 4(4):1339–1368. https://doi.org/10.1002/cphy.c130055

Cámara-Lemarroy CR, Rodriguez-Gutierrez R, Monreal-Robles R, Marfil-Rivera A (2016) Gastrointestinal disorders associated with migraine: a comprehensive review. World J Gastroenterol 22(36):8149–8160. [PMID: 27688656]. https://doi.org/10.3748/wjg.v22.i36.8149

Ceuleers H, Van Spaendonk H, Hanning N, Heirbaut J, Lambeir AM, Joossens J, Augustyns K, De Man JG, De Meester I, De Winter BY (2016) Visceral hypersensitivity in inflammatory bowel diseases and irritable bowel syndrome: the role of proteases. World J Gastroenterol 22(47):10275–10286. https://doi.org/10.3748/wjg.v22.i47.10275

Chandra R, Sokratian A, Chavez KR, King S, Swain SM, Snyder JC, West AB, Liddle RA (2023) Gut mucosal cells transfer α-synuclein to the vagus nerve. JCI Insight 8(23):e172192. https://doi.org/10.1172/jci.insight.172192

Chen J, Wang Q, Wang A, Lin Z (2020) Structural and functional characterization of the gut microbiota in elderly women with migraine. Front Cell Infect Microbiol 9:470. https://doi.org/10.3389/fcimb.2019.00470

Clapp M, Aurora N, Herrera L, Bhatia M, Wilen E, Wakefield S (2017) Gut microbiota's effect on mental health: the gut-brain axis. Clin Pract 7(4):987. https://doi.org/10.4081/cp.2017.987

Cohen F, Yuan H, Silberstein SD (2022) Calcitonin gene-related peptide (CGRP)-targeted monoclonal antibodies and antagonists in migraine: current evidence and rationale. BioDrugs 36(3):341–358. https://doi.org/10.1007/s40259-022-00530-0

Crawford J, Liu S, Tao F (2022) Gut microbiota and migraine. Neurobiol Pain 11:100090. https://doi.org/10.1016/j.ynpai.2022.100090

Cristofori F, Dargenio VN, Dargenio C, Miniello VL, Barone M, Francavilla R (2021) Anti-inflammatory and immunomodulatory effects of probiotics in gut inflammation: a door to the body. Front Immunol 12:578386. https://doi.org/10.3389/fimmu.2021.578386

Cryan JF, Dinan TG (2012) Mind-altering microorganisms: the impact of the gut microbiota on brain and behaviour. Nat Rev Neurosci 13(10):701–712. https://doi.org/10.1038/nrn3346

Dağıdır HG, Topa E, Vuralli D, Bolay H (2023) Medication overuse headache is associated with elevated lipopolysaccharide binding protein and pro-inflammatory molecules in the bloodstream. J Headache Pain 24(1):150. https://doi.org/10.1186/s10194-023-01672-4

Dai YJ, Wang HY, Wang XJ, Kaye AD, Sun YH (2017) Potential beneficial effects of probiotics on human migraine headache: a literature review. Pain Physician 20(2):E251–E255

Diether NE, Hulshof TG, Willing BP, van Kempen TATG (2023) A blend of medium-chain fatty acids, butyrate, organic acids, and a phenolic compound accelerates microbial maturation in newly weaned piglets. PLoS One 18(7):e0289214. https://doi.org/10.1371/journal.pone.0289214

Eastwood J, van Hemert S, Poveda C, Elmore S, Williams C, Lamport D, Walton G (2023) The effect of probiotic bacteria on composition and metabolite production of faecal microbiota using in vitro batch cultures. Nutrients 15(11):2563. https://doi.org/10.3390/nu15112563

Erdener ŞE, Kaya Z, Dalkara T (2021) Parenchymal neuroinflammatory signaling and dural neurogenic inflammation in migraine. J Headache Pain 22:138. https://doi.org/10.1186/s10194-021-01353-0

Fujisaka S, Watanabe Y, Tobe K (2023) The gut microbiome: a core regulator of metabolism. J Endocrinol 256(3):e220111. https://doi.org/10.1530/JOE-22-0111

Gabrielli M, Ojetti V, Candelli M, Santarelli L, Gasbarrini A, Franceschi F (2021) Migraine as a common extra-intestinal presentation of celiac disease. OBM Neurobiol 5:087. https://doi.org/10.21926/obm.neurobiol.2101087

Georgescu D, Reisz D, Gurban CV, Georgescu LA, Ionita I, Ancusa OE, Lighezan D (2017) Migraine in young females with irritable bowel syndrome: still a challenge. Neuropsychiatr Dis Treat 14:21–28. https://doi.org/10.2147/NDT.S144955

Georgescu D, Iurciuc MS, Ionita I, Dragan S, Muntean M, Ancusa OE, Reisz D, Ionita M, Lighezan D (2019) Migraine without aura and subclinical atherosclerosis in young females: is gut microbiota to blame? Medicina (Kaunas) 55(12):786. https://doi.org/10.3390/medicina55120786

Ghavami A, Khorvash F, Heidari Z, Khalesi S, Askari G (2021) Effect of synbiotic supplementation on migraine characteristics and inflammatory biomarkers in women with migraine: results of a randomized controlled trial. Pharmacol Res 169:105668. https://doi.org/10.1016/j.phrs.2021.105668

Guo Y, Luo C, Sun Y, Guo W, Zhang R, Zhang X, Ke X, Wei L (2023) Inhibition of mitochondrial fusion via SIRT1/PDK2/PARL axis breaks mitochondrial metabolic plasticity and sensitizes cancer cells to glucose restriction therapy. Biomed Pharmacother 166:115342. https://doi.org/10.1016/j.biopha.2023.115342

Halaweish HF, Boatman S, Staley C (2022) Encapsulated fecal microbiota transplantation: development, efficacy, and clinical application. Front Cell Infect Microbiol 12:826114. https://doi.org/10.3389/fcimb.2022.826114

Hanci F, Kilinc YB, Kilinc E, Turay S, Dilek M, Kabakus N (2021) Plasma levels of vasoactive neuropeptides in pediatric patients with migraine during attack and attack-free periods. Cephalalgia 41(2):166–175. https://doi.org/10.1177/0333102420957588

Hannun YA, Obeid LM (2018) Sphingolipids and their metabolism in physiology and disease. Nat Rev Mol Cell Biol 19(3):175–191. https://doi.org/10.1038/nrm.2017.107

Hasegawa K, Stewart CJ, Mansbach JM et al (2017) Sphingolipid metabolism potential in fecal microbiome and bronchiolitis in infants: a case–control study. BMC Res Notes 10:325. https://doi.org/10.1186/s13104-017-2659-9

Hassan A, Mehany D, Eldin HG, Abdelghaffar M, Abdelbaky HA, Kamal YS, Hussein M (2023) Helicobacter pylori infection in migraine headache: a true association or an innocent bystander? Int J Neurosci 133(10):1090–1095. https://doi.org/10.1080/00207454.2022.2045291

He Q, Wang W, Xiong Y, Tao C, Ma L, Ma J, You C, International Headache Genetics Consortium (2023) A causal effects of gut microbiota in the development of migraine. J Headache Pain 24(1):90. https://doi.org/10.1186/s10194-023-01609-x

Holzer P, Holzer-Petsche U (2022) Constipation caused by anti-calcitonin gene-related peptide migraine therapeutics explained by antagonism of calcitonin gene-related peptide's motor-stimulating and prosecretory function in the intestine. Front Physiol 12:820006. https://doi.org/10.3389/fphys.2021.820006

Hoyles L, Snelling T, Umlai UK, Nicholson JK, Carding SR, Glen RC, McArthur S (2018) Microbiome-host systems interactions: protective effects of propionate upon the blood-brain barrier. Microbiome 6(1):55. https://doi.org/10.1186/s40168-018-0439-y

Hu Q, Zeng J, Zhang X, He T, Zhang A, Li J, Wei S, Jing M, Li H, Wang X, Chang L, Ma X, Zhao Y (2023) Metabolomics profiles reveal the efficacy of Wuzhuyu decoction on patients with chronic non-atrophic gastritis. Drug Des Devel Ther 17:3269–3280. https://doi.org/10.2147/DDDT.S428783

Izquierdo-Casas J, Comas-Basté O, Latorre-Moratalla ML, Lorente-Gascón M, Duelo A, Vidal-Carou MC, Soler-Singla L (2018) Low serum diamine oxidase (DAO) activity levels in patients with migraine. J Physiol Biochem 74(1):93–99. https://doi.org/10.1007/s13105-017-0571-3

Izquierdo-Casas J, Comas-Basté O, Latorre-Moratalla ML, Lorente-Gascón M, Duelo A, Soler-Singla L, Vidal-Carou MC (2019) Diamine oxidase (DAO) supplement reduces headache in episodic migraine patients with DAO deficiency: a randomized double-blind trial. Clin Nutr 38(1):152–158. https://doi.org/10.1016/j.clnu.2018.01.013

Jiang W, Wang T, Liu C, Deng M, Ren X, Wang F, Zhang Y, Yu X, Yao L, Wang Y (2021) A 16S rRNA gene sequencing-based study of oral microbiota in migraine patients in China. Bioengineered 12(1):2523–2533. https://doi.org/10.1080/21655979.2021.1933840

Johnson EL, Heaver SL, Waters JL et al (2020) Sphingolipids produced by gut bacteria enter host metabolic pathways impacting ceramide levels. Nat Commun 11:2471. https://doi.org/10.1038/s41467-020-16274-w

Kang L, Tang W, Zhang Y, Zhang M, Liu J, Li Y, Kong S, Zhao D, Yu S (2022) The gut microbiome modulates nitroglycerin-induced migraine-related hyperalgesia in mice. Cephalalgia 42(6):490–499. https://doi.org/10.1177/03331024211050036

Kappéter Á, Sipos D, Varga A, Vigvári S, Halda-Kiss B, Péterfi Z (2023) Migraine as a Disease Associated with Dysbiosis and Possible Therapy with Fecal Microbiota Transplantation. Microorganisms 14;11(8):2083. https://doi.org/10.3390/microorganisms11082083. PMID: 37630643; PMCID: PMC10458656.

Kaur H, Bose C, Mande SS (2019) Tryptophan metabolism by gut microbiome and gut-brain-axis: an *in silico* analysis. Front Neurosci 13:1365. https://doi.org/10.3389/fnins.2019.01365

Kilinc E, Firat T, Tore F, Kiyan A, Kukner A, Tunçel N (2015) Vasoactive intestinal peptide modulates c-Fos activity in the trigeminal nucleus and dura mater mast cells in sympathectomized rats. J Neurosci Res 93(4):644–650. https://doi.org/10.1002/jnr.23523

Kilinc YB, Kilinc E, Danis A, Hanci F, Turay S, Ozge A, Bolay H (2023) Mitochondrial metabolism related markers GDF-15, FGF-21, and HIF-1α are elevated in pediatric migraine attacks. Headache 63(8):1076–1086. https://doi.org/10.1111/head.14618

Kim J-H, Lee Y, Kwon Y-S, Sohn J-H (2023) Clinical implications of the association between respiratory and gastrointestinal disorders in migraine and non-migraine headache patients. J Clin Med 12(10):3434. https://doi.org/10.3390/jcm12103434

Lanza M, Filippone A, Ardizzone A, Casili G, Paterniti I, Esposito E, Campolo M (2021) SCFA treatment alleviates pathological signs of migraine and related intestinal alterations in a mouse model of NTG-induced migraine. Cells 10(10):2756. https://doi.org/10.3390/cells10102756

Lanza M, Filippone A, Casili G, Giuffrè L, Scuderi SA, Paterniti I, Campolo M, Cuzzocrea S, Esposito E (2022) Supplementation with SCFAs re-establishes microbiota composition and attenuates hyperalgesia and pain in a mouse model of NTG-induced migraine. Int J Mol Sci 23(9):4847. https://doi.org/10.3390/ijms23094847

Liu J, Xi K, Zhang L, Han M, Wang Q, Liu X (2024) Tryptophan metabolites and gut microbiota play an important role in pediatric migraine diagnosis. J Headache Pain 25(1):2. https://doi.org/10.1186/s10194-023-01708-9

Louis P, Flint HJ (2017) Formation of propionate and butyrate by the human colonic microbiota. Environ Microbiol 19(1):29–41. https://doi.org/10.1111/1462-2920.13589

Lucarini E, Parisio C, Branca JJV, Segnani C, Ippolito C, Pellegrini C, Antonioli L, Fornai M, Micheli L, Pacini A, Bernardini N, Blandizzi C, Ghelardini C, Di Cesare Mannelli L (2020) Deepening the mechanisms of visceral pain persistence: an evaluation of the gut-spinal cord relationship. Cells 9(8):1772. https://doi.org/10.3390/cells9081772

MacEachern SJ, Keenan CM, Papakonstantinou E, Sharkey KA, Patel BA (2018) Alterations in melatonin and 5-HT signalling in the colonic mucosa of mice with dextran-sodium sulfate-induced colitis. Br J Pharmacol 175(9):1535–1547. https://doi.org/10.1111/bph.14163

Maceyka M, Spiegel S (2014) Sphingolipid metabolites in inflammatory disease. Nature 510(7503):58–67. https://doi.org/10.1038/nature13475
Malhotra R (2016) Understanding migraine: potential role of neurogenic inflammation. Ann Indian Acad Neurol 19(2):175–182. https://doi.org/10.4103/0972-2327.182302
Matsuo Y, Komiya S, Yasumizu Y, Yasuoka Y, Mizushima K, Takagi T, Kryukov K, Fukuda A, Morimoto Y, Naito Y, Okada H, Bono H, Nakagawa S, Hirota K (2021) Full-length 16S rRNA gene amplicon analysis of human gut microbiota using MinION™ nanopore sequencing confers species-level resolution. BMC Microbiol 21(1):35. https://doi.org/10.1186/s12866-021-02094-5
Mazziotta C, Tognon M, Martini F, Torreggiani E, Rotondo JC (2023) Probiotics mechanism of action on immune cells and beneficial effects on human health. Cells 12(1):184. https://doi.org/10.3390/cells12010184
Miao S, Tang W, Li H, Li B, Yang C, Xie W, Wang T, Bai W, Gong Z, Dong Z, Yu S (2022) Repeated inflammatory dural stimulation-induced cephalic allodynia causes alteration of gut microbial composition in rats. J Headache Pain 23(1):71. https://doi.org/10.1186/s10194-022-01441-9
Nakaya Y, Tsuboi Y, Okada-Ogawa A, Shinoda M, Kubo A, Chen JY, Noma N, Batbold D, Imamura Y, Sessle BJ, Iwata K (2016) ERK-GluR1 phosphorylation in trigeminal spinal subnucleus caudalis neurons is involved in pain associated with dry tongue. Mol Pain 12:1744806916641680. https://doi.org/10.1177/1744806916641680
Nan N, Wang Q, Li MJ, Xu YS, Guo XM, Xu R, Ma Z, Wang SH, Li J, Zhao H, Gong MX (2023) The microbiota-dependent treatment of Wuzhuyu decoction for chronic migraine model rat associated with anxiety-depression like behavior. Oxidative Med Cell Longev 2023:2302653. https://doi.org/10.1155/2023/2302653
Nezami BG, Srinivasan S (2010) Enteric nervous system in the small intestine: pathophysiology and clinical implications. Curr Gastroenterol Rep 12(5):358–365. https://doi.org/10.1007/s11894-010-0129-9
Nixon GF (2009) Sphingolipids in inflammation: pathological implications and potential therapeutic targets. Br J Pharmacol 158(4):982–993. https://doi.org/10.1111/j.1476-5381.2009.00281.x
Nogal A, Valdes AM, Menni C (2021) The role of short-chain fatty acids in the interplay between gut microbiota and diet in cardio-metabolic health. Gut Microbes 13(1):1–24. https://doi.org/10.1080/19490976.2021.1897212
Parada Venegas D, De la Fuente MK, Landskron G, González MJ, Quera R, Dijkstra G, Harmsen HJM, Faber KN, Hermoso MA (2019) Short chain fatty acids (SCFAs)-mediated gut epithelial and immune regulation and its relevance for inflammatory bowel diseases. Front Immunol 10:277. https://doi.org/10.3389/fimmu.2019.00277
Park J, Wang Q, Wu Q, Mao-Draayer Y, Kim CH (2019) Bidirectional regulatory potentials of short-chain fatty acids and their G-protein-coupled receptors in autoimmune neuroinflammation. Sci Rep 9(1):8837. https://doi.org/10.1038/s41598-019-45311-y
Pekary AE, Sattin A (2022) Rifaximin modulates TRH and TRH-like peptide expression throughout the brain and peripheral tissues of male rats. BMC Neurosci 23(1):9. https://doi.org/10.1186/s12868-022-00694-z
Peterlin BL, Mielke MM, Dickens AM, Chatterjee S, Dash P, Alexander G, Vieira RV, Bandaru VV, Dorskind JM, Tietjen GE, Haughey NH (2015) Interictal, circulating sphingolipids in women with episodic migraine: a case-control study. Neurology 85(14):1214–1223. https://doi.org/10.1212/WNL.0000000000002004
Pham VT, Seifert N, Richard N, Raederstorff D, Steinert RE, Prudence K, Mohajeri MH (2018) The effects of fermentation products of prebiotic fibres on gut barrier and immune functions in vitro. PeerJ 6:e5288. https://doi.org/10.7717/peerj.5288
Prakash S, Patel H, Shah CS (2024) Medication-overuse headache overlapping with serotonin syndrome. BMJ Case Rep 17(1):e256464. https://doi.org/10.1136/bcr-2023-256464
Qin J, Li R, Raes J, Arumugam M, Burgdorf KS, Manichanh C, Nielsen T, Pons N, Levenez F, Yamada T, Mende DR, Li J, Xu J, Li S, Li D, Cao J, Wang B, Liang H, Zheng H, Xie Y et al (2010) A human gut microbial gene catalogue established by metagenomic sequencing. Nature 464(7285):59–65. https://doi.org/10.1038/nature08821

Qu Y, Eguchi A, Ma L, Wan X, Mori C, Hashimoto K (2023) Role of the gut-brain axis via the subdiaphragmatic vagus nerve in stress resilience of 3,4-methylenedioxymethamphetamine in mice exposed to chronic restrain stress. Neurobiol Dis 189:106348. https://doi.org/10.1016/j.nbd.2023.106348

Reigstad CS, Salmonson CE, Rainey JF 3rd, Szurszewski JH, Linden DR, Sonnenburg JL, Farrugia G, Kashyap PC (2015) Gut microbes promote colonic serotonin production through an effect of short-chain fatty acids on enterochromaffin cells. FASEB J 29(4):1395–1403. https://doi.org/10.1096/fj.14-259598

Rohrhofer J, Zwirzitz B, Selberherr E, Untersmayr E (2021) The impact of dietary sphingolipids on intestinal microbiota and gastrointestinal immune homeostasis. Front Immunol 12:635704. https://doi.org/10.3389/fimmu.2021.635704

Roth W, Zadeh K, Vekariya R, Ge Y, Mohamadzadeh M (2021) Tryptophan Metabolism and Gut-Brain Homeostasis. Int J Mol Sci 15;22(6):2973. https://doi.org/10.3390/ijms22062973. PMID: 33804088; PMCID: PMC8000752.

Sasso JM, Ammar RM, Tenchov R, Lemmel S, Kelber O, Grieswelle M, Zhou QA (2023) Gut microbiome-brain alliance: a landscape view into mental and gastrointestinal health and disorders. ACS Chem Neurosci 14(10):1717–1763. https://doi.org/10.1021/acschemneuro.3c00127

Shim JA, Ryu JH, Jo Y, Hong C (2023) The role of gut microbiota in T cell immunity and immune mediated disorders. Int J Biol Sci 19(4):1178–1191. https://doi.org/10.7150/ijbs.79430

Spekker E, Nagy-Grócz G (2023) All roads lead to the gut: the importance of the microbiota and diet in migraine. Neurol Int 15(3):1174–1190. https://doi.org/10.3390/neurolint15030073

Spekker E, Tanaka M, Szabó Á, Vécsei L (2021) Neurogenic inflammation: the participant in migraine and recent advancements in translational research. Biomedicines 10(1):76. https://doi.org/10.3390/biomedicines10010076

Stasiłowicz A, Tykarska E, Rosiak N, Sałat K, Furgała-Wojas A, Plech T, Lewandowska K, Pikosz K, Pawłowicz K, Cielecka-Piontek J (2021) The inclusion of tolfenamic acid into cyclodextrins stimulated by microenvironmental pH modification as a way to increase the anti-migraine effect. J Pain Res 14:981–992. https://doi.org/10.2147/JPR.S295795

Stefanko DP, Barrett RM, Ly AR, Reolon GK, Wood MA (2009) Modulation of long-term memory for object recognition via HDAC inhibition. Proc Natl Acad Sci U S A 106(23):9447–9452. https://doi.org/10.1073/pnas.0903964106

Steidel K, Krause K, Menzler K, Strzelczyk A, Immisch I, Fuest S, Gorny I, Mross P, Hakel L, Schmidt L, Timmermann L, Rosenow F, Bauer S, Knake S (2021) Transcutaneous auricular vagus nerve stimulation influences gastric motility: a randomized, double-blind trial in healthy individuals. Brain Stimul 14(5):1126–1132. https://doi.org/10.1016/j.brs.2021.06.006

Sun W, Guo Y, Zhang S, Chen Z, Wu K, Liu Q, Liu K, Wen L, Wei Y, Wang B, Chen D (2018) Fecal microbiota transplantation can alleviate gastrointestinal transit in rats with high-fat diet-induced obesity via regulation of serotonin biosynthesis. Biomed Res Int 2018:8308671. https://doi.org/10.1155/2018/8308671

Suzuki M, Watanabe A, Huang J, Kobayashi Y, Sakata I (2024) Involvement of the autonomic nervous system in colonic contractions in conscious Suncus murinus. Neurogastroenterol Motil 36:e14716. Advance online publication. https://doi.org/10.1111/nmo.14716

Tang Y, Liu S, Shu H, Xing Y, Tao F (2018) AMPA receptor GluA1 Ser831 phosphorylation is critical for nitroglycerin-induced migraine-like pain. Neuropharmacology 133:462–469. https://doi.org/10.1016/j.neuropharm.2018.02.026

Tang Y, Liu S, Shu H, Yanagisawa L, Tao F (2020) Gut microbiota dysbiosis enhances migraine-like pain via TNFα upregulation. Mol Neurobiol 57(1):461–468. https://doi.org/10.1007/s12035-019-01721-7

Terry N, Margolis KG (2017) Serotonergic mechanisms regulating the GI tract: experimental evidence and therapeutic relevance. Handb Exp Pharmacol 239:319–342. https://doi.org/10.1007/164_2016_103

Tham YY, Choo QC, Muhammad TST, Chew CH (2020) Lauric acid alleviates insulin resistance by improving mitochondrial biogenesis in THP-1 macrophages. Mol Biol Rep 47(12):9595–9607. https://doi.org/10.1007/s11033-020-06019-9

Vajdi M, Khorvash F, Rouhani MH, Ghavami A, Clark CCT, Askari G (2023) Effect of inulin supplementation on clinical symptoms, inflammatory and oxidative stress markers in women with migraine: study protocol for a randomized clinical trial. Trials 24(1):722. https://doi.org/10.1186/s13063-023-07765-4

Vecsey CG, Hawk JD, Lattal KM, Stein JM, Fabian SA, Attner MA, Cabrera SM, McDonough CB, Brindle PK, Abel T, Wood MA (2007) Histone deacetylase inhibitors enhance memory and synaptic plasticity via CREB:CBP-dependent transcriptional activation. J Neurosci 27(23):6128–6140. https://doi.org/10.1523/JNEUROSCI.0296-07.2007

Vinolo MA, Rodrigues HG, Nachbar RT, Curi R (2011) Regulation of inflammation by short chain fatty acids. Nutrients 3(10):858–876. https://doi.org/10.3390/nu3100858

Waschek JA, Baca SM, Akerman S (2018) PACAP and migraine headache: immunomodulation of neural circuits in autonomic ganglia and brain parenchyma. J Headache Pain 19:23. https://doi.org/10.1186/s10194-018-0850-6

Wu B, Wang S, Qin G, Xie J, Tan G, Zhou J, Chen L (2017) Protein kinase C γ contributes to central sensitization in a rat model of chronic migraine. J Mol Neurosci 63(2):131–141. https://doi.org/10.1007/s12031-017-0960-7

Yang W, Ip SP, Liu L, Xian YF, Lin ZX (2020) *Uncaria rhynchophylla* and its major constituents on central nervous system: a review on their pharmacological actions. Curr Vasc Pharmacol 18(4):346–357. https://doi.org/10.2174/1570161117666190704092841

Yong D, Lee H, Min HG, Kim K, Oh HS, Chu MK (2023) Altered gut microbiota in individuals with episodic and chronic migraine. Sci Rep 13(1):626. https://doi.org/10.1038/s41598-023-27586-4

Zhang W, Lei M, Wen Q, Zhang D, Qin G, Zhou J, Chen L (2022) Dopamine receptor D2 regulates GLUA1-containing AMPA receptor trafficking and central sensitization through the PI3K signaling pathway in a male rat model of chronic migraine. J Headache Pain 23(1):98. https://doi.org/10.1186/s10194-022-01469-x

Zhernakova A, Kurilshikov A, Bonder MJ, Tigchelaar EF, Schirmer M, Vatanen T, Mujagic Z, Vila AV, Falony G, Vieira-Silva S, Wang J, Imhann F, Brandsma E, Jankipersadsing SA, Joossens M, Cenit MC, Deelen P, Swertz MA, LifeLines Cohort Study, Weersma RK et al (2016) Population-based metagenomics analysis reveals markers for gut microbiome composition and diversity. Science 352(6285):565–569. https://doi.org/10.1126/science.aad3369

Zhao Z, Jin H, Yin Y, Hou Y, Wang J, Tang C, Fu J (2021) Association of Migraine with Its Comorbidities and Food Specific Immunoglobulin G Antibodies and Inflammatory Cytokines: Cross-Sectional Clinical Research. J Pain Res 5;14:2359–2368. https://doi.org/10.2147/JPR.S316619. PMID: 34385841; PMCID: PMC8352645.

Zhou Q, Xie Z, Wu D, Liu L, Shi Y, Li P, Gu Q (2022) The effect of indole-3-lactic acid from *Lactiplantibacillus plantarum* ZJ316 on human intestinal microbiota in vitro. Foods 11(20):3302. https://doi.org/10.3390/foods11203302

Role of Antioxidants in the Management of Migraine Pain

7

Arghya Kusum Dhar and Sidhartha Sankar Kar

7.1 Introduction

Migraine is a complicated neurological condition that causes significant disability. It is a combination of various factors and affects around 16% of the global population. The condition typically starts in infancy and becomes more common in the ages between 22 and 55. Women are more likely to be affected by migraines compared to males, with a ratio of 3 to 1. There is also a genetic predisposition for migraines to occur within families (Noseda and Burstein 2013; Burstein et al. 2015).

A headache that lasts for fewer than 15 days in a given month is called an episodic migraine (EM), while chronic migraine (CM) is defined as more than 15 days of headache per month with at least 8 days of migraine characteristics or response to triptans, for at least 3 months (Silberstein et al. 2005; Serrano et al. 2013). Headaches caused by EM can develop into CM if not treated properly (Katsarava et al. 2004). Subjects with CM are more severely disabled by headaches than those with EM (Buse et al. 2012). Despite numerous attempts, the precise mechanism behind migraine headache etiology is still unknown. Many hypotheses have been advanced on the underlying process, with some suggesting vascular dysfunction, neurogenic inflammation, and trigeminovascular circuit activation (Dodick 2018).

Because traditional pharmacologic treatments for migraine pain are ineffective and involve unwanted side effects, patients are looking for alternative pain relief options. The hunt for new treatments for migraine, a debilitating condition, has led researchers and clinicians to investigate plant-based therapies and their potential mechanisms of action (Katsarava et al. 2004). An increasing number of people are showing interest in nutraceutical and herbal remedies for migraine.

A. K. Dhar (✉)
School of Pharmacy, The Neotia University, Sarisha, West Bengal, India

S. S. Kar
Faculty of Pharmacy, C.V. Raman Global University, Bhubaneswar, Odisha, India

S. Bhatt et al. (eds.), *Management of Migraine Pain*,
https://doi.org/10.1007/978-981-97-4529-6_7

7.1.1 Migraine Headache

Migraine symptoms include a throbbing headache on one side of the head; sensitivity to light, sound, and smell; nausea; and vomiting. Movement can worsen the headache, which can last from a few hours to several days (Headache Classification Committee of the International Headache Society 2018). One-third of migraines begin with a negative or positive aura, including dazzling lights or scotomas, paresthesia or numbness, dysarthria or weakening, or cortical spreading depression. Research suggests that pial, arachnoid, dural, and major cerebral artery and sinus nociceptors activate migraine headaches (Noseda and Burstein 2013; Olesen et al. 2009). The meninges and intracranial vasculature receive nociceptive innervation from thin, myelinated Aδ fibers and nonmyelinated C-fibers, which release neuropeptides including substance P (SP) and calcitonin gene-related peptide (CGRP). From the trigeminal ganglion (TG), these fibers proceed to the dura via the ophthalmic branch and, to a lesser extent, the maxillary and mandibular divisions (Noseda and Burstein 2013; Levy and Strassman 2002). Extra dura innervation comes from the higher cervical dorsal root ganglia. Axon terminals of nociceptive nerve fibers of the dura mater release the vasoactive neuropeptides CGRP, SP, neurokinin A, and pituitary adenylate cyclase-activating peptide (Noseda and Burstein 2013; Goadsby et al. 2017).

7.1.2 Pathophysiology of Migraine

The cortical spreading depression (CSD) that causes migraine aura is a series of slow depolarization waves that travel throughout the cortex. CSD has been linked to headache pathophysiology by certain researchers (Yarnell 2017; Magis and Schoenen 2011).

Migraines are brain disorders caused by disruptions in brain homeostasis, activating the trigeminovascular structure, stimulating nociceptors of cerebral vessels, and affecting brain signal transmission. Further signal transmission stimulates brain areas important for migraine symptoms, including pain and neurological symptoms (Fig. 7.1) (Evers et al. 2006).

However, the processes behind migraine pathogenesis remain unclear (Evers et al. 2006; Ferrari et al. 2001). Migraine sufferers experience increased brain activity known as cortical hyper-reactivity. Possible causes include reduced thalamic or brain stem neuronal suppression or preactivation (Takaku et al. 2013). Trigeminal ganglion stimulation, which innervates the meninges, may induce CSD accompanied by aura and migraine. SP, neurokinin A, or CGRP release is a possible outcome of activation (Goadsby and Sprenger 2010). The release of cytokines leads to endothelial and platelet activation, leading to increased nitrogen oxide (NO) production and vasodilation, causing migraine symptoms (Goadsby and Sprenger 2010).

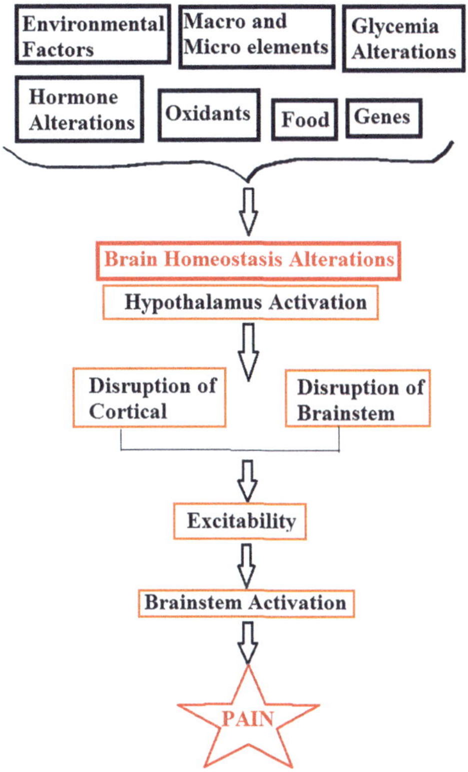

Fig. 7.1 Neurobiological steps involved in the development of migraine pain (Patil et al. 2021)

Gamma-aminobutyric acid receptors and amino acids such as aspartic acid and glutamic and may contribute to migraine etiopathogenesis (D'Andrea et al. 2014; Arulmozhi et al. 2007; Welch et al. 1990). Head examinations of migraine patients using 31P-MRS revealed decreased energy and oxygen metabolism of the brain (Hobbs 1989; Hayes and Foreman 1987). Researchers found that migraines could be associated with energy deficits in the brain, which could lead to pain from hypometabolism and oxidative stress. These factors can be caused by things like fasting, intense aromas, lack of sleep, physical exertion, or photosensitivity, all of which impact mitochondrial activity and energy metabolism (Heptinstall et al. 1985).

There are a lot of factors that can affect the development of migraines, such as heredity, food, oxidants, environmental factors, both micro- and macro-elements, metabolic problems (glycemia fluctuations), and hormonal changes. The balance of the brain is impacted by each of these components. Thus, the hypothalamus is stimulated. As a result, the excitability of the brainstem and the cortex is altered and disrupted. The result is the stimulation of the brainstem. Migraine pain is caused by the neurological processes that have already been covered.

7.1.3 Role of Reactive Oxygen Species and Oxidative Stress in Pathological Progression of Migraine

Hypothesis suggests mitochondria may contribute to migraine pathophysiology, with known links between mitochondria abnormalities and migraine (Sparaco et al. 2006; Montagna et al. 1988; Stuart and Griffiths 2012; Schaefer et al. 2008). Small organelles called mitochondria generate energy, produce reactive oxygen species (ROS), regulate apoptosis, and maintain Ca^{2+} homeostasis (Sparaco et al. 2006). Since mitochondria are essential for cell Ca^{2+} sequestration, their malfunction may cause pain hypersensitization (Kim et al. 2004). Vasoconstriction during CSD is also mediated by astrocyte Ca^{2+} accumulation. This mechanism is mediated by arachidonic acid metabolite phospholipase A2 (Metea and Newman 2006). Neuronal function depends on mitochondrial Ca^{2+} homeostasis. Thus, Ca^{2+} abnormalities can cause downstream imbalances, increasing migraine risk (Stuart and Griffiths 2012). Cytochrome-c oxidase (COX), commonly known as complex IV, in the mitochondria respiratory transport chain binds nitrogen oxide (NO) (Leao 1947). COX-rich arteries may sequester NO, preventing vasodilation. Some migraine patients have COX-negative fibers with increased fat storage (Sparaco et al. 2006). Only one occurrence of cluster headache has been linked to mitochondrial DNA defects (Shimomura et al. 1994). Additionally, two frequent mitochondrial DNA polymorphisms (16519C → T and 3010G → A) are linked to pediatric cyclic vomiting syndrome and migraine (Zaki et al. 2009).

Some researchers believe that oxidative stress has a role in the development of headaches (Gupta et al. 2009). Levels of catalase, coenzyme Q10 (CoQ10), and lipid peroxidation (LPO) were used to identify oxidative stress. Some researchers think that oxidative stress and LPO are involved in migraine pathophysiology because they control energy metabolism and cerebral blood flow, two factors that could be migraine triggers (Tuncel et al. 2008; Ciancarelli et al. 2003). Since ROS are thought to have a role in the development of pain, it follows that LPO levels may be seen as an indirect measure of the production of ROS within cells due to oxidative stress. Thus, it is believed that antioxidants, whether they be vitamins or cofactors, might assist lower free radical levels, which are crucial in mitochondrial illnesses.

During the absence of migraine attacks, Tozzi-Ciancarelli et al. (1997) discovered an increase in the number of compounds interacting with thiobarbituric acid in migraineurs. Alp et al. (2010) found variations in cell redox status indicators in migraine sufferers, including Total Antioxidant Status (TAS), Total Oxidant Status (TOS), and Oxidative Stress Index (OSI). TAS levels were lower and TOS values were significantly greater in migraine patients without aura compared to controls (Alp et al. 2010). Researchers examined oxidative stress parameters and DNA damage, as determined by 8-hydroxy-20-deoxyguanosine (8-OHdG) levels in migraine patients with and without aura (Geyik et al. 2016). There was no statistically significant difference in TOS, TAS, or OSI values between the migraine-free control group and the study population. Patients experiencing migraines with or without aura did not differ in terms of TOS, TAS, and OSI values (Geyik et al. 2016). Migraine sufferers, both with and without auras, had significantly higher 8-OHdG plasma concentrations than controls and migraine sufferers with auras than those without (Geyik et al. 2016).

Yigit et al. studied migraine patients in 2018 to determine the impact of oxidative stress on plasma lymphocyte destruction. The authors assessed the amount of a peptide present in the blood and heart, UTS2R, by measuring its levels in the serum. High amounts of the peptide, which has powerful vasoconstrictive properties, are observed in migraine sufferers who do not experience aura (Yigit et al. 2018). The study also included an assessment of plasma MDA levels, TOS, TAS, OSI, and CAT activity (Yigit et al. 2018). In this study, migraine patients had significantly higher TOS, OSI, lymphocyte DNA damage, and MDA concentration compared to controls. Lower levels of TAS, UTS2R, and CAT activity were seen in the study group compared to the control group (Yigit et al. 2018).

A study by Aytaç et al. found that migraine patients with high oxidative stress have a higher risk of developing white matter hyperintensities (WHM) (Aytaç et al. 2014). The results demonstrated that the study group had higher plasma MDA concentration and lower CAT activity in comparison to the control group. Researchers found that CAT activity was lower and MDA levels were higher in patients with migraine having WHM-type lesions compared to patients without these lesions (Aytaç et al. 2014).

7.2 Antioxidants in the Therapy of Migraine Pain

In recent years, nutraceuticals have garnered attention as migraine therapeutic options (D'Onofrio et al. 2017). Riboflavin, CoQ10, and alpha lipoic acid are examples of this category (D'Onofrio et al. 2017; Ferroni et al. 2018). A diet rich in antioxidant-rich foods can protect cells from oxidative stress by slowing its start, progression, and cascade. Antioxidants in food also reduce oxidative processes, quench molecular oxygen, and scavenge free radicals (Baiano and del Nobile 2015). Continued migraine etiology research has advanced treatment research. Current research suggests that the complicated pathophysiology of migraine is influenced by genetic, epigenetic, and environmental variables (Ferroni et al. 2018). As discussed above, oxidative stress contributes to migraine etiology. Oxidative stress can be reduced using antioxidants (Ferroni et al. 2018). Migraine-prevention medications also have antioxidant properties (Ferroni et al. 2018). In addition to supplements, natural antioxidants can be found in plant-based foods such as vegetables, fruits, flowers, and spices (Deng et al. 2012; Carlsen et al. 2010). The primary antioxidants found in plants are carotenoids, polyphenols, xantophylls, anthocyanins, lignans, flavonoids, and vitamin C and E (Chiaiese et al. 2019; Xu et al. 2017; Schepetkin et al. 2016).

7.3 Potential of Nutraceutics in Treating Migraine

It is possible to reduce the frequency and severity of migraines by dietary interventions that address mitochondrial malfunction, reduced energy generation, and oxidative stress. Milder migraine symptoms and enhanced mitochondrial metabolism have been associated with the low-carbohydrate, high-fat ketogenic diet (Batch et al. 2020; Caminha et al. 2021; Haslam et al. 2021; Kovács et al. 2021; Papetti et al. 2021; Qu et al. 2021). Each diet, including the ketogenic diet, has several versions with distinct components that may interact and produce unexpected results. It is thus dangerous to prescribe a complicated diet for a particular illness; however, one possible alternative is to tailor a broad "healthy diet" to a particular disease by removing foodstuffs that contain nutrients that are known to exacerbate the condition or by adding nutrients that may alleviate symptoms or reduce the risk of the disease. This section delves into specific nutrients that might enhance a balanced diet and potentially alleviate migraine symptoms. The topic includes mitochondria, oxidative stress, and energy production.

7.3.1 Vitamins

7.3.1.1 Thiamine

Thiamine (vitamin B1) insufficiency is linked to various human illnesses, including the notable beriberi condition (Wiley and Gupta 2021). Thiamine has been used to treat migraines since 1949 (Braaf 1949, 1951). In mammals, thiamine is absorbed from the environment and converted to thiamine pyrophosphate (TPP) in the cytoplasm (Subramanian et al. 2013). The mitochondrial thiamine pyrophosphate transporter (MTPPT) transports most TPP into mitochondria. Thiamine may contribute to migraine etiology by promoting neuronal impulse propagation and myelin sheath maintenance, although the mechanisms are unknown (Wiley and Gupta 2021; Henssen et al. 2022). A case study revealed that thiamine reduced symptoms of chronic cluster headache, suggesting that it modulates pain (Antonio et al. 2018). Another study found decreased thiamine levels in two female migraine patients (Prakash et al. 2016). These instances were found to exhibit clinical signs that are indicative of Wernicke's encephalopathy (WE). Thiamine used intravenously alleviated headaches and WE. The authors speculated that a vicious cycle of thiamine shortage due to nausea, vomiting, and anorexia would exacerbate migraine headaches, as WE mimics migraine symptoms. Supplementation may break this cycle, offering potential for migraine treatment.

Migraine sufferers consumed less thiamine than healthy controls, according to a case-control study by Faraji et al. (2018). Additionally, thiamine consumption was adversely linked with migraine attack severity in patients. These relationships were not significant after adjusting for calorie intake. The research concluded that thiamine may contribute to the onset of migraine by influencing brain energy metabolism and that maintaining an adequate energy balance is critical for the condition. The brains and livers of rats on a low-thiamine diet showed no damage to the mitochondria, according to Parker et al. (1984). In the group that did not get enough thiamine, mitochondrial ATP synthesis was abnormal, and the pyruvate dehydrogenase complex and alpha-ketoglutarate dehydrogenase complex in the brain and liver were less active. This suggests that mitochondrial defects caused by thiamine shortage may be due to decreased enzyme activity. In conclusion, thiamine is a crucial human vitamin that is mostly delivered to mitochondria and may play a role in mitochondrial functioning. Thiamine may modulate pain, and its lack may be associated to migraines.

7.3.1.2 Riboflavin

Riboflavin (vitamin B2) is essential for myelin formation in the neurological system, and its absence may damage myelin lamellae (Saedisomeolia and Ashoori 2018). Due of its function in migraine etiology and mitochondria maintenance, riboflavin was connected to mitochondria. In a recent study, Yamanaka et al. found that riboflavin may prevent migraine attacks, although the evidence is insufficient to establish a judgment (Yamanaka et al. 2021). There is no evidence that riboflavin shortage causes migraine headaches, which is relevant given the large number of people worldwide with genetically determined riboflavin absorption and utilization impairment (Kennedy 2016). The American Academy of Neurology classifies riboflavin as a Level B drug for migraine prophylaxis, having been proven effective in clinical trials (Plantone et al. 2021).

Because defects in mitochondrial energy metabolism are involved in the etiology of migraines, riboflavin has potential uses in the prevention of migraines in both adults and adolescents (Schoenen et al. 1998; Yee 1999). Although the precise method by which riboflavin protects against oxidative stress remains unclear, it is believed to do so by modulating the mitochondria-energy production-oxidative stress circuit. Patients suffering from migraines who had mitochondrial DNA haplotypes other than H reaped greater benefits from riboflavin, according to research by Di Lorenzo et al. (2009). Results were explained by linking haplotype H to enhanced activity in complex I, a key riboflavin target. Migraines may have ethnic consequences, as haplotype H is primarily found in European populations. Riboflavin also protects against migraine by lowering neuroinflammation, a key component of the neurogenic underpinning of the condition (Cutrer and Charles 2008; Marashly and Bohlega 2017).

Riboflavin influences mitochondrial energy generation through the triacylglycerol (TCA) cycle; oxidative phosphorylation, and nucleotide, amino acid, and fatty acid metabolism (Tolomeo et al. 2021; Udhayabanu et al. 2017). These processes require flavoenzymes like dehydrogenases, reductases, and oxidases. Efficacy of flavoenzymes relies on their redox cofactors, FAD or FMN, which are derived from riboflavin (Giancaspero et al. 2015). Riboflavin is crucial for flavoenzyme-mediated antioxidant defense, preventing cell stress and death. Riboflavin can reduce mitochondrial ROS generation and mtDNA release, which stimulates NLRP3 and non-canonical inflammasomes, improving the mitochondria/energy production/oxidative stress axis (Yamanaka et al. 2021). It inhibits caspase-1, NLRC4, and AIM2, which are crucial for cell proliferation and other biological functions (missing in melanoma) (Ahn and Lee 2020). Riboflavin may prevent migraines by promoting antioxidant and anti-inflammatory effects in mitochondrial dysfunctions. It has a crucial role in mitochondrial homeostasis, production of energy, and protection against oxidative stress in the brain. Clinical researches have verified its potential to reduce migraines.

7.3.1.3 Niacin

Vitamin B3, sometimes known as niacin, is a combination of the compounds nicotinic acid and nicotinamide, which are building blocks for niacin and NADPH. In their roles as cofactors or cosubstrates, they are indispensable to numerous enzymes that contribute to cellular homeostasis and the etiology of human diseases (Kirkland and Meyer-Ficca 2018). The NAD/NADH ratio governs enzyme activity in key metabolic processes such as fatty acid synthesis and oxidation, TCA cycle and glycolysis. Niacin insufficiency impairs mitochondrial respiration and reduces oxidative phosphorylation (Pirinen et al. 2020). Niacin affects mitochondrial functioning by many pathways, such as mitochondrial sirtuins, PARPs, polymerases, and mitophagy (Kirkland and Meyer-Ficca 2018).

A case study found that sustained-release niacin considerably alleviated migraine headache attacks in a patient (Velling et al. 2003). The authors proposed that niacin negatively regulated the kynurenine pathway. Migraines are associated with low amounts of tryptophan, which is directed to the serotonin pathway (Giniatullin 2022). Recently, riboflavin has been shown to be useful in migraines and may transform tryptophan into niacin (Fukuwatari and Shibata 2013). Studies have indicated that a combination of small doses of caffeine, niacin, tryptophan, and acetylsalicylic acid (ASA) may alleviate migraine symptoms following an episode (Gedye 2001). The finding that a combined treatment was helpful for 9 out of 12 migraine patients validates this notion.

For the treatment of migraines and tension headaches, Prousky and Seely combed through the medical literature on niacin up until 2004 (Prousky and Seely 2005). The search for niacin as a treatment for migraines turned up no randomized or controlled trials. Inadequate control, missing patient and therapy data, and inadequate ways for assessing treatment efficacy were among the major issues the authors identified in the majority of the publications. Randomized trials are necessary to establish the clinical utility of niacin for migraine headache treatment; however, preliminary evidence suggests it may have some effect (2005 authors). The mechanism of niacin's possible benefits in migraine was poorly understood and only hypothesized. Prousky and Seely suggested that niacin may have potential as a peripheral vasodilator, potentially affecting cerebral vasodilation and contributing to migraine pathophysiology (Mason and Russo 2018). Moreover, niacin's migraine preventive effectiveness stems from its enhancement of brain energy shortage via complex I of electron transport chain (ETC).

Despite calls for randomized clinical trials, more research on the efficacy of niacin in treating migraines has not been done. Niacin indirectly affects brain energy production through glycolysis, and its lack inhibits mitochondrial functioning. Study indicates potential of niacin for migraine prevention, but further study and clinical trials are needed to support its usage in migraine treatment.

7.3.1.4 Vitamin C

Water-soluble vitamin C is a key nonenzymatic antioxidant. Humans require food to synthesize vitamin C, as they lack the enzyme L-glucono-gamma-lactone-oxidase (Ravetti et al. 2019). Fruits and vegetables contain vitamin C (Du et al. 2012). Vitamin C is active as L-ascorbic acid (LAA), a weak acid. This unstable organic molecule has a high decomposition rate related to temperature, light, heavy metal activity, and pH (Ravetti et al. 2019). Ascorbic acid plays a role in detoxification and neuronal metabolism (Covarrubias-Pinto et al. 2015). LAA has a dual impact on human redox balance: Reducing agent and enzyme cofactor (Mandl et al. 2009). Although LAA has reducing properties, it can also increase the oxidizing properties of metals like iron and copper (Mandl et al. 2009).

Complex regional pain syndrome (CRPS) in the limbs is more common in those with migraines, inflammatory bowel disease, and asthma, and it is thought to be caused by nerve fiber degeneration (Shaik and Gan 2015). A prophylactic treatment that was approved was daily vitamin C supplementation (200–1500 g) for a maximum of 50 days after a wrist injury, which lowered the likelihood of CRPS (Shaik and Gan 2015). According to the study, vitamin C may also reduce neuroinflammation and ROS activity during migraines (Shaik and Gan 2015). Vitamin C, when administered intravenously at a dosage of 5 g/day on days 1 and 3 of discomfort, successfully reduces neuralgia that can occur after shingles, according to research by Kim et al. Vitamin C supplementation, on the other hand, did not reduce herpes zoster pain (Kim et al. 2016).

In 2006, Chayasirisobhon conducted a study to investigate the impact of a daily regimen of a combination medication consisting of ten antioxidant capsules on individuals suffering from migraines, both with and without auras. Each capsule in this unblinded, open-label trial contained 120 mg of pink bark extract, 60 mg of vitamin C, and 30 international units of vitamin E. Patients reported less frequent and less severe headaches after finishing treatment (Chayasirisobohon 2006).

The effect of 150 mg of vitamin C and extract from *Pinus radiata* bark on migraine symptoms was investigated in an open-label trial by Chayasirisobhon. Over a 3-month period, patients were given the aforementioned dosages of vitamin C and *P. radiata* bark. Significant improvements were observed in patients following therapy: the occurrence and intensity of headaches diminished (Chayasirisobhon 2013).

7.3.1.5 Vitamin E

Reduced magnesium levels, platelet disorder, central serotonin dysmodulation, and progesterone (PG) generation are among the several mechanisms associated with menstrual migraine (Chayasirisobhon 2013). In the follicular and luteal phases, PG levels in the endometrium increase thrice; during menstruation, they rise even further. An increase in PG levels in the blood during the luteal phase prompted the use of anti-PG drugs (Granella et al. 1997). Enzymes phospholipase A2 and cyclo-oxygenase are responsible for converting arachidonic acid to PG, although vitamin E blocks this process (Ziaei et al. 2005).

Vitamin E, a medication that inhibits prostaglandins and has minimal adverse effects, is useful in relieving headache and symptoms of migraine (Ziaei et al. 2009; MacGregor 2010). This significantly reduced functional impairment and the need for rescue drugs (Ziaei et al. 2009; MacGregor 2010). Short-term use of preventive medicines like nonsteroidal anti-inflammatory medications and triptans can alleviate headache symptoms, but many patients have increasing symptoms after discontinuing treatment. There is no evidence that the recommended dosage of 400 IU of vitamin E daily for three cycles of menstruation can induce breakthrough headaches. Women suffering from severe and persistent migraines may get relief with this medication (Ziaei et al. 2009; MacGregor 2010). Vitamin E supplementation for 5 days during menstruation alleviated migraine pain and functional impairment in women, according to research by Ziaei et al. (2009). Further research in a larger population could determine the effectiveness of vitamin E in treating menstrual migraines in women.

7.3.1.6 Coenzyme Q10

CoQ10 (ubiquinone) plays a crucial role in bioenergetics and antioxidant defense through significant cellular redox reactions (Alcázar-Fabra et al. 2016). CoQ10's main function is ETC electron transport. CoQ accepts electrons from NADH or succinate to move electrons from complex I or II to complex III (Shults et al. 1997).

ROS generation involves CoQ10. Accumulating reduced CoQ10 leads to superoxide production via retrograde electron transfer from complex I to NAD+ (Pryde and Hirst 2011). CoQ10 is renowned for its antioxidant properties, including inhibiting lipid peroxidation and regenerating vitamin E (Littarru and Tiano 2007). Oxidative DNA damage is another threat that CoQ10 wards off (Alahmar et al. 2021). CoQ10 connects mitochondrial activity with oxidative stress and energy generation. Researchers discovered that pancreatic beta cells with intact mitochondria and tacrolimus-induced oxidative stress were both ameliorated by CoQ10 supplementation (Luo et al. 2019). Recent controlled randomized trials indicate CoQ10 may be a safe and effective supplementary medication for migraine therapy, as summarised by Paraohan et al. (2020). The primary effect of CoQ10 is to decrease migraine frequency. Sazali et al. conducted a meta-analysis of six randomized control studies comparing CoQ10 to placebo or as an additional treatment (Sazali et al. 2021). Although CoQ10 supplementation did not significantly reduce migraine headache severity, it did reduce the duration and frequency of headache attacks.

Researchers noted a significant enhancement in migraine severity, duration, frequency, high dynamic range (HDR), and lactate levels among individuals with migraines who consumed CoQ10 and L-carnitine supplements for duration of 8 weeks, as compared to those who were administered a placebo (Hajihashemi et al. 2019). Researchers observed that migraineurs who took a supplement containing nano-curcumin and CoQ10 for 8 weeks had a considerable improvement in their headache frequency, intensity, and duration (HDR) compared to those who took either supplement alone or a placebo (Parohan et al. 2021). Following the conclusion of the trial, the collective group achieved superior scores on migraine-specific surveys compared to the other groups. The subjects in the trial reported no negative effects.

Among a sample of 32 patients, a significant reduction in the number of headache days by more than 50% was observed in 61.3% of individuals who received a daily dosage of 150 mg of CoQ10 for the treatment of migraines (Rozen et al. 2002). The mean frequency of headache episodes fell from 7.34 days at the onset of treatment to 2.95 days after a 3-month period. The frequency of migraine attacks decreased without any adverse consequences. Another study analyzed five papers on CoQ10 supplementation in migraineurs and reached a similar finding (Zeng et al. 2019). Numerous randomized clinical researches indicate that supplementing with CoQ10 may reduce headache severity, frequency, and duration in migraineurs without side effects. It is interesting that there are little investigations on the mechanisms of CoQ10's positive effects in migraine, despite its well-established involvement in mitochondrial energy production and oxidative stress avoidance.

7.4 Potential of Plant Derived Antioxidants in the Treatment of Migraine

Most antioxidants present in plants are carotenoids; polyphenols; vitamins C and E; and anthocyanins, lignans, flavonoids, and phenolic acid. Preliminary data indicate that curcumin, citron, and coriander show promise as preventative treatments for migraines, while menthol and chamomile appear to be effective for acute treatment. Nevertheless, numerous kinds of research exhibited a significant likelihood of bias. Scientific assessments demonstrate that antioxidants obtained from different plants have the capacity to enhance the management of migraines due to their diverse physiological impacts (Lopresti et al. 2020).

7.4.1 Terpenoids

7.4.1.1 Curcumin

A sesquiterpene, curcumin (**1**) is a yellow dye with a specific chemical formula. This compound was first isolated in the nineteenth century by Vogel and Pelletier from the turmeric plant (*Curcuma longa*). One of the most well-known polyphenols found in plants, curcumin, is mostly obtained from the turmeric root (Aggarwal et al. 2003). The curcumin level in turmeric rhizome is thought to be around 50% (Xu et al. 2018). Various sources indicate that curcumin, derived from isolated turmeric curcuminoids, makes up approximately 60–85% of the composition. It is then followed by demethoxycurcumin, which accounts for 2% to 15–27%, and bisdemethoxycurcumin, which makes up 3% to 5–15% (Lee et al. 2013). Curcumin is extensively utilized as a condiment and in the textile dyeing sector. Curcumin is associated with a multitude of advantageous characteristics, including antioxidative, antiatherosclerotic, antibacterial, immunomodulatory, and antiaging effects (Wang et al. 2018). As a result, curcumin has garnered attention as a subject of research for its potential application in treating various conditions (Cole et al. 2007). These include a broad spectrum of disorders that impact various parts of the neurological system, including the central nervous system, spinal cord, peripheral and cranial nerves, nerve roots, autonomic nervous system, neuromuscular junction, and myopathies. Curcumin positively affects the neurological system by inhibiting the production of ROS. It also restores the decreased activity of superoxide dismutase (SOD) and the lowered level of glutathione (GSH), which are both antioxidants (Novak Kujundžić et al. 2019). In addition, studies on the neurological system have shown that it inhibits the expression of cyclooxygenase (COX-s), interleukin (IL)-1, and interleukin (IL-6), which in turn reduces inflammation. It also has antiapoptic properties, meaning it can stop cells from dying (Goschorska et al. 2020) (Fig. 7.2).

Potential of curcumin in treating migraines is receiving increased attention from scientists. The effectiveness of sumatriptan (ST) when administered alone versus in combination with curcumin was examined in a migraine treatment trial by Bulboacă et al. A nitroglycerine-induced migraine model in rats was used for the study. The gastrointestinal system has low absorption of curcumin, so it was administered intravenously as either an alcoholic solution diluted in saline or as liposomes (Pulido-Moran et al. 2016). The authors documented the antioxidant properties of curcumin, including its ability to decrease MDA concentration, reduce generation of reactive nitrogen species, lower TOS levels, and enhance TAS levels. These effects were observed for each dose and form of curcumin given, whether in solution or liposomes. Nevertheless, curcumin exhibited enhanced antioxidative effects when delivered in liposomal form. The authors attribute the desired results to the potential of curcumin to neutralize hydroxyl or peroxyl radicals, directly interact with peroxide radical anion, and inhibit the activity of NF κB nuclear transcription factors (NF κBs), which play a role in the transcription of proinflammatory proteins (Pulido-Moran et al. 2016). The researchers then conducted further experiments using a nitroglycerine-induced migraine model in rats. The objective was to evaluate the effects of naproxen and curcumin solution on pain perception and markers of oxidative stress (Bulboacă et al. 2018). Subsequent to administering liposomal curcumin, nociception, NO generation, TOS values, and MDA concentrations were all reduced. Improved antioxidant mechanisms, including increased thiol levels and total antioxidant capacity (TAC), were observed in the combination of curcumin and naproxen as compared to naproxen alone. Nevertheless, the scientists noted that patients who received curcumin in the form of liposomes exhibited a more

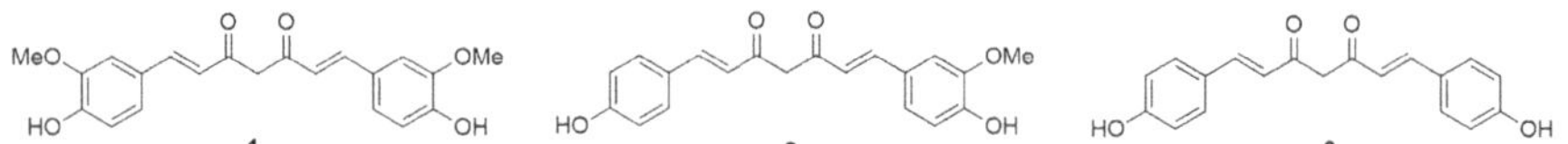

1 2 3

Fig. 7.2 Chemical structures of curcumin (**1**), demethoxycurcumin (**2**), and bisdemethoxycurcumin (**3**)

pronounced antioxidative impact. Another study has shown that curcumin has a beneficial antioxidative and analgesic impact, surpassing the effects of propranolol and indomethacin treatment (Bulboacă et al. 2019).

A study was conducted on persons to explore the possible therapeutic benefits of curcumin in the management of migraines. Tumor necrosis factor α (TNF-α) is a probable mechanism in the development of migraines. It induces migraine symptoms by heightening neuronal hyper-excitability, stimulating pain receptors, and facilitating the generation of prostanoids. Neuroinflammation then starts as a result of this (Bulboacă et al. 2017). The study conducted by Abdolahi et al. focused on individuals who experience sporadic or episodic migraines. The experiment aimed to find out how curcumin and ω-3 acids affected the TNF-α gene expression when used together. A decrease in TNF-α expression was indicated by a decrease in plasma mRNA levels in patients who were given both medications. Patients who took just one of the tested medications did not experience this (Abdolahi et al. 2017). The study investigated the antioxidant properties of nanocurcumin, in combination with coQ10, a commonly used anti-inflammatory medication, in individuals suffering from sporadic migraines. The findings of the study indicate that the concurrent use of curcumin and coQ10 effectively reduces the length, intensity, and occurrence rate of migraine episodes. On additional to that, the antioxidants used in the trial were well-tolerated by the patients (Parohan et al. 2019).

7.4.1.2 Ginkgolides

Ginkgolides are bioactive terpenoid lactones. Ginkgolide B is a unique type of diterpenoid trilactone that consists of six five-membered rings. The compound features a spiro [4,4]-nonane carbocyclic ring, a tetrahydrofuran ring, and a highly distinctive tert-butyl group attached to one of the rings. The ginkgolides were initially extracted from the *Ginkgo biloba* (GB) tree in 1932 (Stromgaard and Nakanishi 2004). The process of determining the structure was successfully completed in 1967 by Maruyama et al. (1967)

Research has demonstrated that an extract derived from GB leaves has been shown to decrease the death of hippocampus neurons, which occurs as a result of temporary global ischemia (TGA), due to its antioxidative and anti-inflammatory properties (Tulsulkar and Shah 2013). Flavonoids and terpene lactones are the main bioactive components discovered in GB leaf extract (Choi et al. 2019). Quercetin is credited by several authors for its antioxidative effects, as opposed to ginkgolide B (Shi et al. 2009). Among nutraceutical agents used in migraine treatment, ginkgolide B is the most frequently mentioned (D'Andrea et al. 2014). While the primary way it affects migraines is by adjusting the transmission of glutamate in the brain and acting as an antagonist to platelet-activating factor (PAF) receptors, the literature also mentions its antioxidative characteristics (Li et al. 2009; Ma et al. 2013). This appears to be crucial considering the fluctuations in the redox state during the progression of the migraine. Ginkgolide B is considered one of the primary active components of GB leaf extract, with ginkgolide A, C, and terpenes. It has been found to have positive effects on memory (Shah et al. 2011).

Ginkgolide B was reported to be beneficial in treating migraine with aura in patients who experienced at least one migraine episode per month by D'Andrea et al. (2009) during 6 months open-label multicenter study. The study participants were individuals who were in good health, without any other medical conditions, specifically without any cerebral focal activation. During the trial period, which was divided into two stages lasting 2 months each, the patients consumed the Migrasoll® preparation twice daily. The composition of this formulation includes 60 mg of GB terpenes phytosome, 11 mg of CoQ10, with 8.7 mg of vitamin B2. The participants in the trial group experienced a significant reduction in the frequency of migraine episodes. More precisely, 42.2% of patients experienced a total cessation of migraine occurrences following the conclusion of the experiment. Additionally, five patients did not respond to the preventative treatment using Migrasoll®. Furthermore, the introduction of supplements led to a reduction in the length of aura across the full duration of the trial (D'Andrea et al. 2009). Participants in the study were instructed to record the symptoms of aura and the level of pain experienced during the first instance of a migraine. Patients were advised to take two capsules of Migrasoll® orally at the onset of aura for future migraine attacks. The patients persisted in using analgesics. The results indicated that the use of Migrasoll® resulted in a significant decrease in the length of aura, as measured in minutes.

7.4.1.3 Feverfew

Feverfew, scientifically known as *Tanacetum parthenium* (L.) Schultz Bip., is a member of the Asteraceae family. It is a promising species within the *Tanacetum* L. genus but has not been thoroughly studied. Pharmacopoeias of many countries such as United States, United Kingdom, and Ukraine recognize feverfew herb as a raw material (British Pharmacopoeia Commission 2009; Council of Europe 2008). Parthenolide, artecanin, and germacranolide are sesquiterpene lactones that mainly determine the chemical composition of source components of feverfew. The phenolic compounds are another important group (Upton 2011; United States Pharmacopeia 2007). Hydroxycinnamic acids (e.g., chlorogenic acid, dicaffeoylquinic acid, and chicoric acid) and flavonoids (e.g., tanetin, quercetagetin-3,6-dimethyl ether, quercetagetin-3,6,3′-trimethyl ether, quercetin, apigenin, luteolin, luteolin-7-glucuronide, chrysoeriol, and santin) make up the bulk of these substances (Williams et al. 1999; Akpulat et al. 2005).

The butterbur rhizome and/or leaves are used to make preparations. Based on the chemical analysis, the main ingredients include petasins (**5**), essential oils, and pyrrolizidine alkaloids (PAs) (Borlak et al. 2022). Butterbur contains sesquiterpenes called petasins, which have pharmacological effects. To be more precise, they are petasol and angelic acid esters. The medical usage of sesquiterpenes is affected by differences in quantity and composition among species, despite these variances (Thomet et al. 2001a). Due in part to their ability to block the production of prostaglandin and leukotriene, petasins add to butterbur's anti-inflammatory characteristics (Bickel et al. 1994). The anti-inflammatory effects of petasins make butterbur a suitable option for treating migraine symptoms, considering the involvement of inflammation in causing migraine pain (Edvinsson et al. 2019). The generation of leukotrienes in human blood cells is inhibited by petasin, isopetasin, and neopetasin (Thomet et al. 2001b). Petasin stops platelet-activating factor and complement factor 5a from causing a rise in intracellular Ca^{2+} by blocking the release of eosinophil cationic protein. Eosinophils that have been activated are unable of moving 5-lipoxygenase out of the cytosol into the nucleus due to the inhibitory effect of petasin on phospholipase A2 activity. This provides more evidence that petasins can block a variety of signaling molecules within cells. This suggests that petasins can block phospholipase C (beta) and other early signaling pathways initiated by G protein-coupled receptors of granulocytes (Thomet et al. 2001a). In migraine attacks, the inflammatory response is significant, and the increased phospholipase C activity in CSF fluid of migraineurs supports this theory (Fonteh et al. 2011). Recent research strongly suggests that petasins can inhibit CGRP nociceptive pathways, and butterbur is an excellent migraine preventative.

7.4.2 Phenolic Compounds

Research has substantiated the preventive and defensive properties of phenolic chemicals, specifically caffeoylquinic acids (**6**; Fig. 7.3), against neurodegenerative disorders (Murai and Matsuda 2023). Extracts of *T. parthenium* and *Salix alba* were found by Giacomo et al. (2019) to have neuromodulatory and anti-inflammatory effects, which they attribute to their capability to eradicate damaging free radicals. Further, changes in the rat brain were observed after *T. parthenium* water extract administration, with quantities of neurofilament and myelin-associated glycoprotein being reduced in particular. A thorough understanding of the phenolic profile and specific mechanisms of various phytochemical markers is essential for addressing concerns regarding potential neurotoxic implications. This is significant because antioxidant chemicals may potentially have pro-oxidant effects (Hordiei et al. 2023). It is essential to ascertain the overall quantities of phenolic compounds obtained from phenolic sources, as they play a vital function as antioxidants. These molecules are crucial in removing harmful free radicals and ROS, therefore, ensuring the equilibrium of redox reactions (Goschorska et al. 2020). Oxidative stress is strongly linked to other inflammatory processes, including neuroinflammation, which is believed to play a role in the development of migraines and similar diseases (Kursun et al. 2021). Inflammation and the equilibrium between oxidants and antioxidants play a significant role in the etiopathogenesis of migraines (Borkum 2016).

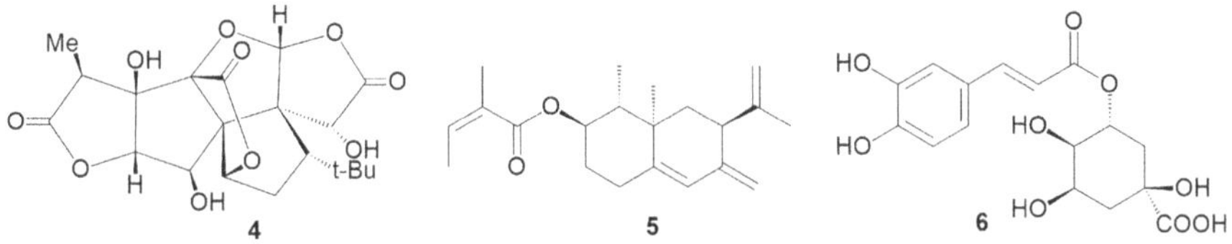

Fig. 7.3 Chemical structures of compounds **4**, **5**, and **6**

Migraine attacks may actually be a preventive mechanism for the nervous system against oxidative stress, according to one theory. The pain of migraines can be reduced by using phenolic antioxidants (Bakirhan et al. 2022).

7.4.3 Flavonoids

Tanacetum species have been found to contain significant phenolic compounds, specifically hydroxycinnamic acids and flavonoids, according to scientific studies (Pareek et al. 2011). Studies provide valuable data on the total concentrations of these compounds, enabling a comprehensive understanding of their presence and role in the *T. parthenium* components. The polyphenols catechin, tannin, caffeic acid, and ferulic acid are significant components of areca nuts. The popular traditional medicine for migraine in the Indian states of Tamil Nadu and Kerala is areca nut. This may be because procyanidins, which are polyphenolic chemicals found in areca nuts, have an inhibitory effect on macrophages' inducible nitric oxide synthase (iNOS) (Huang et al. 2010).

Proanthocyanidins, or condensed tannins, are conjugation products containing two or more flavan-3-ol subunits, such as catechin (**7**) and epicatechin (**8**) (Fig. 7.4). Procyanidins fall under this category also.

Procyanidins are generally conjugated with single-bond links C4–C8 or C4–C6 (B type, **9**), and with double-bond linkages C4–C8/C2–O7 (A type, **10**; Fig. 7.4) (Xie and Dixon 2005). The various linking types, subunit compositions, and chain lengths of proanthocyanidins result in a variety of structures with distinct biological functions (Pierini et al. 2008). According to a prior study, procyanidins prevent the inflammatory effects of lipopolysaccharide (LPS) with interferon-γ in Raw 264.7 macrophages. These effects include the creation of nitric oxide, the stimulation of nitric oxide synthase (iNOS) protein expression, and the synthesis of prostaglandin

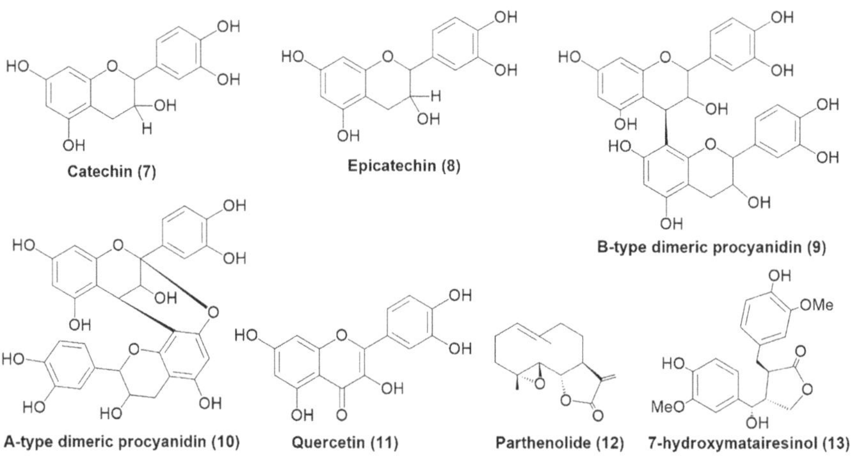

Fig. 7.4 Chemical structures of compounds **7**–**12**

E2 (PGE2) (Terra et al. 2007). Procyanidins may therefore have the ability to upregulate endogenous antioxidant enzymes and lower oxidative stress, which underlies their antioxidant and anti-inflammatory properties.

One of the abundant flavonols in plants, fruits, vegetables, bark, tea, and other items is quercetin (**11**; Fig. 7.4). Because of its strong antioxidant activity, quercetin demonstrated effectiveness in reducing inflammation. In an in vitro investigation, quercetin inhibited calcium channels, disrupted cyclooxygenase and nitric oxide pathways, and caused basilar artery relaxation (Yuan et al. 2018). For this reason, quercetin was chosen for the current study to look into its impact on migraines. Additionally, it has the potential to be a migraine therapy agent. The study findings indicate that there was an increase in the quantity of immunoreactive CGRP and C-fos cells in the TSN fibers of the rat model of migraines induced by nitroglycerine. However, the injection of quercetin slowed down this cell multiplication. Even more encouraging is the finding that tecagepant and olcegepant, which are CGRP receptor antagonists, provide relief and protection from migraines (Ho et al. 2016). The rat model of nitroglycerine-induced migraine shows that allodynia, freezing, resting, exploratory, and locomotor behavior can be alleviated by quercetin pretreatment (Zhang et al. 2017). Based on these results, quercetin may be useful in treating behavioral problems and migraine pain. One way quercetin helps migraine sufferers deal with their migraines is by lowering inflammatory factors like CGRP and immunoreactive C-fos cells (Foudah et al. 2022).

Tanacetum species have been found to contain significant phenolic compounds, specifically hydroxycinnamic acids and flavonoids, according to scientific studies (Pareek et al. 2011). Their research provides valuable data on the total concentrations of these compounds, enabling a comprehensive understanding of their presence and role in the *T. parthenium* materials. One of the well-studied herbs for migraine prophylaxis that has demonstrated effectiveness in treating both adult and pediatric migraineurs is *T. parthenium* (Moscano et al. 2019). A member of the Asteraceae family, *T. parthenium* has long been used as an analgesic, anti-inflammatory, and antipyretic (Pareek et al. 2011). The chemicals that make up the active phytocomplex are parthenolide (**12**; Fig. 7.4), flavonoids, and volatile oils. They have an anti-inflammatory effect by inhibiting the formation of prostaglandin and nitric oxide (Materazzi et al. 2013). It's interesting to note that oral administration of *S. alba* reduced 5-HT turnover in the rat hippocampal region and shown modulatory effect on the 5-HT pathway (Ulrich-Merzenich et al. 2012). Given the antioxidant and reducing properties of the *S. alba* and *T. parthenium* extracts, phenolic compounds—namely, gallic acid, epicatechin, resveratrol, and catechin—that are well-known for their antioxidant and anti-inflammatory properties were the focus of a fingerprint HPLC-fluorimetric analysis. Consistent with what Esatbeyoglu and colleagues found (2010), *S. alba* extract has larger levels of epicatechin and catechin compared to *T. parthenium* extract. Conversely, the *T. parthenium* extract exhibits higher quantities of gallic acid and resveratrol, with resveratrol being undetectable in the *S. alba* extract (Close et al. 2019). By activating the trigeminal nociceptive system, CSD may also function as a migraine initiator. Instead, low 5-HT levels may impact migraine attacks by reducing pain perception and disrupting the balance in the regulation of cerebrovascular nociception (Supornsilpchai et al. 2006). According to the findings, K^+ (60 mM)-induced 5HIIA/5-HT could be entirely inhibited by *S. alba* and *T. parthenium* extracts. This is significant since brain 5-HT catabolism has long been thought to be accurately indicated by these markers (Brunetti et al. 2014). If the results from *S. alba* match the data that Ulrich-Merzenich et al. (2012) have already published, then the results from *T. parthenium* seem to contradict the previously shown inhibition of 5-HT release in the rat hippocampus (Mittra et al. 2000). The neuroprotective properties of these extracts are also consistent with the phenolic compounds—catechin, epicatechin, gallic acid, and resveratrol—that they contain (Shanan et al. 2019).

7.4.4 Lignans

Due to their diverse biological actions, plant lignans are being studied for possible application as medicinal agents. They offer a variety of pharmacophores that occur naturally and can be produced chemically (Giuliano et al. 2020). Numerous studies have shown that lignans with various structural groups can often exhibit both antioxidant and anti-inflammatory properties at the same time (Jeong et al. 2013). One type of polyphenol that is frequently included in whole grains, seeds, nuts, legumes, and vegetables is called lignans. These plant metabolites have been the focus of intense research over the last 10 years as possible treatments for a variety of neuropathologic disorders and neurodegenerative diseases, including Parkinson's disease (PD) and Alzheimer's disease (AD) (Yang et al. 2018). It has been demonstrated that lignans possess significant neuroprotective, anti-inflammatory, antioxidant, and immunomodulatory properties (Park et al. 2015). Treatment with lignans extracted from *Schisandra chinensis* fruits was shown to lower the levels of amyloid β (Aβ) peptide 1–42, attenuate Aβ1–42-induced neuronal cell injury by increasing the activities of glutathione peroxidase and SOD and lowering the levels of reactive oxygen species, lactate dehydrogenase, and malondialdehyde (Jeong et al. 2013; Yang et al. 2018). Research has demonstrated that administering 7-hydroxymatairesinol (**13**; Fig. 7.4) over an extended period of time can impede the degeneration of striatal dopaminergic terminals in a rat model of PD, leading to an enhancement in motor function (Cosentino et al. 2007).

7.5 Preclinical and Clinical Studies of Antioxidants in the Treatment of Migraine

The only butterbur extract that has undergone clinical evaluation for therapeutic efficacy in the prevention of migraines is Petadolex® (Borlak et al. 2022). The course of treatment lasted between 3 and 4 months. In a randomized, double-blind, placebo-controlled trial, the pivotal study by Lipton et al. assessed the efficacy of

4 months of treatment with Petadolex® at dosages of 50 mg bid or 75 mg bid. About 233 participants with episodic migraine were included in the study (Lipton et al. 2004). Study participants using 150 mg of Petadolex® daily experienced fewer migraine attacks than those on a placebo. The reduction was 48% at 150 mg, 36% at 100 mg, and 26% in the placebo group; however, there was no statistically significant difference when using 100 mg as compared to the placebo. At the 4-month mark, there was a significant decrease in migraine episode occurrence ($p < 0.05$ for 150 mg of Petadolex® vs. placebo) in 68% of patients taking 150 mg, 56% of patients taking 100 mg, and 49% of patients using a placebo. Comprehensive reexamination of the data by Diener et al. demonstrated that the 100 mg dose of Petadolex® was effective, lending credence to the clinical investigators' claims of high efficacy and excellent tolerability (Grossman and Schmidramsl 2001; Diener et al. 2004). Compared to the placebo group, which had a rate of 15% (45 vs.), the Petadolex® study arm had a responder rate of 50% or higher for migraine frequency. The efficacy of Petadolex® in alleviating episodic migraine in 63 children in primary school was the subject of the third study. The design of the study included a parallel-group design with random assignment and partial double-blinding. Attack frequency decreased significantly across all groups relative to baseline (Oelkers-Ax et al. 2008).

7.6 Conclusion and Future Prospects

There is hope for the prevention and treatment of migraines by the targeting of key pathways associated with migraine etiology by terpenoids, chlorogenic acids, phenolic compounds, lignan, phenolics coupled with terpene-origin compounds, and so forth. Specifically, there is promising evidence that these compounds, when consumed regularly through food, can alleviate migraines. At the intersection of pharmaceuticals and nutrition, functional foods and their bioactive components will pave the way for the development of novel therapeutic interventions aimed at preventing migraines. Furthermore, knowledge of the phytogeographical profiles of this family of chemicals can help with cultivation techniques and resource management. Finding areas with ideal environmental conditions can help to ensure the desired uniform phytochemical makeup (Murai and Matsuda 2023). The available research on the topic supports the use of these plant-derived chemicals; consequently, if they show promise in additional studies including both humans and animals, they may be administered as a stand-alone treatment or in conjunction with other preventive medical interventions to enhance outcomes. Results suggested that a lower incidence of headaches may be linked to increased adherence to diets high in phytochemicals. Additionally, it offered compelling proof that led compounds found in phytoconstituents might be used to treat or prevent migraine headaches. These substances can be further examined to see if they work as medications to prevent migraines.

References

Abdolahi M, Tafakhori A, Togha M, Okhovat AA, Siassi F, Eshraghian MR, Sedighiyan M, Djalali M, Mohammadzadeh Honarvar N, Djalali M (2017) The synergistic effects of ω-3 fatty acids and nano-curcumin supplementation on tumor necrosis factor (TNF)-α gene expression and serum level in migraine patients. Immunogenetics 69:371–378. https://doi.org/10.1007/s00251-017-0992-8

Aggarwal BB, Kumar A, Bharti AC (2003) Anticancer potential of curcumin: preclinical and clinical studies. Anticancer Res 23:363–398

Ahn H, Lee GS (2020) Riboflavin, vitamin B2, attenuates NLRP3, NLRC4, AIM2, and non-canonical inflammasomes by the inhibition of caspase-1 activity. Sci Rep 10:19091

Akpulat H, Tepe B, Sokmen A, Daferera D, Polissiou M (2005) Composition of the essential oils of Tanacetum argyrophyllum (C. Koch) Tvzel. var. argyrophyllum and Tanacetum parthenium (L.) Schultz Bip. (Asteraceae) from Turkey. Biochem Syst Ecol 33:511–516

Alahmar AT, Sengupta P, Dutta S, Calogero AE (2021) Coenzyme Q10, oxidative stress markers, and sperm DNA damage in men with idiopathic oligoasthenoteratospermia. Clin Exp Reprod Med 48:150–155

Alcázar-Fabra M, Navas P, Brea-Calvo G (2016) Coenzyme Q biosynthesis and its role in the respiratory chain structure. Biochim Biophys Acta 1857:1073–1078

Alp R, Selek S, Alp SI, Taşkin A, Koçyiğit A (2010) Oxidative and antioxidative balance in patients with migraine. Eur Rev Med Pharmacol Sci 14:877–882

Antonio C, Massimo T, Gianpaolo Z, Immacolata PM, Erika T (2018) Oral high-dose thiamine improves the symptoms of chronic cluster headache. Case Rep Neurol Med 2018:3901619

Arulmozhi DK, Veeranjaneyulu A, Bodhankar SL (2007) The herbal approach for the treatment of migraine. Pharmacol Online 1:19–37

Aytaç B, Coşkun Ö, Alioğlu B, Durak ZE, Büber S, Tapçi E, Ocal R, Inan LE, Durak İ, Yoldaş TK (2014) Decreased antioxidant status in migraine patients with brain white matter hyperintensities. Neurol Sci 35:1925–1929

Baiano A, del Nobile MA (2015) Antioxidant compounds from vegetable matrices: biosynthesis, occurrence, and extraction systems. Crit Rev Food Sci Nutr 56:2053–2068

Bakirhan H, Pehlivan M, Cankay TU, Kocak M (2022) Migraine severity, disability and duration: is a good diet quality, high intake of phytochemicals and polyphenols important? Front Nutr 9:1041907

Batch JT, Lamsal SP, Adkins M, Sultan S, Ramirez MN (2020) Advantages and disadvantages of the ketogenic diet: a review article. Cureus 12:e9639

Bickel D, Roder T, Bestmann HJ, Brune K (1994) Identification and characterization of inhibitors of peptido-leukotriene—synthesis from Petasites hybridus. Planta Med 60:318–322

Borkum JM (2016) Migraine triggers and oxidative stress: a narrative review and synthesis. Headache 56:12–35

Borlak J, Diener H-C, Kleeberg-Hartmann J, Messlinger K, Silberstein S (2022) Petasites for migraine prevention: new data on mode of action, pharmacology and safety. A narrative review. Front Neurol 13:864689

Braaf MM (1949) Migraine headache treated successfully by head-traction manipulation and thiamin chloride. N Y State J Med 49:1812–1816

Braaf MM (1951) Migraine headache: an analysis of 124 cases treated by head-traction manipulation and thiamin chloride. N Y State J Med 51:528–533

British Pharmacopoeia Commission (2009) British pharmacopoeia, vol 3. The Stationery Office, London, UK

Brunetti L, Orlando G, Ferrante C, Recinella L, Leone S, Chiavaroli A, Di Nisio C, Shohreh R, Manippa F, Ricciuti A, Vacca M (2014) Peripheral chemerin administration modulates hypothalamic control of feeding. Peptides 51:115–121

Bulboacă AE, Bolboacă SD, Stănescu IC, Sfrângeu CA, Bulboacă AC (2017) Preemptive analgesic and antioxidative effect of curcumin for experimental migraine. Biomed Res Int. https://doi.org/10.1155/2017/4754701

Bulboacă AE, Bolboacă SD, Stănescu IC, Sfrângeu CA, Porfire A, Tefas L, Bulboacă AC (2018) The effect of intravenous administration of liposomal curcumin in addition to sumatriptan treatment in an experimental migraine model in rats. Int J Nanomedicine 13:3093–3103. https://doi.org/10.2147/IJN.S162087

Bulboacă AE, Bolboacă SD, Bulboacă AC, Porfire AS, Tefas LR, Suciu SM, Dogaru G, Stănescu IC (2019) Liposomal curcumin enhances the effect of naproxen in a rat model of migraine. Med Sci Monit 25:5087–5097. https://doi.org/10.12659/MSM.915607

Burstein R, Noseda R, Borsook D (2015) Migraine: multiple processes, complex pathophysiology. J Neurosci 35:6619–6629

Buse DC, Manack AN, Fanning KM, Serrano D, Reed ML, Turkel CC et al (2012) Chronic migraine prevalence, disability, and sociodemographic factors: results from the American Migraine Prevalence and Prevention study. Headache 52(10):1456–1470

Caminha MC, Moreira AB, Matheus FC, Rieger DK, Moreira JD, Dalmarco EM, Demarchi IG, Lin K (2021) Efficacy and tolerability of the ketogenic diet and its variations for preventing migraine in adolescents and adults: a systematic review. Nutr Rev 80(6):1634–1647

Carlsen MH, Halversen BL, Holte K, Bohn SK, Dragland S, Sampson L, Willey C, Senoo H, Umezono Y, Sanada C et al (2010) The total antioxidant content of more than 3100 foods, beverages, spices, herbs and supplements used worldwide. Nutr J 9:3

Chayasirisobhon S (2013) Efficacy of Pinus radiata bark extract and vitamin combination product as a prophylactic therapy for recalcitrant migraine and long-term results. Acta Neurol 22:13–21

Chayasirisobohon S (2006) Use of a pink bark extract and antioxidant vitamin combination product as therapy for migraine in patients refractory to pharmacologic medication. Headache 46:788–793

Chiaiese P, Corrado G, Minutolo M, Barone A, Errico A (2019) Transcriptional regulation of ascorbic acid during fruit ripening in pepper (Capsicum annuum) varieties with low and high antioxidants content. Plants 8:206

Choi MS, Kim JK, Kim DH, Yoo HH (2019) Effects of gut microbiota on the bioavailability of bioactive compounds from ginkgo leaf extracts. Metabolites 9:132. https://doi.org/10.3390/metabo9070132

Ciancarelli I, Tozzi-Ciancarelli MG, Di Massimo C, Marini C, Carolei A (2003) Urinary nitric oxide metabolites and lipid peroxidation by-products in migraine. Cephalalgia 23(1):39–42

Close LN, Eftekhari S, Wang M, Charles AC, Russo AF (2019) Cortical spreading depression as a site of origin for migraine: role of CGRP. Cephalalgia 39(3):428–434

Cole GM, Teter B, Frautschy SA (2007) Neuroprotective effects of curcumin. In: Aggarwal BB, Surh YJ, Shishodia S (eds) The molecular targets and therapeutic uses of curcumin in health and disease, Advances in experimental medicine and biology, vol 595. Springer, Boston, MA

Cosentino M, Marino F, Ferrari M, Rasini E, Bombelli R, Luini A, Legnaro M, Delle Canne MG, Luzzani M, Crema F, Paracchini S, Lecchini S (2007) Estrogenic activity of 7-hydroxymatairesinol potassium acetate (HMR/lignan) from Norway spruce (Picea abies) knots and of its active metabolite enterolactone in MCF-7 cells. Pharmacol Res 56(2):140–147

Council of Europe (ed) (2008) European pharmacopoeia, 6th edn. Council of Europe, Strasbourg, France

Covarrubias-Pinto A, Acuña AI, Beltrán FA, Torres-Díaz L, Castro MA (2015) Old things new view: ascorbic acid protects the brain in neurodegenerative disorders. Int J Mol Sci 16:28194–28217

Cutrer FM, Charles A (2008) The neurogenic basis of migraine. Headache 48:1411–1414

D'Andrea G, Bussone G, Allais G, Aggugia M, D'Onofrio F, Maggio M, Moschiano F, Saracco MG, Terzi MG, Petretta V et al (2009) Efficiacy of ginkgolide B in the prophylaxis of migraine with aura. Neurol Sci 30(Suppl 1):S121–S124

D'Andrea G, Cevoli S, Cologno D (2014) Herbal therapy in migraine. Neurol Sci 35(Suppl 1):S135–S140. https://doi.org/10.1007/s10072-014-1757-x

D'Onofrio F, Raimo S, Spitaleri D, Casucci G, Bussone G (2017) Usefulness of nutraceuticals in migraine prophylaxis. Neurol Sci 38:S117–S120

Deng GF, Shen C, Xu XR, Kuang RD, Guo YJ, Zeng LS, Gao LL, Lin X, Xie JF, Xia EQ (2012) Potential of fruit wastes as natural resources of bioactive compounds. Int J Mol Sci 13:8308–8323

Di Lorenzo C, Pierelli F, Coppola G, Grieco GS, Rengo C, Ciccolella M, Magis D, Bolla M, Casali C, Santorelli FM et al (2009) Mitochondrial DNA haplogroups influence the therapeutic response to riboflavin in migraineurs. Neurology 72:1588–1594

Diener HC, Rahlfs VW, Danesch U (2004) The first placebo-controlled trial of a special butterbur root extract for the prevention of migraine: reanalysis of efficacy criteria. Eur Neurol 51:89–97

Dodick DW (2018) A phase-by-phase review of migraine pathophysiology. Headache 58(Suppl 1):4–16

Du J, Cullen JJ, Buettner GR (2012) Ascorbic acid: chemistry, biology and the treatment of cancer. Biochim Biophys Acta 1826:443–457

Edvinsson L, Haanes KA, Warfvinge K (2019) Does inflammation have a role in migraine? Nat Rev Neurol 15:483–490

Esatbeyoglu T, Wray V, Winterhalter P (2010) Dimeric procyanidins: screening for B1 to B8 and semisynthetic preparation of B3, B4, B6, and B8 from a polymeric procyanidin fraction of white willow bark (Salix alba). J Agric Food Chem 58(13):7820–7830

Evers S, Afra J, Frese A, Goadsby PJ, Linde M, May A et al (2006) EFNS guideline on the drug treatment of migraine–report of an EFNS task force. Eur J Neurol 13(6):560–572

Faraji H, Paknahad Z, Chitsaz A (2018) Dietary intake of thiamine in migraine patients and healthy subjects: a case-control study. Clin Nutr Res 7:40–47

Ferrari MD, Roon KI, Lipton RB, Goadsby PJ (2001) Oral triptans (serotonin 5-HT1B/1D agonists) in acute migraine treatment: a meta-analysis of 53 trials. Lancet 358(9294):1668–1675

Ferroni P, Barbanti P, Della-Morte D, Palmirotta R, Jirillo E, Guadagni F (2018) Redox mechanisms in migraine: novel therapeutics and dietary interventions. Antioxid Redox Signal 28:1144–1183

Fonteh AN, Chung R, Sharma TL, Fisher RD, Pogoda JM, Cowan R et al (2011) Cerebrospinal fluid phospholipase C activity increases in migraine. Cephalalgia 31:456–462

Foudah AI, Devi S, Alqarni MH, Alam A, Salkini MA, Kumar M, Almalki HS (2022) Quercetin attenuates nitroglycerin-induced migraine headaches by inhibiting oxidative stress and inflammatory mediators. Nutrients 14:4871

Fukuwatari T, Shibata K (2013) Nutritional aspect of tryptophan metabolism. Int J Tryptophan Res 6:3–8

Gedye A (2001) Hypothesized treatment for migraines using low doses of tryptophan, niacin, calcium, caffeine, and acetylsalicylic acid. Med Hypotheses 56:91–94

Geyik S, Altunisik E, Neyal AM, Taysi S (2016) Oxidative stress and DNA damage in patients with migraine. J Headache Pain 17:10

Giacomo V, Ferrante C, Ronci M, Cataldi A, Valerio V, Rapino M, Recinella L, Chiavaroli A, Leone S, Vladimir-Knezevic S et al (2019) Multiple pharmacological and toxicological investigations of Tanacetum parthenium and Salix alba extracts: focus on potential application as antimigraine agents. Food Chem Toxicol 133:110783

Giancaspero TA, Colella M, Brizio C, Difonzo G, Fiorino GM, Leone P, Brandsch R, Bonomi F, Iametti S, Barile M (2015) Remaining challenges in cellular flavin cofactor homeostasis and flavoprotein biogenesis. Front Chem 3:30

Giniatullin R (2022) Serotonin in migraine: the puzzling role of ionotropic 5-HT(3) receptor in the context of established therapeutic effect of metabotropic 5-HT(1) subtypes. Br J Pharmacol 179(3):400–415

Giuliano C, Siani F, Mus L, Ghezzi C, Cerri S, Pacchetti B, Bigogno C, Blandini F (2020) Neuroprotective effects of lignan 7-hydroxymatairesinol (HMR/lignan) in a rodent model of Parkinson's disease. Nutrition 69:110494

Goadsby PJ, Sprenger T (2010) Current practice and future directions in the prevention and acute management of migraine. Lancet Neurol 9(3):285–298

Goadsby PJ, Holland PR, Martins-Oliveira M, Hoffmann J, Schankin C, Akerman S (2017) Pathophysiology of migraine: a disorder of sensory processing. Physiol Rev 97:553–622

Goschorska M, Gutowska I, Baranowska-Bosiacka I, Barczak K, Chlubek D (2020) The use of antioxidants in the treatment of migraine. Antioxidants (Basel) 9(2):116. https://doi.org/10.3390/antiox9020116. PMID: 32012936; PMCID: PMC7070237

Granella F, Sances G, Messa G, de Marinis M, Manzoni GC (1997) Treatment of menstrual migraine. Cephalalgia 17(Suppl 20):35–38

Grossman W, Schmidramsl H (2001) An extract of Petasites hybridus is effective in the prophylaxis of migraine. Altern Med Rev 6:303–310

Gupta R, Pathak R, Bhatia MS, Banerjee BD (2009) Comparison of oxidative stress among migraineurs, tension-type headache subjects, and a control group. Ann Indian Acad Neurol 12(3):167–172

Hajihashemi P, Askari G, Khorvash F, Reza Maracy M, Nourian M (2019) The effects of concurrent Coenzyme Q10, L-carnitine supplementation in migraine prophylaxis: a randomized, placebo-controlled, double-blind trial. Cephalalgia 39:648–654

Haslam RL, Bezzina A, Herbert J, Spratt N, Rollo ME, Collins CE (2021) Can ketogenic diet therapy improve migraine frequency, severity and duration? Healthcare 9:1105

Hayes NA, Foreman JC (1987) The activity of compounds extracted from feverfew on histamine release from rat mast cells. J Pharm Pharmacol 39(6):466–470

Headache Classification Committee of the International Headache Society (2018) The International Classification of Headache Disorders, 3rd edition. Cephalalgia 38:1–211

Henssen D, Kluin SJP, Kleerebezem J, Cappellen V, van Walsum AM, Mulleners WM, Vissers K (2022) White matter changes in the trigeminal spinal tract in chronic migraineurs: an ex vivo study combining ultra-high field diffusion tensor imaging and polarized light imaging microscopy. Pain 163(4):779–785

Heptinstall S, Williamson L, White A, Mitchell JR (1985) Extracts of feverfew inhibit granule secretion in blood platelets and polymorphonuclear leucocytes. Lancet 325(8437):1071–1074

Ho TW, Ho AP, Ge YJ, Assaid C, Gottwald R, MacGregor EA, Mannix LK, van Oosterhout WP, Koppenhaver J, Lines C et al (2016) Randomized controlled trial of the CGRP receptor antagonist telcagepant for prevention of headache in women with perimenstrual migraine. Cephalalgia 36:148–161

Hobbs C (1989) Feverfew. Herbalgram 20:26–35

Hordiei K, Gontova T, Trumbeckaite S, Yaremenko M, Raudone L (2023) Phenolic composition and antioxidant activity of Tanacetum parthenium cultivated in different regions of Ukraine: insights into the flavonoids and hydroxycinnamic acids profile. Plants 12:2940

Huang P-L, Chi C-W, Liu T-Y (2010) Effects of Areca catechu L. containing procyanidins on cyclooxygenase-2 expression in vitro and in vivo. Food Chem Toxicol 48(1):306–313

Jeong EJ, Lee HK, Lee KY, Jeon BJ, Kim DH, Park JH, Song JH, Huh J, Lee JH, Sung SH (2013) Effects of lignan-riched extract of Shisandra chinensis on amyloid-β-induced cognitive impairment and neurotoxicity in the cortex and hippocampus of mouse. J Ethnopharmacol 146(1):347–354

Katsarava Z, Schneeweiss S, Kurth T, Kroener U, Fritsche G, Eikermann A et al (2004) Incidence and predictors for chronicity of headache in patients with episodic migraine. Neurology 62(5):788–790

Kennedy DO (2016) B vitamins and the brain: mechanisms, dose and efficacy—a review. Nutrients 8:68

Kim HK, Park SK, Zhou J-L et al (2004) Reactive oxygen species (ROS) play an important role in a rat model of neuropathic pain. Pain 111(1–2):116–124

Kim MS, Kim DJ, Na CH, Shin BS (2016) A study of intravenous administration of vitamin C in the treatment of acute herpetic pain and postherpetic neuralgia. Ann Dermatol 28:677–683

Kirkland JB, Meyer-Ficca ML (2018) Niacin. Adv Food Nutr Res 83:83–149

Kovács Z, Brunner B, Ari C (2021) Beneficial effects of exogenous ketogenic supplements on aging processes and age-related neurodegenerative diseases. Nutrients 13:2197

Kursun O, Yemisci M, van den Maagdenberg AMJM, Karatas H (2021) Migraine and neuroinflammation: the inflammasome perspective. J Headache Pain 22:55

Leao AA (1947) Further observations on the spreading depression of activity in the cerebral cortex. J Neurophysiol 10(6):409–414
Lee WH, Loo CY, Bebawy M, Luk F, Mason RS, Rohanizadeh R (2013) Curcumin and its derivatives: their application in neuropharmacology and neuroscience in the 21st century. Curr Neuropharmacol 11:338–378. https://doi.org/10.2174/1570159X11311040002
Levy D, Strassman AM (2002) Mechanical response properties of A and C primary afferent neurons innervating the rat intracranial dura. J Neurophysiol 88:3021–3031
Li R, Chen B, Wu W, Bao L, Li J, Qi R (2009) Ginkgolide B suppresses intercellular adhesion molecule-1 expression via blocking nuclear factor-kB activation in human vascular endothelial cells stimulated by oxidized low-density lipoprotein. J Pharmacol Sci 110:362–369. https://doi.org/10.1254/jphs.08275FP
Lipton RB, Gobel H, Einhaupl KM, Wilks K, Mauskop A (2004) Petasites hybridus root (butterbur) is an effective preventive treatment for migraine. Neurology 63:2240–2244
Littarru GP, Tiano L (2007) Bioenergetic and antioxidant properties of coenzyme Q10: recent developments. Mol Biotechnol 37:31–37
Lopresti AL, Smith SJ, Drummond PD (2020) Herbal treatments for migraine: a systematic review of randomised-controlled studies. Phytother Res 34(10):2493–2517
Luo K, Yu JH, Quan Y, Shin YJ, Lee KE, Kim HL, Ko EJ, Chung BH, Lim SW, Yang CW (2019) Therapeutic potential of coenzyme Q(10) in mitochondrial dysfunction during tacrolimus-induced beta cell injury. Sci Rep 9:7995
Ma L, Liu X, Zhao Y, Chen B, Li X, Qi R (2013) Ginkgolide B reduces LOX-1 expression by inhibiting Akt phosphorylation and increasing Sirt1 expression in oxidized LDL-stimulated human umbilical vein endothelial cells. PLoS One 8:e74769
MacGregor EA (2010) Prevention and treatment of menstrual migraine. Drugs 70(14):1799–1818
Magis D, Schoenen J (2011) Treatment of migraine: update on new therapies. Curr Opin Neurol 24(3):203–210
Mandl J, Szarka A, Banhegyi E (2009) Vitamin C: update on physiology and pharmacology. Br J Pharmacol 157:1097–1110
Marashly ET, Bohlega SA (2017) Riboflavin has neuroprotective potential: focus on Parkinson's disease and migraine. Front Neurol 8:333
Maruyama M, Terahara A, Itagaki Y, Nakanishi K (1967) The ginkgolides. I. Isolation and characterization of the various groups. Tetrahedron Lett 8(4):299–302. https://doi.org/10.1016/S0040-4039(00)71538-3
Mason BN, Russo AF (2018) Vascular contributions to migraine: time to revisit? Front Cell Neurosci 12:233
Materazzi S, Benemei S, Fusi C, Gualdani R, De Siena G, Vastani N, Andersson DA, Trevisan G, Moncelli MR, Wei X, Dussor G, Pollastro F, Patacchini R, Appendino G, Geppetti P, Nassini R (2013) Parthenolide inhibits nociception and neurogenic vasodilatation in the trigeminovascular system by targeting the TRPA1 channel. Pain 154(12):2750–2758
Metea MR, Newman EA (2006) Glial cells dilate and constrict blood vessels: a mechanism of neurovascular coupling. J Neurosci 26(11):2862–2870
Mittra S, Datta A, Singh SK, Singh A (2000) 5-Hydroxytryptamine-inhibiting property of Feverfew: role of parthenolide content. Acta Pharmacol Sin 21(12):1106–1114
Montagna P, Sacquegna T, Martinelli P et al (1988) Mitochondrial abnormalities in migraine. Preliminary findings. Headache 28(7):477–480
Moscano F, Guiducci M, Maltoni L, Striano P, Ledda MG, Zoroddu F, Raucci U, Villa MP, Parisi P (2019) An observational study of fixed-dose Tanacetum parthenium nutraceutical preparation for prophylaxis of pediatric headache. Ital J Pediatr 45(1):36
Murai T, Matsuda S (2023) The chemopreventive effects of chlorogenic acids, phenolic compounds in coffee, against inflammation, cancer, and neurological diseases. Molecules 28:2381
Noseda R, Burstein R (2013) Migraine pathophysiology: anatomy of the trigeminovascular pathway and associated neurological symptoms, cortical spreading depression, sensitization, and modulation of pain. Pain 154(Suppl 1):S44–S53

Novak Kujundžić R, Stepanić V, Milković L, Čipak Gašparović A, Tomljanović M, Gall Trošelj K (2019) Curcumin and its potential for systemic targeting of inflamm-aging and metabolic reprogramming in cancer. Int J Mol Sci 20:1180. https://doi.org/10.3390/ijms20051180

Oelkers-Ax R, Leins A, Parzer P, Hillecke T, Bolay HV, Fischer J et al (2008) Butterbur root extract and music therapy in the prevention of childhood migraine: an explorative study. Eur J Pain 12:301–313

Olesen J, Burstein R, Ashina M, Tfelt-Hansen P (2009) Origin of pain in migraine: evidence for peripheral sensitization. Lancet Neurol 8:679–690

Papetti L, Moavero R, Ferilli MAN, Sforza G, Tarantino S, Ursitti F, Ruscitto C, Vigevano F, Valeriani M (2021) Truths and myths in pediatric migraine and nutrition. Nutrients 13:2714

Pareek A, Suthar M, Rathore G, Bansal V (2011) Feverfew (Tanacetum parthenium L. Schultz Bip.): a systematic review. Pharmacogn Rev 5:103–110

Park HJ, Zhao TT, Lee KS, Lee SH, Shin KS, Park KH, Choi HS, Lee MK (2015) Effects of (−)-sesamin on 6-hydroxydopamine-induced neurotoxicity in PC12 cells and dopaminergic neuronal cells of Parkinson's disease rat models. Neurochem Int 83–84:19–27

Parker WD Jr, Haas R, Stumpf DA, Parks J, Eguren LA, Jackson C (1984) Brain mitochondrial metabolism in experimental thiamine deficiency. Neurology 34:1477–1481

Parohan M, Sarraf P, Javanbakht MH, Foroushani AR, Ranji-Burachaloo S, Djalali M (2019) The synergistic effects of nano-curcumin and coenzyme Q10 supplementation in migraine prophylaxis: a randomized, placebo-controlled, double-blind trial. Nutr Neurosci 26:1–10. https://doi.org/10.1080/1028415X.2019.1627770

Parohan M, Sarraf P, Javanbakht MH, Ranji-Burachaloo S, Djalali M (2020) Effect of coenzyme Q10 supplementation on clinical features of migraine: a systematic review and dose-response meta-analysis of randomized controlled trials. Nutr Neurosci 23:868–875

Parohan M, Sarraf P, Javanbakht MH, Foroushani AR, Ranji-Burachaloo S, Djalali M (2021) The synergistic effects of nanocurcumin and coenzyme Q10 supplementation in migraine prophylaxis: a randomized, placebo-controlled, double-blind trial. Nutr Neurosci 24:317–326

Patil RJ, Nikam SM, Milmile PS, Awandekar NB, Umekar MJ (2021) Herbal treatment for migraine. Int J Pharmacogn Life Sci 2(1):61–66

Pierini R, Kroon PA, Guyot S, Ivory K, Johnson IT, Belshaw NJ (2008) Procyanidin effects on oesophageal adenocarcinoma cells strongly depend on flavan-3-ol degree of polymerization. Mol Nutr Food Res 52:1399–1407

Pirinen E, Auranen M, Khan NA, Brilhante V, Urho N, Pessia A, Hakkarainen A, Kuula J, Heinonen U, Schmidt MS et al (2020) Niacin cures systemic NAD(+) deficiency and improves muscle performance in adult-onset mitochondrial myopathy. Cell Metab 31:1078–1090.e1075

Plantone D, Pardini M, Rinaldi G (2021) Riboflavin in neurological diseases: a narrative review. Clin Drug Investig 41:513–527

Prakash S, Kumar Singh A, Rathore C (2016) Chronic migraine responding to intravenous thiamine: a report of two cases. Headache 56:1204–1209

Prousky J, Seely D (2005) The treatment of migraines and tension-type headaches with intravenous and oral niacin (nicotinic acid): systematic review of the literature. Nutr J 4:3

Pryde KR, Hirst J (2011) Superoxide is produced by the reduced flavin in mitochondrial complex I: a single, unified mechanism that applies during both forward and reverse electron transfer. J Biol Chem 286:18056–18065

Pulido-Moran M, Moreno-Fernandez J, Ramirez-Tortosa C, Ramirez-Tortosa C (2016) Curcumin and health. Molecules 21:264. https://doi.org/10.3390/molecules21030264

Qu C, Keijer J, Adjobo-Hermans MJW, van deWal M, Schirris T, van Karnebeek C, Pan Y, Koopman WJH (2021) The ketogenic diet as a therapeutic intervention strategy in mitochondrial disease. Int J Biochem Cell Biol 138:106050

Ravetti S, Clemente C, Brignone S, Hergert L, Allemandi D, Palma S (2019) Ascorbic acid in skin health. Cosmetics 6:58

Rozen TD, Oshinsky ML, Gebeline CA, Bradley KC, Young WB, Shechter AL, Silberstein SD (2002) Open label trial of coenzyme Q10 as a migraine preventive. Cephalalgia 22:137–141

Saedisomeolia A, Ashoori M (2018) Riboflavin in human health: a review of current evidences. Adv Food Nutr Res 83:57–81
Sazali S, Badrin S, Norhayati MN, Idris NS (2021) Coenzyme Q10 supplementation for prophylaxis in adult patients with migraine—a meta-analysis. BMJ Open 11:e039358
Schaefer AM, McFarland R, Blakely EL et al (2008) Prevalence of mitochondrial DNA disease in adults. Ann Neurol 63(1):35–39
Schepetkin IA, Ramstead AG, Kirpotina LN, Voyich JM, Jutila MA, Quinn MT (2016) Therapeutic potential of polyphenols from Epilobium angustifolium (Fireweed). Phytother Res 30:1287–1297
Schoenen J, Jacquy J, Lenaerts M (1998) Effectiveness of high-dose riboflavin in migraine prophylaxis: a randomized controlled trial. Neurology 50:466–470
Serrano D, Manack AN, Reed ML, Buse DC, Varon SF, Lipton RB (2013) Cost and predictors of lost productive time in chronic migraine and episodic migraine: results from the American Migraine Prevalence and Prevention (AMPP) study. Value Health 16(1):31–38
Shah ZA, Nada SE, Dore S (2011) Heme oxygenase 1, beneficial role in permanent ischemic stroke and in Ginkgo Biloba (EGB761) neuroprotection. Neuroscience 180:248–255
Shaik MM, Gan SH (2015) Vitamin supplementation as possible prophylactic treatment against migraine with aura and menstrual migraine. Biomed Res Int 2015:469529
Shanan N, GhasemiGharagoz A, Abdel-Kader R, Breitinger HG (2019) The effect of Pyrroloquinoline quinone and resveratrol on the survival and regeneration of cerebellar granular neurons. Neurosci Lett 694:192–197
Shi C, Zhao L, Zhu B, Li Q, Yew DT, Yao Z, Xu J (2009) Protective effects of Ginkgo biloba extract (EGb761) and its constituent's quercetin and ginkgolide B against β-amyloid peptide-induced toxicity in SH-SY5Y cells. Chem Biol Interact 181:115–123. https://doi.org/10.1016/j.cbi.2009.05.010
Shimomura T, Kitano A, Marukawa H et al (1994) Point mutation in platelet mitochondrial tRNA(Leu(UUR)) in a patient with cluster headache. Lancet 344(8922):625
Shults CW, Haas RH, Passov D, Beal MF (1997) Coenzyme Q10 levels correlate with the activities of complexes I and II/III in mitochondria from parkinsonian and nonparkinsonian subjects. Ann Neurol 42:261–264
Silberstein S, Olesen J, Bousser M, Diener H, Dodick D, First M, et al. International Headache Society (2005) The International Classification of Headache Disorders, 2nd Edition (ICHD-II)—revision of criteria for 8.2 medication-overuse headache. Headache 45(10):1424–1425
Sparaco M, Feleppa M, Lipton RB, Rapoport AM, Bigal ME (2006) Mitochondrial dysfunction and migraine: evidence and hypotheses. Cephalalgia 26(4):361–372
Stromgaard K, Nakanishi K (2004) Chemistry and biology of terpene trilactones from Ginkgo biloba. Angew Chem Int Ed 43(13):1640–1658. https://doi.org/10.1002/anie.200300601
Stuart S, Griffiths LR (2012) A possible role for mitochondrial dysfunction in migraine. Mol Genet Genomics 287(11–12):837–844
Subramanian VS, Nabokina SM, Lin-Moshier Y, Marchant JS, Said HM (2013) Mitochondrial uptake of thiamin pyrophosphate: physiological and cell biological aspects. PLoS One 8:e73503
Supornsilpchai W, Sanguanrangsirikul S, Maneesri S, Srikiatkhachorn A (2006) Serotonin depletion, cortical spreading depression, and trigeminal nociception. Headache 46(1):34–39
Takaku S, Osono E, Kuribayashi H, Takaku C, Hirama N, Takahashi H (2013) A case of migraine without aura that was successfully treated with an herbal medicine. J Altern Complement Med 19(12):970–972
Terra X, Valls J, Vitrac X, Merrillon JM, Arola L, Ardevol A, Blade C, Fernandez-Larrea J, Pujadas G, Salvado J, Blay M (2007) Grape-seed procyanidins act as antiinflammatory agents in endotoxin-stimulated RAW 264.7 macrophages by inhibiting NFjB signaling pathway. J Agric Food Chem 55:4357–4365
Thomet OA, Wiesmann UN, Schapowal A, Bizer C, Simon HU (2001a) Role of petasin in the potential anti-inflammatory activity of a plant extract of Petasites hybridus. Biochem Pharmacol 61:1041–1047

Thomet OA, Wiesmann UN, Blaser K, Simon HU (2001b) Differential inhibition of inflammatory effector functions by petasin, isopetasin and neopetasin in human eosinophils. Clin Exp Allergy 31:1310–1320

Tolomeo M, Nisco A, Barile M (2021) Alteration of flavin cofactor homeostasis in human neuromuscular pathologies. Methods Mol Biol (Clifton NJ) 2280:275–295

Tozzi-Ciancarelli MG, De Matteis G, Di Massimo C, Marini C, Ciancarelli I, Carolei A (1997) Oxidative stress and platelet responsiveness in migraine. Cephalalgia 17:580–584

Tulsulkar J, Shah ZA (2013) Ginkgo biloba prevents transient global ischemia-induced delayed hippocampal neuronal death through antioxidant and anti-inflammatory mechanism. Neurochem Int 62:189–197. https://doi.org/10.1016/j.neuint.2012.11.017

Tuncel D, Tolun FI, Gokce M, Imrek S, Ekerbiçer H (2008) Oxidative stress in migraine with and without aura. Biol Trace Elem Res 126(1–3):92–97

Udhayabanu T, Manole A, Rajeshwari M, Varalakshmi P, Houlden H, Ashokkumar B (2017) Riboflavin responsive mitochondrial dysfunction in neurodegenerative diseases. J Clin Med 6:52

Ulrich-Merzenich G, Kelber O, Koptina A, Freischmidt A, Heilmann J, Müller J, Zeitler H, Seidel MF, Ludwig M, Heinrich EU, Winterhoff H (2012) Novel neurological and immunological targets for salicylate-based phytopharmaceuticals and for the anti-depressant imipramine. Phytomedicine 19(10):930–939

United States Pharmacopeia (2007) The United States pharmacopeial convention, 30th edn. United States Pharmacopeia, North Bethesda, MD. Suppl USP30-NF25. https://www.uspnf.com/. Accessed 18 May 2023

Upton R (2011) American Herbal Pharmacopoeia: botanical pharmacognosy-microscopic characterization of botanical medicines. CRC Press, Boca Raton, FL

Velling DA, Dodick DW, Muir JJ (2003) Sustained-release niacin for prevention of migraine headache. Mayo Clin Proc 78:770–771

Wang R, Li JB, Zhao YL, Li YP, Yin L (2018) Investigating the therapeutic potential and mechanism of curcumin in breast cancer based on RNA sequencing and bioinformatics analysis. Breast Cancer 25:206–212. https://doi.org/10.1007/s12282-017-0816-6

Welch KMA, D'Andrea G, Tepley N, Barkley GL, Ramadan NM (1990) The concept of migraine as a state of central neuronal hyperexcitability. Neurol. Clin. 8:817–828

Wiley KD, Gupta M (2021) Vitamin B1 thiamine deficiency. In: StatPearls. StatPearls Publishing, Treasure Island, FL

Williams CA, Harborne JB, Eagles J (1999) Variations in lipophilic and polar flavonoids in the genus Tanacetum. Phytochemistry 52:1301–1306

Xie DY, Dixon RA (2005) Proanthocyanidin biosynthesis—still more questions than answers? Phytochemistry 66:2127–2144

Xu DP, Li Y, Meng X, Zhou T, Zhou Y, Zheng J, Zhang JJ, Li HB (2017) Natural antioxidants in foods and medicinal plants: extraction, assessment and resources. Int J Mol Sci 18:96

Xu XY, Meng X, Li S, Gan RY, Li Y, Li HB (2018) Bioactivity, health benefits, and related molecular mechanisms of curcumin: current progress, challenges, and perspectives. Nutrients 10:1553. https://doi.org/10.3390/nu10101553

Yamanaka G, Suzuki S, Morishita N, Takeshita M, Kanou K, Takamatsu T, Morichi S, Ishida Y, Watanabe Y, Go S et al (2021) Experimental and clinical evidence of the effectiveness of riboflavin on migraines. Nutrients 13:2612

Yang BY, Han W, Han H, Liu Y, Guan W, Li XM, Kuang HX (2018) Effects of lignans from *Schisandra chinensis* rattan stems against Aβ(1–42)-induced memory impairment in rats and neurotoxicity in primary neuronal cells. Molecules 23(4):E870

Yarnell E (2017) Herbal medicine and migraine. Altern Complement Therapies 23(5):192–201

Yee AJ (1999) Effectiveness of high-dose riboflavin in migraine prophylaxis. Neurology 52:431–432

Yigit M, Sogut O, Tataroglu Ö, Yamanoglu A, Yigit E, Güler EM, Ozer OF, Kocyigit A (2018) Oxidative/antioxidative status, lymphocyte DNA damage, and urotensin-2 receptor level in patients in migraine attacks. Neuropsychiatr Dis Treat 14:367–374

Yuan T-Y, Niu Z-R, Chen D, Chen Y-C, Zhang H-F, Fang L-H, Du G-H (2018) Vasorelaxant effect of quercetin on cerebral basilar artery in vitro and the underlying mechanisms study. J Asian Nat Prod Res 20:477–487

Zaki EA, Freilinger T, Klopstock T et al (2009) Two common mitochondrial DNA polymorphisms are highly associated with migraine headache and cyclic vomiting syndrome. Cephalalgia 29(7):719–728

Zeng Z, Li Y, Lu S, Huang W, Di W (2019) Efficacy of CoQ10 as supplementation for migraine: a meta-analysis. Acta Neurol Scand 139:284–293

Zhang M, Liu Y, Zhao M, Tang W, Wang X, Dong Z, Yu S (2017) Depression and anxiety behaviour in a rat model of chronic migraine. J Headache Pain 18:27

Ziaei S, Zakeri M, Kazemnejad A (2005) A randomised controlled trial of vitamin E in the treatment of primary dysmenorrhoea. Br J Obstet Gynaecol 112(4):466–469

Ziaei S, Kazemnejad A, Sedighi A (2009) The effect of vitamin E on the treatment of menstrual migraine. Med Sci Monit 15(1):CR16–CR19

8 Herbal Drugs for Management of Migraine Pain

Neha Mattoo, Yashdeep Mukheja, and Anurag Kuhad

8.1 Introduction

Migraine is ranked amongst the top ten disabling medical conditions and is an intricate disorder marked by recurring headache episodes coupled with nausea. Often unilateral, these episodes may involve visual or sensory symptoms which significantly impact daily life. Migraine comprises of two main subcategories: episodic migraine and chronic migraine. The migraine episode unfolds in specific phases starting with the Premonitory Phase affecting 30–50% of patients. It lasts from hours to days. This phase encompasses symptoms like fatigue, cognitive difficulties, mood changes, food cravings, neck pain, and yawning. Following this phase is the Aura Phase which is experienced by 90% of patients and it is characterized by neurological symptoms lasting from 5 min to an hour. The Headache Phase, which affects almost 80–90% of the population, lasts 4–74 h and involves symptoms like headaches, nausea, photophobia, and phonophobia. Last is the Postdrome Phase that affects 10–12% of these individuals and lasts from 12 to 24 h involving symptoms such as fatigue, difficulty concentrating, and neck stiffness (Goadsby 2012). Existing pharmacotherapies for migraine encompass acute interventions such as the use of triptans and ergots as well as preventive/prophylactic approaches like calcium channel blockers (CCBs), beta-blockers, and calcitonin gene-related peptide (CGRP) antibodies (Chaplin 2018; Chinese Medical Association Group 2016). Patient dissatisfaction with these medications often arises from their limited efficacy and potential to induce comorbidities as well as due to the emergence of undesirable side effects (Lipton et al. 2019). Additionally, the frequent overuse of acute pain-relief medications by patients suffering from migraine may contribute to the

N. Mattoo · Y. Mukheja · A. Kuhad (✉)
Pharmacology Research Laboratory, University Institute of Pharmaceutical Sciences, UGC-Centre of Advanced Study, Panjab University, Chandigarh, India
e-mail: anurag.kuhad@pu.ac.in

S. Bhatt et al. (eds.), *Management of Migraine Pain*,
https://doi.org/10.1007/978-981-97-4529-6_8

transition from episodic migraine to chronic migraine. This overreliance on medication is associated with an elevated risk of psychological comorbidities, hence amplifying the overall disease burden (Schwedt et al. 2018; Radat et al. 2005). Globally, migraine affects 1.1 billion people, with prevalence remaining relatively stable. More prevalent in women, migraine exerts a substantial economic impact, costing Europe €18–27 billion annually. The prevalence of migraine increased by 1.7% (0.7–2.8) from 1990 to 2019, and in 2019 there were 1.1 billion (0.98–1.3) prevalent cases and 525.5 (78.8–1194) years lived with disability (YLDs) per 100,000 population (Safiri et al. 2021). Despite healthcare advancements, 30–40% of the patients express dissatisfaction with current treatments and seem to fall short of optimal outcomes. Herbal drugs especially in patients with gastrointestinal diseases present an opportunity for exploration. Identifying migraine triggers, including certain foods and components is crucial. Phytomedicines offer a dual approach by serving both as a traditional and modern choice for effectively managing migraine symptoms. Phytomedicines involving medicinal formulations derived from plants hold a significant potential in discovering new therapeutic avenues for migraine. Growing clinical and preclinical evidence support their effectiveness and prompt many patients to explore herbal and plant-based therapies as alternative or complementary options to conventional pharmacological treatments, hence seeking an effective and well-tolerable remedy (Rajapakse and Davenport 2019).

8.2 History

The historical trajectory of headache treatment encompasses both general headaches and migraines which extends across the epochs ranging from the Neanderthal era to the Space Age. The origins of headache treatment can be traced back to a magical hypothesis which subsequently gave rise to mystical therapeutic practices. Over time, the evolution of rational treatment methods for this age-old ailment has been gradual and complex.

In the present era of molecular medicine, delving into the origins and historical progression of headache treatment becomes a valuable as well as essential endeavour. Understanding and pinpointing the site from where the journey began and comprehending the transformative stages it underwent provides crucial insights for the on-going pursuit of safe and effective migraine treatments. From ancient mystical beliefs to the current advancements in molecular medicine, the historical narrative of headache treatment serves as a guidepost for refining and advancing therapeutic approaches. As we navigate the complexities of migraines, this historical perspective informs as well as lights our path to develop cutting-edge, evidence-based treatments that can enhance the well-being of individuals grappling with this pervasive and ancient ailment (Edmeads 1999).

Artifacts dating back over 9000 years indicate treatment of headaches and migraines through trephination, which is a procedure aimed to release perceived evil spirits from the head. This practice persisted through the ancient civilizations like in Egypt (Collado-Vázquez and Carrillo 2014). Initial written records providing details about migraine appeared in Mesopotamia around 4000 BCE and the Egyptian

Ebers Papyrus from 1550 BCE contained similar descriptions. Hippocrates later proposed a theory which linked migraines to vapours ascending from the stomach to the head (Collado-Vázquez and Carrillo 2014; Zargaran et al. 2016).

However, during the medieval period the Persian physician Ibn Sina who was also known as Avicenna had extensively documented herbal and plant-based therapies for migraines in his Canon of Medicine. Table 8.1 summarises Avicenna's

Table 8.1 The medicinal plants that Avicenna investigated for migraine treatment along with the comprehension of their pharmacological effects and active constituents

Family	Scientific name	Common name	Pharmacological effect	Active components
Asteraceae	*Cichorium intybus* L.	Chicory extract	It acts as a COX-2 inhibitor	Sesquiterpene lactones (Fig. 8.1) esp. 8-deoxylactucin (Cavin et al. 2005)
	Matricaria chamomilla L.	Chamomile	It reduces NO levels as well as has anti-inflammatory action	Sesquiterpenes (Fig. 8.1), flavonoids (Singh et al. 2011)
Iridaceae	*Crocus sativus*	Safron	It reduces NO levels, has anti-inflammatory action, also causes serotonergic modulation	Crocin safranal
Lamiaceae	*Mentha piperita*	Peppermint	Causes 2-h pain relief from migraine superior to placebo	Menthol
Rosaceae	*Rosa damascena* mill L.	Damask rose	Analgesic and anti-inflammatory action	Polyphenol flavonoids
Lauraceae	*Laurus nobilis* L.	Bay laurel	It reduces NO levels	Sesquiterpene lactones (Fig. 8.1)

NO nitric oxide, *COX* cyclooxygenase

Fig. 8.1 Scheme illustrates several subclasses of sesquiterpene lactones

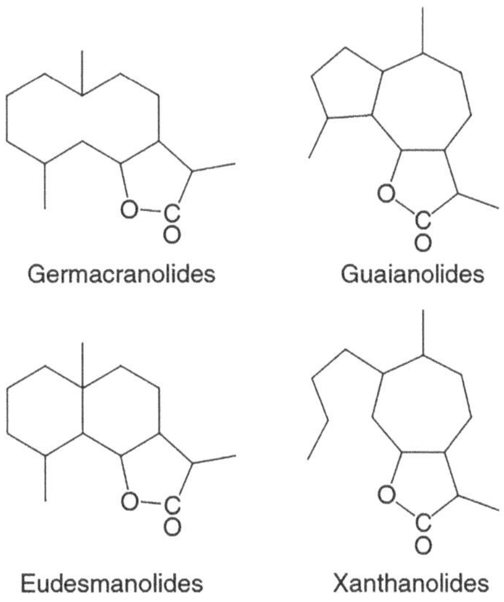

studies on medicinal plants for migraine treatment, alongside comprehension of their pharmacological effects based on in vivo and in vitro research (Zargaran et al. 2016; Abokrysha 2009).

8.3 Herbal Drugs for Migraine Pain Treatment

8.3.1 Feverfew (*Tanacetum parthenium*)

Feverfew, scientifically known as *Tanacetum parthenium* (L.) Schultz Bip. [formerly *Chrysanthemum parthenium* (L.) Bernh.], belongs to the Asteraceae/Compositae family which holds a rich medicinal heritage. Its probable origins lie in South-eastern Europe, Asia Minor, and the Caucasus but it has since spread across Europe and North and South America, thriving in roadside areas and neglected spaces (Evans 2009). Derived from the Latin term febrifugia meaning "fever reducer," feverfew was historically prescribed by the first-century Greek physician Dioscorides for "all hot inflammations." However, in the context of migraine treatment, current scientific understanding projects feverfew's anti-migraine properties primarily to parthenolide, a sesquiterpene lactone (Fig. 8.2) (Rajapakse and Davenport 2019; Rajapakse and Pringsheim 2016). Parthenolide exerts its effects through antiplatelet actions, relaxation of vascular smooth muscles, and anti-inflammatory mechanisms. It is Harvested during its flowering stage. Feverfew's active ingredients concentration, such as parthenolide and other compounds vary based on factors like geographic location, harvest condition, plant parts used, and storage duration and conditions. The whole-plant extracts contain a compound that inhibits prostaglandin biosynthesis, differing from salicylates by not impeding prostaglandin synthase cyclooxygenation, these extracts inhibit platelet aggregation by hampering thromboxane synthesis through cellular phospholipase inactivation (Heptinstall et al. 1992; Collier et al. 1980). Feverfew extracts also curtail secretory activity in blood platelets and polymorphonuclear leukocytes (PMNs) with a distinct pattern from other platelet aggregation inhibitors, exerting a more pronounced effect on PMNs than even high concentrations of nonsteroidal anti-inflammatory agents (Groenewegen and Heptinstall 1990). Parthenolide from feverfew is also investigated for its serotonin (5HT) receptor-blocking capabilities. Studies demonstrate that parthenolide inhibits serotonin release in a dose-dependent manner (Rajapakse and Davenport 2019; Marles et al. 1992).

Feverfew is not advisable during pregnancy or breastfeeding due to its potential menstrual flow-increasing effects in early pregnancy. It's not recommended for

Fig. 8.2 Parthenolide

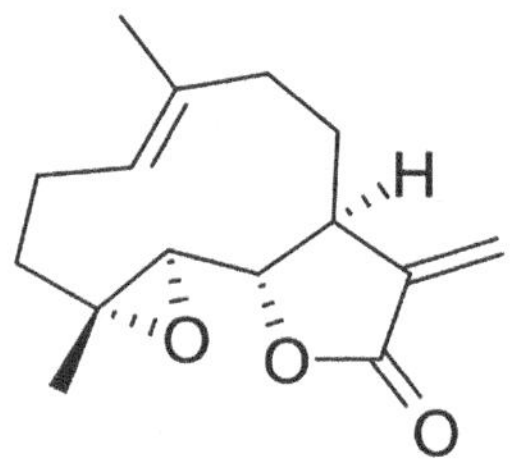

children under 2 years old. It's contraindicated for individuals that are allergic to feverfew or other Asteraceae family members like ragweed, chrysanthemums, marigolds, chamomile, yarrow, and daisies. Individuals (6–15%) may experience mouth ulcers and gastric disturbances and contact dermatitis while handling fresh feverfew, often occurring within the first week of use. Though rare symptoms like diarrhoea, flatulence, nausea, and vomiting may cause treatment discontinuation in some individuals. Long-term toxicity data is lacking and there is no documented occurrence of serious adverse effects in individuals consuming feverfew products (Ernst and Pittler 2000; Pfaffenrath et al. 2002).

8.3.2 Butterbur (*Petasites hybridus*)

Butterbur (*Petasites hybridus*) is a perennial herb that has a longstanding history of medicinal use due to its pharmacological properties. It belongs to the Asteraceae plant family and thrives in various global regions with around 18 identified species in its genus. Traditionally, it has been employed for treating migraines, tension, headache-related pain, stomach discomfort, allergic rhinitis, asthma, and gastrointestinal spasms. The active constituent of butterbur is petasins (Fig. 8.3), which fall under the category of sesquiterpenes, specifically esters of petasol and angelic acid. Different species within the genus exhibit variations in sesquiterpene amounts and compositions, impacting their therapeutic efficacy (Ernst and Pittler 2000). Petasins contribute significantly to butterbur's anti-inflammatory effect by inhibiting the synthesis of prostaglandins and leukotrienes (Borlak et al. 2022). Petasins block calcium influx in neutrophils and eosinophils and inhibits leukotriene production, hence exhibiting anti-inflammatory effects by reducing prostaglandin E2 (PGE2), leukotriene B4 (LTB4), and cysteinyl-leukotrienes (LTs) in both animal and human cellular systems (Lee et al. 2004). Asthmatic patients on steroids also benefit from its anti-inflammatory action. Studies in rats show petasins' calcium-blocking effects on L-type voltage-dependent calcium channels, leading to vasorelaxation (Jiang et al. 2014). Several experimental investigations have affirmed the neuroprotective potential of petasins. Bakkenolides, a type of sesquiterpene, exhibited neuron safeguarding properties by hindering the activation of nuclear factor-κB (NF-κB), a pivotal mediator in inflammatory responses. For instance, bakkenolide-VI isolated from *Petasites tatewakianus* displayed neuroprotection in primary neuron cultures subjected to oxygen-glucose deprivation and oxidative stress. *Petasites tricholobus*-derived bakkenolides also demonstrated neuroprotective effects linked to their antioxidant capabilities (NCCIH n.d.-a). Butterbur extracts work by diminishing CGRP release from trigeminal nerve fibers through the desensitization of TRPA1 and TRPV1 receptor channels (Lee et al. 2004).

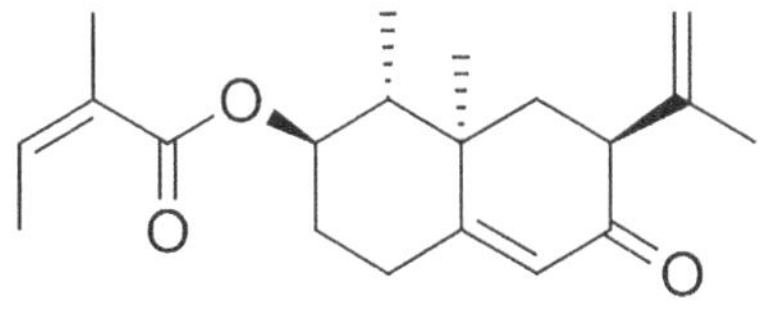

Fig. 8.3 Petasin (Adapted from: https://pubchem.ncbi.nlm.nih.gov/compound/Petasin)

Some butterbur variants contain pyrrolizidine alkaloids (PAs), posing risks of liver, lung, and circulatory damage, potentially leading to cancer. Products labelled or certified as PA-free are advisable for use. Some nations have withdrawn butterbur due to liver toxicity concerns. Studies suggest PA-free butterbur is safe orally for up to 16 weeks, but some marketed as PA-free may still contain Pas like Petadolex, and may be still linked to liver damage. Long-term safety for these extracts and products isn't confirmed. PA-free butterbur must be avoided during pregnancy and breast-feeding due to birth defects and liver damage risks. Limited data exists on the safety of PA-free butterbur in these cases. Though it is generally well-tolerated, it can cause belching, headache, itching, diarrhoea, breathing difficulties, fatigue, an upset stomach, and drowsiness. Allergic reactions may occur in individuals who are sensitive to ragweed, chrysanthemums, marigolds, and daisies (NCCIH n.d.-a).

8.3.3 *Cannabis sativa*

Cannabis sativa, also known as marijuana, has a long-standing history for both medicinal and recreational purposes for over 2500 years. Throughout ancient times, it was utilized to alleviate various conditions like acute pain, anxiety, cancer pain, chronic pain, depression, headaches, and migraines. Extracts of leaves and flowers from varieties such as *Cannabis indica*, *Cannabis ruderalis*, and *Cannabis sativa*, which encompass around 400 compounds, are majorly constituted by Δ9-tetrahydrocannabinol (THC) (Fig. 8.4), cannabidiol (CBD) (Fig. 8.5), flavonoids, and terpenes. THC and CBD are primary components in various medical cannabis preparations (Okusanya et al. 2022; Baron 2018). A recent study suggests that Cannabinoid systems impact various pathways (glutamine, inflammatory, opiate, and serotonin) both centrally and peripherally to mitigate migraine pain. Anandamide (AEA) heightens the effectiveness of 5-HT1A and blocks 5-HT2A receptors, which offers therapeutic potential in both acute and preventive migraine treatment as it has an impact on the periaqueductal grey matter which is recognised as a migraine trigger. Cannabinoids also have dopamine-blocking properties and anti-inflammatory effects. The endocannabinoid system consists of CB1 (central/peripheral nervous system) and CB2 (peripheral/immune tissues) receptors. The activation of CB1 receptors leads to decreased neurotransmission of dopamine, γ-aminobutyric acid (GABA), and glutamate whereas the activation of CB2 leads to analgesia and decreased immune system function. They prevent migraines by inhibiting serotonin release from platelets and inducing peripheral vasoconstriction. CB1 receptors play a role in reducing pain sensation through a serotonin-related process, while CB2 receptors contribute to pain relief without inducing tolerance or side effects. Ongoing research suggests that the endocannabinoid system contributes significantly to mitigating migraines, although more investigation is needed (Del Giudice et al. 2007).

Exogenous cannabinoids' safety both in immediate and prolonged usage has been studied regarding its impact on cardiovascular, respiratory, and immune systems and cognition. Acute doses may lead to dose-dependent effects on the cardiovascular system and may lead to increased blood pressure and causing adverse

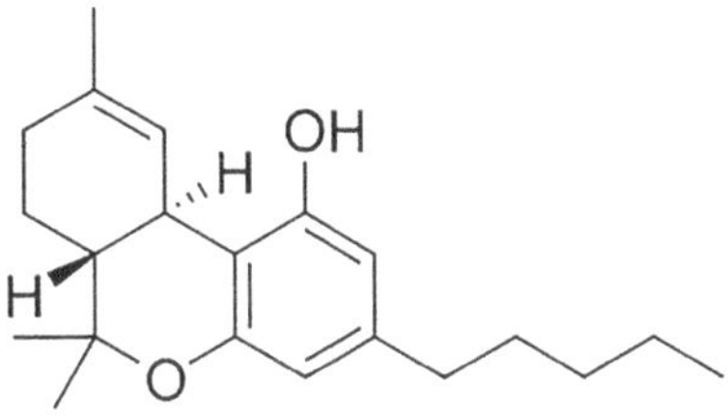

Fig. 8.4 Δ9-Tetrahydrocannabinol

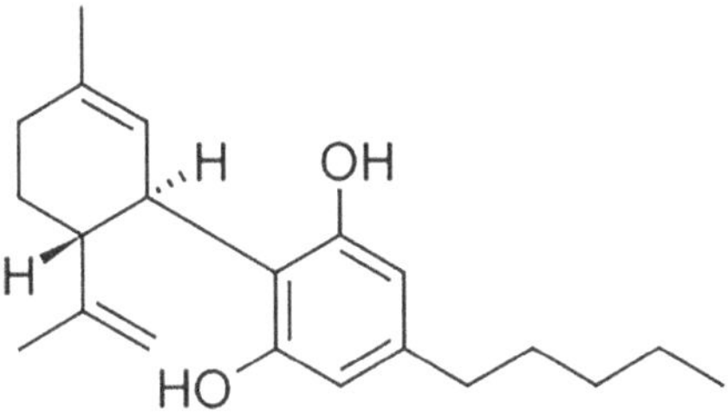

Fig. 8.5 Cannabidiol

events like angina, myocardial infarction, and cardiomyopathy in individuals who already have existing heart conditions (Hall and Degenhardt 2009). Cannabis is contraindicated in these patients. In the respiratory system, acute cannabis use can induce airway inflammation leading to increased airway resistance and damage to lung tissue over time, which can potentially lead to chronic bronchitis and an increased risk of emphysema (Sachs et al. 2015). Acute use of cannabis in patients with existing cognitive difficulties during acute migraine episodes may exacerbate acute and chronic cognitive impairment. It affects recall, memory, attention, and executive function although evidences vary in various studies. While chronic use of cannabis might not cause persistent neuropsychological deficits in most cases, individuals using it from at an early age may face long-term cognitive challenges which specifically can affect visual attention, verbal fluency, impulsivity, executive functioning and discontinuing use of cannabis in adolescent users might not fully restore cognitive deficits which can potentially lead to loss of IQ. The above discussion suggests that the use of cannabis, especially in young children and adolescence, is not advised for treatment of migraine pain (Sachs et al. 2015; Vurallı et al. 2018).

8.4 Potential Phytomedicines for Pain Treatment in Migraine

8.4.1 Saint John's Wort (*Hypericum perforatum*)

Hypericum perforatum also known as Saint John's Wort (SJW) has been used as a medicinal plant for centuries. Hypericum consists of dried aerial parts of *Hypericum perforatum*, family Hypericaceae (Clusiaceae), gathered usually at the time of flowering or shortly before. Hypericum contains a variety of constituents that cause its biological activity, specifically *Anthraquinones*, mainly hypericin and pseudohypericin (Fig. 8.6).

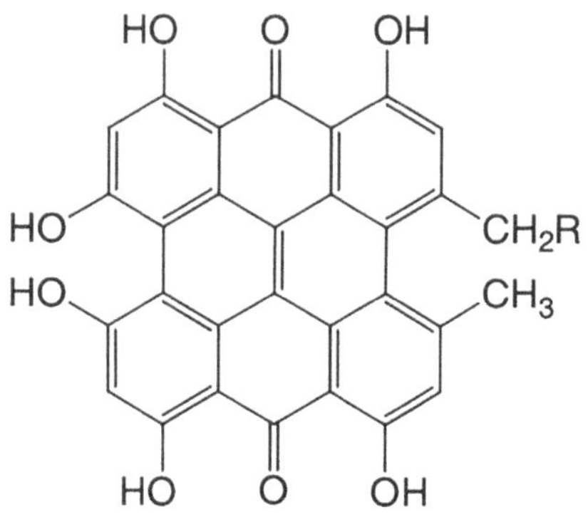

Hypericin, R = H
Pseudohypericin, R = OH

Fig. 8.6 Hypericin and Pseudohypericin

In a study conducted in 2013 (Okusanya et al. 2022), researchers investigated the efficacy of a dried extract of St. John's Wort (SJW) in an animal model of meningeal nociception, induced by nitric oxide (NO) donors. These donors triggered delayed meningeal inflammation, leading to the expression of inflammatory markers within the dura mater. The study revealed that the SJW extract may have the potential to alleviate the pro-nociceptive effects, induced by NO donors GTN and SNP. These donors prompted meningeal inflammation, indicated by increased interleukin (IL)-1B and iNOS levels, along with prolonged cold sensitivity and heat-induced pain, resembling NO-induced migraine attacks. The administration of a single oral dose of 5 mg/kg SJW dried extract countered these nociceptive behaviours and lowered IL-1B and iNOS expression levels. This effect occurred due to the blocking of protein kinase C-mediated pathways involving NF-κB, cAMP response element-binding protein (CREB), and signal transducer and activator of transcription (STAT1), consistent with both SJW extract and its active compound hypericin (Galeotti and Ghelardini 2013a). The findings suggest that SJW or its extracts may offer therapeutic potential in treating migraines. Notably, SJW demonstrates a favourable safety and tolerability profile even though it has the ability to induce cytochrome P450 (CYP) enzymes which lead to the necessity for extensive testing (Galeotti and Ghelardini 2013b).

Frequently reported adverse effects of SJW include GIT irritation, allergic reactions, dizziness, confusion, tiredness, sedation, and a dry mouth. Generally, these reactions are mild, moderate, or transient. Photosensitivity reactions affecting the skin are serious adverse effects which are associated with SJW. Recent data indicates that the occurrence of these photosensitivity reactions are dose-dependent, implying that higher doses cause increased sensitivity. It is essential to note that the primary evidence backing up this observation is derived from anecdotal reports in animals, particularly pale-skinned cattle. These reports suggest that animals consuming *Hypericum perforatum* (SJW) may experience photosensitivity reactions resembling severe sunburn (Henderson et al. 2002).

8.4.2 Damask Rose (*Rosa damascena* mill L.)

Damask rose (*Rosa damascena* mill L.), known for its beautiful blooms and fragrant essential oil, boasts a complex chemical profile. Its oil is rich in monoterpenoid alcohols like citronellol, geraniol, and nerol, which contribute significantly to its characteristic rose scent. Figure 8.7 shows the main constituents of Damask Rose. Additionally, phenylethyl alcohol adds another layer of floral aroma. The exact composition of the oil can vary depending on the rose variety, climate, and harvesting methods.

In a study conducted in 2017 a placebo-controlled crossover trial, 40 migraine patients were divided into two groups and treated with either topical *R. damascena* oil or a placebo for their first two consecutive migraine attacks. After a week the groups were switched. Pain intensity during the attacks was measured at various time points up to 24 h, alongside secondary outcomes such as photosensitivity and sound sensitivity and nausea/vomiting. The analysis was based on syndrome differentiation and aimed to evaluate *R. damascena* oil's impact on different migraine types (hot and cold). When syndrome differentiation was applied, individuals with the "hot" type of migraine experienced notably lower pain intensity scores at certain intervals (30, 45, 60, 90, and 120 min) after using *R. damascena* oil compared to those with the "cold" type. This suggests that syndrome differentiation could potentially guide the selection of patients who might benefit from short-term relief using topical *R. damascena* oil for migraine pain.

However, further research with larger populations and longer follow-up periods is necessary for a more comprehensive evaluation of *R. damascena* oil's efficacy in migraine treatment (Niazi et al. 2017).

8.4.3 Ginkgo (*Ginkgo biloba*)

Ginkgo leaves belong to the dioecious *Ginkgo biloba* tree, commonly known as the Maidenhair tree (Ginkgoaceae). This tree is the only surviving species from the pre-Ice Age fossil family. Among the various compounds extracted from ginkgo, it is the diterpene lactones and flavonoids that exhibit therapeutic properties such as five identified diterpene lactones, ginkgolides A, B, C, J, M (Fig. 8.8) (Evans 2009). Ginkgolide B has been used widely for treatment of diseases like depression, anxiety, headaches, and memory issues. It has been proposed to have neuroprotective

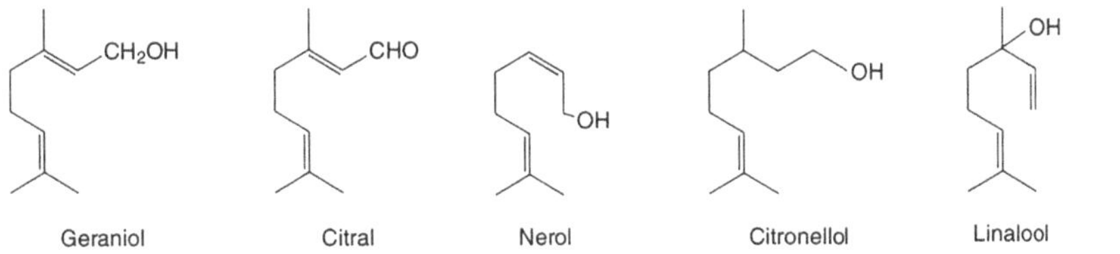

Fig. 8.7 The main constituents of Damask Rose

Fig. 8.8 Ginkgolide structures

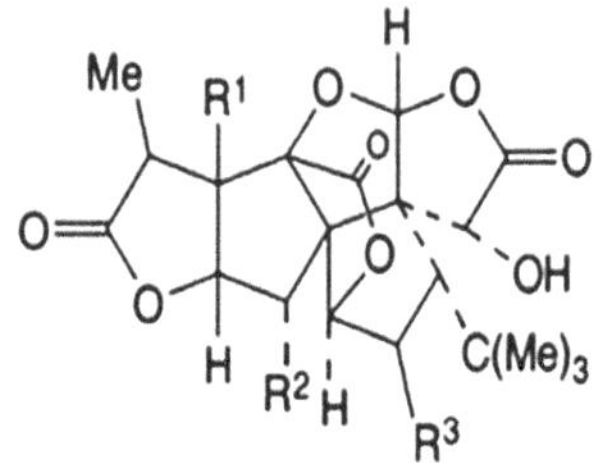

Ginkgolide structures

	R1	R2	R3
Ginkgolide A:	OH	H	H
Ginkgolide B:	OH	OH	H
Ginkgolide C:	OH	OH	OH
Ginkgolide J:	OH	H	OH
Ginkgolide M:	H	OH	OH

qualities but the exact mechanisms behind its actions need more exploration. Studies indicate that the extract which is primarily composed of flavonoids and terpene lactones contributes to the lessening of neuronal damage caused by transient global ischemia through its antioxidative and anti-inflammatory attributes which makes it a prominent nutraceutical in migraine treatment as it primarily modulates glutamatergic transmission and platelet-activating factor receptor antagonism (Tulsulkar and Shah 2013; Ma et al. 2013). Its antioxidative properties are crucial due to changes in redox status during migraine episodes. Alongside ginkgolide A, C, and terpenes, ginkgolide B stands out as a key active compound in *Ginkgo biloba* extract, positively affecting memory functions (Shah et al. 2011).

Ginkgo biloba is usually considered safe and well-tolerated. The recommended maximum dose is 240 mg/day. Some mild adverse effects may include headaches, heart palpitations, GI upset, constipation, and allergic skin reactions. It is contraindicated or cautious use is advised for individuals with bleeding disorders or individuals on NSAIDs, antiplatelet, or anticoagulant therapies as drug interactions may pose risks of spontaneous bleeding. Limited evidence exists regarding the safety and efficacy of ginkgo in pregnant women, nursing mothers, or infants, hence the use of ginkgo is usually contraindicated in such populations. In epileptic patients or those prone to seizures, due to the potential seizure threshold-lowering effect of the ginkgo toxin present in both seeds and leaves, it should be given with caution and constant monitoring should be done (Nguyen and Alzahrani 2023).

8.4.4 Peppermint (*Mentha piperita*)

Peppermint also known as *Mentha piperita*, possesses a complex chemical composition that contributes to its therapeutic properties in the treatment of migraine pain.

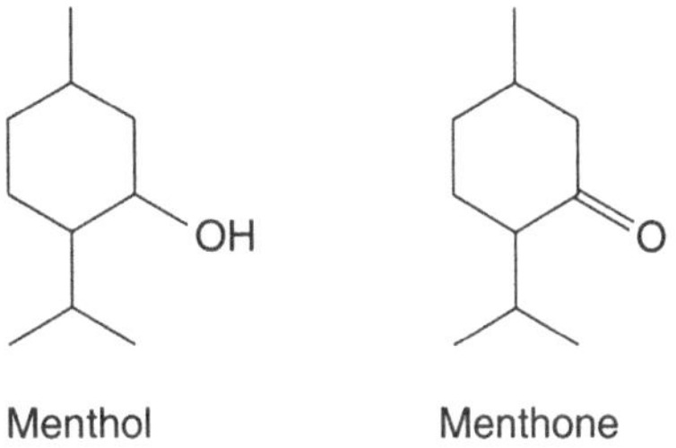

Fig. 8.9 (**a**) Menthol; (**b**) Menthone

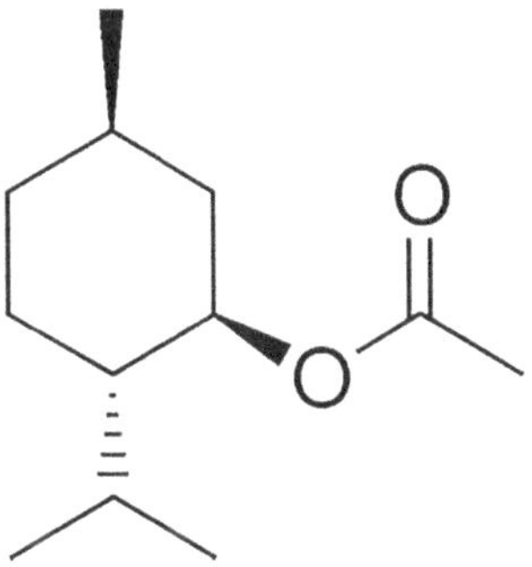

Fig. 8.10 Menthyl acetate (Adapted from: https://pubchem.ncbi.nlm.nih.gov/compound/Menthyl-acetate)

The primary active compounds found in peppermint include menthol, menthone (Fig. 8.9a, b), and menthyl acetate (Fig. 8.10) (Evans 2009). Menthol is responsible for the characteristic cooling sensation and analgesic effects of peppermint. It acts as a natural vasodilator, helping to widen blood vessels and improve blood flow, which can alleviate the constricted blood vessels often associated with migraines. Menthone exhibits anti-inflammatory properties that can help reduce swelling and inflammation in the head and thereby provide relief from migraine symptoms. Menthyl acetate contributes to the calming and soothing effects of peppermint, which can help relax the mind and reduce stress, a common trigger for migraines (Lopresti et al. 2020). Peppermint has been found to exert its effects on migraine pain through various mechanisms. One key mechanism involved is its ability to modulate pain perception. The menthol component in peppermint acts as a natural analgesic, providing relief from pain by desensitizing the nerve receptors that transmit pain signals. Another way in which peppermint may alleviate migraine pain is through its muscle relaxant effects. By relaxing the muscles in the head and neck, peppermint can help relieve tension and alleviate the symptoms of migraines. Furthermore, peppermint has been shown to possess vasodilatory properties. Overall, the mechanisms of action of peppermint involve pain modulation, anti-inflammatory effects, muscle relaxation, and vasodilation (Lopresti et al. 2020; Kumar Paswan et al. 2021). It has potential side effects, although rare, but some individuals may experience allergic reactions such as rashes, itching, and difficulty breathing when exposed to peppermint. It is recommended to perform a patch test before applying peppermint oil topically to ensure no adverse reactions occur. Peppermint oil can also cause skin irritation or sensitization in some individuals, hence it is advised to dilute the oil with carrier oil before use. Ingesting excessive amounts of peppermint oil may lead to digestive issues such as heartburn, nausea, or diarrhoea. Pregnant or breastfeeding women, as well as individuals with gastroesophageal reflux disease or gallbladder disorders should consult with a healthcare professional before using peppermint for migraine pain relief (NCCIH n.d.-b). A 2017 study explored the TRPM8 receptor's role which is also known as the cold and menthol receptor, in migraine management. TRPM8 activation reversed lowered facial pain thresholds caused by meningeal inflammation. The study suggested that triggering TRPM8 in the face might alleviate migraines by reducing TRPV1 activity, indicating a potential therapeutic target for migraine treatments, particularly those with a menthol component (Kayama et al. 2017). A small randomized trial with 35 patients treating 118 migraine attacks found that a 10% menthol solution was significantly better than placebo in achieving pain relief at 2 h and sustained pain relief. The menthol solution also effectively relieved associated migraine symptoms like nausea, vomiting, and sensitivity to light and sound. This study highlighted menthol's safety, tolerability, and potential effectiveness in acute migraine treatment, especially in children due to its safety and topical application. However, more research is needed for validation (Borhani-Haghighi et al. 2010).

Menthol has been observed to induce hepatocellular changes in rodents. Adverse responses to enteric-coated peppermint oil capsules are uncommon but may include sensitivity reactions, contact dermatitis, stomach pain, acid reflux, perianal burning, bradycardia, and muscle tremors. Inhalation of menthol can lead to apnea and laryngoconstriction. There are reported cases of patients exhibiting contact sensitivity to menthol and *Mentha piperita*, with oral symptoms like burning oral discomfort and intermittent oral ulceration. Excessive exposure to concentrated menthol vapor in rodent studies has shown potentially lethal effects. Pregnant women are contraindicated menthol use, even within recommended doses (Mughal 2022).

8.4.5 Coriander (*Coriandrum sativum*)

Coriander, also known as cilantro, has been valued for its medicinal properties for centuries. It is rich in antioxidants, which helps reduce inflammation and oxidative stress, both of which can trigger migraines. Coriander contains several chemical compounds that have shown potential in relieving pain. One key compound is linalool, which has been found to possess analgesic properties. Linalool structure (Fig. 8.11) acts as a natural sedative and has been shown to reduce inflammation which can contribute to migraine pain. Another compound found in coriander is terpinene-4-ol, which exhibits anti-inflammatory effects and may help alleviate migraine symptoms. It contains flavonoids such as quercetin (Fig. 8.12) and kaempferol (Fig. 8.13), which have been studied for their potential ability to reduce pain and inflammation. These chemical compounds present in coriander make it a promising candidate for the treatment of migraine pain. The presence of vitamins like vitamin C and vitamin K in coriander aids in strengthening blood vessels and reducing the risk of migraines. Additionally, it also possesses analgesic and anti-inflammatory properties which can provide relief from migraine pain. By inhibiting the release of certain chemicals that trigger pain and inflammation, it hence acts as a natural painkiller. It also possesses calming properties which can help alleviate the anxiety and stress often associated with migraines.

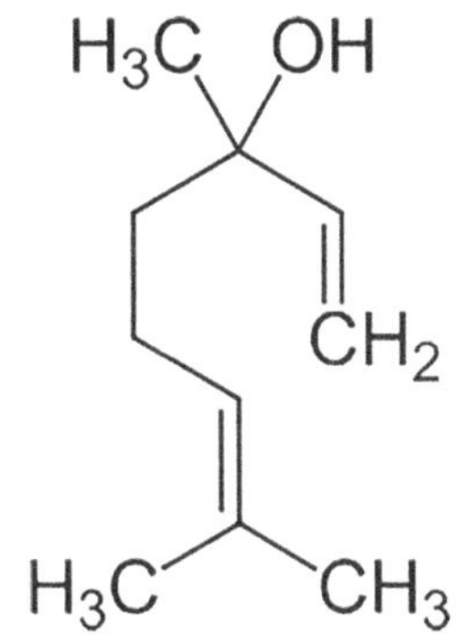

Fig. 8.11 Linalool

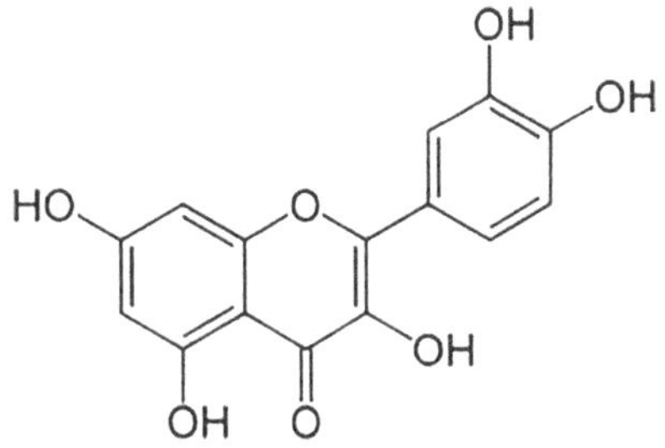

Fig. 8.12 Quercetin

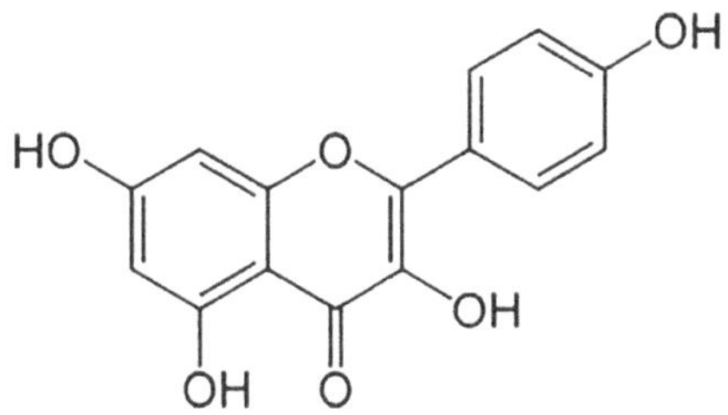

Fig. 8.13 Kaempferol (Adapted from: https://pubchem.ncbi.nlm.nih.gov/compound/Kaempferol)

Whether consumed as a spice, in the form of coriander oil, or brewed in tea, coriander can be used as a natural remedy to manage and prevent migraine pain effectively (Lopresti et al. 2020). Coriander, a commonly used herb, offers various methods for migraine relief. One method is by consuming coriander seeds. These seeds can be ground into a powder and then mixed with water or juice to create a drink. Alternatively, coriander leaves can be crushed and added to salads, soups, or other dishes to incorporate their migraine-relieving properties. Another way to utilize coriander is through its essential oil. When diluted with carrier oil, such as coconut or olive oil, it can be massaged onto the temples, forehead, and neck for alleviating migraine pain. Additionally, coriander tea can be brewed by steeping crushed coriander seeds or leaves in hot water. This herbal tea can be consumed multiple times a day to promote headache relief. Experimenting with these different methods can help individuals find the most effective way to utilize coriander for their migraine pain treatment (Lopresti et al. 2020; Kumar Paswan et al. 2021).

Studies confirm the general safety of coriander in context to migraine pain treatment. Coriander is commonly used both as a culinary herb and a medicinal plant and it has shown a favourable safety profile. Research indicates that its consumption and topical application are well-tolerated by a majority of individuals and in the majority of the population. The herb has been a staple in traditional medicine which suggests that there is a historical acceptance of its safety. While coriander is generally deemed safe, it's crucial to acknowledge that there is potential for allergic reactions in certain individuals. Due to its high allergenic potential, some people may experience adverse effects such as skin rashes or respiratory issues. This highlights the importance of individual variations in response to herbal treatments and emphasizing the necessity for caution and more aggressive testing, specifically for patients prone to or those with known allergies. Understanding and monitoring potential side effects contributes to a more informed and personalized approach in incorporating coriander into migraine pain management strategies (Kern et al. 2022).

8.4.6 Chamomile (*Matricaria chamomilla*)

Chamomile is a medicinal herb known for its calming properties and has been used for centuries in traditional medicine. It contains various beneficial compounds including chamazulene, bisabolol, and flavonoids, that have anti-inflammatory and analgesic effects. Chamomile, a traditional herbal remedy, contains various active compounds that have the potential to alleviate migraine symptoms. One of the key compounds found in chamomile is bisabolol (Fig. 8.14), which possesses anti-inflammatory and analgesic properties. These properties contribute to the reduction of pain and inflammation associated with migraines. Additionally, chamazulene (Fig. 8.15), another compound in chamomile, demonstrates antioxidant and anti-inflammatory effects that may help in mitigating migraine symptoms. Matricin and apigenin are also present in chamomile and possess anxiolytic properties, potentially aiding in the relief of stress and anxiety commonly associated with migraines. The combined action of these active compounds in chamomile makes it a promising natural option for managing migraine pain. These properties make chamomile an effective remedy for migraine pain. Chamomile can help reduce the severity and frequency of migraines by inhibiting the release of inflammatory substances and promoting muscle relaxation. Additionally, chamomile has sedative properties, which can help alleviate the associated anxiety and stress often experienced during migraines. This herb can be consumed as a tea or applied topically in the form of essential oil for migraine relief. However, it is important to note that chamomile may interact with certain medications, so it is advisable to consult a healthcare professional before using it as a treatment for migraine pain (Galeotti and Ghelardini 2013b). Chamomile has been recognized for its efficacy in alleviating migraine pain. It contains compounds that possess anti-inflammatory and analgesic properties, which help in reducing the intensity and duration of migraine attacks. The active ingredients in chamomile, such as bisabolol and chamazulene, work to relax the blood vessels and soothe the nervous system, providing relief from the throbbing sensation and associated symptoms. Several studies have shown that

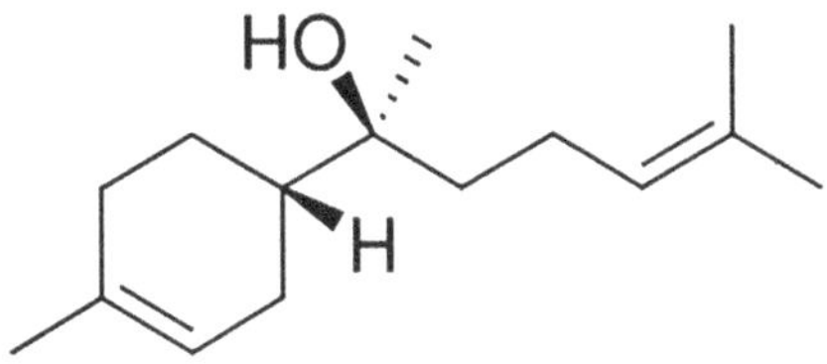

Fig. 8.14 Bisabolol (Adapted from: https://pubchem.ncbi.nlm.nih.gov/compound/Bisabolol)

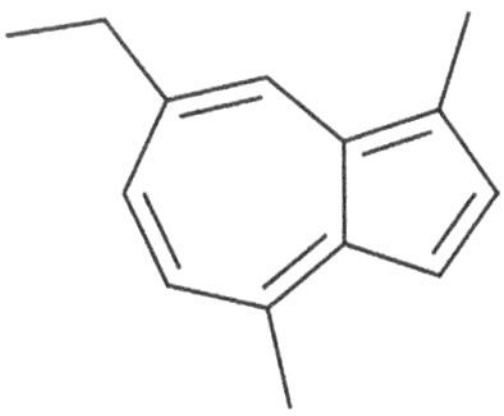

Fig. 8.15 Chamazulene (Adapted from: https://pubchem.ncbi.nlm.nih.gov/compound/Chamazulene)

chamomile tea, when consumed during a migraine episode, can significantly reduce pain severity and frequency. Additionally, chamomile essential oil can be applied topically on the temples and forehead to alleviate headache symptoms. The soothing aroma of chamomile also aids in relaxation and stress reduction, which are common triggers for migraines. Chamomile proves to be a natural and effective herbal remedy for individuals seeking relief from migraine pain (Lopresti et al. 2020; Ahmadifard et al. 2020).

Chamomile, widely recognized for its potential therapeutic benefits, exhibits a generally favourable safety profile in the context of migraine pain treatment. It is well-tolerated by most individuals and adverse effects are infrequent. However, it is essential to acknowledge that it has potential for allergic reactions, especially in individuals with pre-existing allergies to plants specifically in the Asteraceae family, to which chamomile belongs. Individuals with known sensitivities to these plants should be prescribed chamomile with caution as well as constant monitoring should to be done to avoid any adverse drug reactions and consider consulting a healthcare professional before incorporating chamomile into their migraine treatment regimen. While chamomile is generally considered safe, the prolonged and excessive use may pose a risk of skin inflammation or irritation. This emphasizes the importance of moderation and appropriate usage to handle, as well as avoid, potential adverse effects. Consulting a healthcare professional becomes particularly crucial for those with a history of skin sensitivities or conditions to ensure the safe integration of chamomile into their migraine management plan (Kern et al. 2022).

References

Abokrysha NT (2009) Ibn Sina (Avicenna) on pathogenesis of migraine compared with the recent theories. Headache 49(6):923–927. https://doi.org/10.1111/j.1526-4610.2009.01394.x

Ahmadifard M et al (2020) The efficacy of topical basil essential oil on relieving migraine headaches: a randomized triple-blind study. Complement Med Res 27(5):310–318. https://doi.org/10.1159/000506349

Baron EP (2018) Medicinal properties of cannabinoids, terpenes, and flavonoids in cannabis, and benefits in migraine, headache, and pain: an update on current evidence and cannabis science. Headache 58(7):1139–1186. https://doi.org/10.1111/head.13345

Borhani-Haghighi A et al (2010) Cutaneous application of menthol 10% solution as an abortive treatment of migraine without aura: a randomised, double-blind, placebo-controlled, crossed-over study. Int J Clin Pract 64(4):451–456. https://doi.org/10.1111/j.1742-1241.2009.02215.x

Borlak J et al (2022) Petasites for migraine prevention: new data on mode of action, pharmacology and safety. A narrative review. Front Neurol 13:864689. https://doi.org/10.3389/fneur.2022.864689

Cavin C et al (2005) Inhibition of the expression and activity of cyclooxygenase-2 by chicory extract. Biochem Biophys Res Commun 327(3):742–749. https://doi.org/10.1016/j.bbrc.2004.12.061

Chaplin S (2018) SIGN on the pharmacological management of migraine. Prescriber 29:27–31

Chinese Medical Association Group (2016) Guide to the prevention and treatment of migraine in China [in Chinese]. Chin J Pain Med 22:721–727

Collado-Vázquez S, Carrillo JM (2014) Cranial trepanation in the Egyptian. Neurologia 29(7):433–440. https://doi.org/10.1016/j.nrl.2011.05.012

Collier HOJ et al (1980) Extract of feverfew inhibits prostaglandin biosynthesis. Lancet 316(8200):922–923. https://doi.org/10.1016/s0140-6736(80)92084-x

Del Giudice E et al (2007) Cannabidiol, unlike synthetic cannabinoids, triggers activation of RBL-2H3 mast cells. J Leukoc Biol 81(6):1512–1522. https://doi.org/10.1189/jlb.1206738

Edmeads J (1999) History of migraine treatment. J Can Pharmacol Clin 6 Suppl A:5A–8A

Ernst E, Pittler MH (2000) The efficacy and safety of feverfew (Tanacetum parthenium L.): an update of a systematic review. Public Health Nutr 3(4a):509–514. https://doi.org/10.1017/s1368980000000598

Evans WC (2009) Trease and Evans' pharmacognosy. Elsevier Health Sciences

Galeotti N, Ghelardini C (2013a) St. John's wort reversal of meningeal nociception: a natural therapeutic perspective for migraine pain. Phytomedicine 20(10):930–938. https://doi.org/10.1016/j.phymed.2013.03.007

Galeotti N, Ghelardini C (2013b) St. John's wort relieves pain in an animal model of migraine. Eur J Pain (London, England) 17(3):369–381. https://doi.org/10.1002/j.1532-2149.2012.00196.x

Goadsby PJ (2012) Pathophysiology of migraine. Ann Indian Acad Neurol 15(5):15. https://doi.org/10.4103/0972-2327.99993

Groenewegen WA, Heptinstall S (1990) A comparison of the effects of an extract of feverfew and parthenolide, a component of feverfew, on human platelet activity in-vitro. J Pharm Pharmacol 42(15):553–557. https://doi.org/10.1111/j.2042-7158.1990.tb07057.x

Hall W, Degenhardt L (2009) Adverse health effects of non-medical cannabis use. Lancet 374(9698):1383–1391. https://doi.org/10.1016/s0140-6736(09)61037-0

Henderson LM et al (2002) St John's wort (Hypericum perforatum): drug interactions and clinical outcomes. Br J Clin Pharmacol 54(4):349–356. https://doi.org/10.1046/j.1365-2125.2002.01683.x

Heptinstall S et al (1992) Parthenolide content and bioactivity of feverfew (Tanacetum parthenium (L.) Schultz-Bip.). Estimation of commercial and authenticated feverfew products. J Pharm Pharmacol 44(5):391–395. https://doi.org/10.1111/j.2042-7158.1992.tb03631.x

Jiang Q et al (2014) Total bakkenolides protects neurons against cerebral ischemic injury through inhibition of nuclear factor-κB activation. CNS Neurol Disord Drug Targets 13(5):874–884. https://doi.org/10.2174/18715273113129990104

Kayama Y et al (2017) Functional interactions between transient receptor potential M8 and transient receptor potential V1 in the trigeminal system: relevance to migraine pathophysiology. Cephalalgia 38(5):833–845. https://doi.org/10.1177/0333102417712719

Kern C, Gombert C, Roso A, Garcia C (2022) Effect of the supplementation of virgin coriander seed oil on reducing reactivity in healthy women with sensitive skin: a randomized double-blind placebo-controlled pilot clinical study. Food Nutr Res 66. https://doi.org/10.29219/fnr.v66.7730

Kumar Paswan V, Shekhar Singh C, Kukreja G et al (2021) Health benefits and functional and medicinal properties of some common Indian spices. In: Herbs and spices—new processing technologies. IntechOpen. https://doi.org/10.5772/intechopen.98676

Lee DKC et al (2004) Butterbur, a herbal remedy, confers complementary anti-inflammatory activity in asthmatic patients receiving inhaled corticosteroids. Clin Exp Allergy 34(1):110–114. https://doi.org/10.1111/j.1365-2222.2004.01838.x

Lipton RB et al (2019) Unmet acute treatment needs from the 2017 migraine in America symptoms and treatment study. Headache 59(15):1310–1323. https://doi.org/10.1111/head.13588

Lopresti AL, Smith SJ, Drummond PD (2020) Herbal treatments for migraine: a systematic review of randomised-controlled studies. Phytother Res 34(10):2493–2517. https://doi.org/10.1002/ptr.6701

Ma L et al (2013) Ginkgolide B reduces LOX-1 expression by inhibiting AKT phosphorylation and increasing SIRT1 expression in oxidized LDL-stimulated human umbilical vein endothelial cells. PLoS One 8(9):e74769. https://doi.org/10.1371/journal.pone.0074769

Marles RJ et al (1992) A bioassay for inhibition of serotonin release from bovine platelets. J Nat Prod 55(15):1044–1056. https://doi.org/10.1021/np50086a003

Mughal SS (2022) Peppermint oil, its useful, and adverse effects on human health: a review. Authorea. https://doi.org/10.22541/au.166401168.86712355/v1

NCCIH (n.d.-a). https://www.nccih.nih.gov/health/butterbur#:~:text=PA%2Dfree%20butterbur%20is%20generally,chrysanthemums%2C%20marigolds%2C%20and%20daisies

NCCIH (n.d.-b). https://www.nccih.nih.gov/health/peppermint-oil#:~:text=Peppermint%20oil%20appears%20to%20be,abdominal%20pain%2C%20and%20dry%20mouth

Nguyen T, Alzahrani T (2023) Ginkgo biloba. In: StatPearls [Internet]. StatPearls Publishing, Treasure Island, FL. https://www.ncbi.nlm.nih.gov/books/NBK541024/

Niazi M et al (2017) Efficacy of topical Rose (Rosa damascena Mill.) oil for migraine headache: a randomized double-blinded placebo-controlled cross-over trial. Complement Ther Med 34:35–41. https://doi.org/10.1016/j.ctim.2017.07.009

Okusanya BO et al (2022) Medical cannabis for the treatment of migraine in adults: a review of the evidence. Front Neurol 13:871187. https://doi.org/10.3389/fneur.2022.871187

Pfaffenrath V et al (2002) The efficacy and safety of Tanacetum parthenium (Feverfew) in migraine prophylaxis—a double-blind, multicentre, randomized placebo-controlled dose-response study. Cephalalgia 22(7):523–532. https://doi.org/10.1046/j.1468-2982.2002.00396.x

Radat F et al (2005) Psychiatric comorbidity in the evolution from migraine to medication overuse headache. Cephalalgia 25(7):519–522. https://doi.org/10.1111/j.1468-2982.2005.00910.x

Rajapakse T, Davenport WJ (2019) Phytomedicines in the treatment of migraine. CNS Drugs 33(5):399–415. https://doi.org/10.1007/s40263-018-0597-2

Rajapakse T, Pringsheim T (2016) Nutraceuticals in migraine: a summary of existing guidelines for use. Headache 56(4):808–816. https://doi.org/10.1111/head.12789

Sachs J, McGlade E, Yurgelun-Todd D (2015) Safety and toxicology of cannabinoids. Neurotherapeutics 12(4):735–746. https://doi.org/10.1007/s13311-015-0380-8

Safiri S et al (2021) Global, regional, and national burden of migraine in 204 countries and territories, 1990 to 2019. Pain 163(2):e293–e309. https://doi.org/10.1097/j.pain.0000000000002275

Schwedt TJ et al (2018) Factors associated with acute medication overuse in people with migraine: results from the 2017 migraine in America symptoms and treatment (MAST) study. J Headache Pain 19(1):38. https://doi.org/10.1186/s10194-018-0865-z

Shah ZA, Nada SE, Doré S (2011) Heme oxygenase 1, beneficial role in permanent ischemic stroke and in Gingko biloba (EGb 761) neuroprotection. Neuroscience 180:248–255. https://doi.org/10.1016/j.neuroscience.2011.02.031

Singh O et al (2011) Chamomile (Matricaria chamomilla L.): an overview. Pharmacogn Rev 5(9):82. https://doi.org/10.4103/0973-7847.79103

Tulsulkar J, Shah ZA (2013) Ginkgo biloba prevents transient global ischemia-induced delayed hippocampal neuronal death through antioxidant and anti-inflammatory mechanism. Neurochem Int 62(2):189–197. https://doi.org/10.1016/j.neuint.2012.11.017

Vurallı D, Ayata C, Bolay H (2018) Cognitive dysfunction and migraine. J Headache Pain 19(1):109. https://doi.org/10.1186/s10194-018-0933-4

Zargaran A et al (2016) A review on the management of migraine in the Avicenna's Canon of Medicine. Neurol Sci 37(3):471–478. https://doi.org/10.1007/s10072-016-2498-9

9 Neuromodulation and Neurostimulation Approaches for the Management of Migraine Pain

Mithun Singh Rajput, Surabhi Patil, Viral Patel, and Jigna Shah

9.1 Introduction

A prevalent and extremely incapacitating primary headache, migraines are generally more common and have a significant socioeconomic impact. According to Lipton et al. (2001), America has the highest incidence of migraine, followed by Europe, whereas Africa and Asia have a lesser prevalence. There have been several notable changes made to the classification of migraine in the most recent version of the International Headache Society's International Classification of Headache Disorders, ICHD-III (beta) (HIS 2013). For example, chronic migraine, which was formerly categorised as a migraine complication, is now regarded as a migraine subcategory. Additionally, according to the revised version, patients must be diagnosed as having chronic migraine if they experience headache (tension-type or migraine-like) on 15 or more days per month for a period longer than 3 months and exhibit migraine-like symptoms on at least 8 of those days. By classifying migraines based on the frequency of attacks, the categorization highlights the most severe types of headaches within the migraine spectrum.

Research on the underlying pathophysiology of chronic migraine is still in its early stages and our understanding of the same is in its infancy stage. Although there has been no agreement on the pathophysiology of migraines over time, there are currently two main pools of thought regarding the underlying mechanism of migraines in general: one holds that changes in the brain play a major role in the

M. S. Rajput · S. Patil · J. Shah (✉)
Department of Pharmacology, Institute of Pharmacy, Nirma University, Ahmedabad, Gujarat, India
e-mail: jigna.shah@nirmauni.ac.in

V. Patel
Department of Pharmaceutics, Ramanbhai Patel College of Pharmacy, Charotar University of Science and Technology (CHARUSAT), Anand, Gujarat, India

S. Bhatt et al. (eds.), *Management of Migraine Pain*,
https://doi.org/10.1007/978-981-97-4529-6_9

development of migraines, while the other holds that external triggers are the primary cause of migraines. The perivascular nerves that innervate the major cerebral vessels were thought to be the source of activation for migraine attacks, which gave rise to the early vascular theory of migraines (Mungoven et al. 2021). Although cerebrovascular changes were once thought to be the basis of migraine pathophysiology, new research has shown that neuronal changes in subcortical areas play a crucial role in the development of migraines (Akerman et al. 2011). This evidence includes reports of fatigue and impaired concentration occurring hours prior to the onset of migraines, as well as evidence from experimental animals showing that cortical spreading depression can activate brainstem trigeminovascular neurons without the need for peripheral input (Lambert et al. 2011). In fact, a recent theory suggests that there is an oscillation between sensitivity and threshold, and reduced neuronal tone in brainstem activity. Reduced brainstem tone impairs ongoing endogenous analgesic circuits' ability to modulate painful inputs, and increases the likelihood that an external stimulus would activate trigeminal pathways and cause headaches. It seems that the peripheral and central neural systems work in concert to generate migraines, making migraine disorder a complicated condition (Puledda et al. 2017). Although it's possible that peripheral and central mechanisms work together, each migraineur may have a different ratio of these causes. Chronic migraine transformation and onset may be influenced by changes in the structure and function of the brain in pain-related regions causing atypical pain processing, cortical hyperexcitability, central sensitization, and neurogenic inflammation (Mungoven et al. 2021) (Fig. 9.1).

Moreover, research on individuals with familial hemiplegic migraine (FHM) demonstrates the genetic component of migraine. It has been demonstrated that this condition is associated with mutations in the genes for voltage-gated calcium channel (CACNA1A), voltage-gated sodium channel (SCN1A), and Na^+/K^+ ATPase (ATP1A2) (Bussone 2004).

Currently, medication, psychoeducation, lifestyle control, and, if available, psychotherapy interventions are functional in migraine management (Bonfert et al. 2013). Acute therapeutic options have grown in number over time. Currently, there are two categories of medications that have demonstrated effectiveness in treating migraine attacks: non-specific medications (such as analgesics and non-steroidal

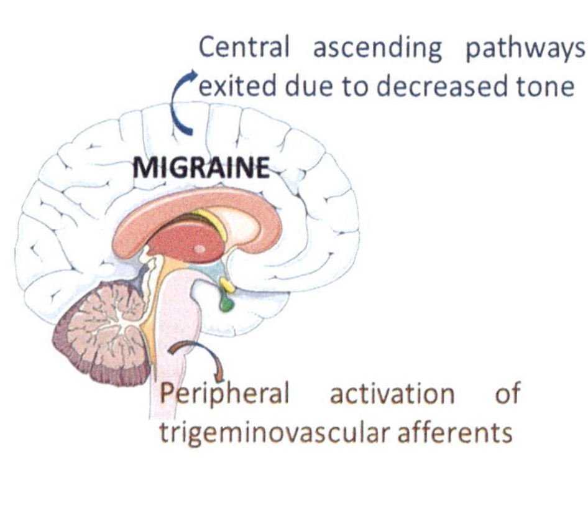

Fig. 9.1 Chief theories of migraine attack

anti-inflammatory drugs, or NSAIDs) and specific medications (such as triptans and ergot derivatives) (Antonaci et al. 2016). Every specific migraine medication has been assigned with a level of evidence, based on the calibre of available studies: level A indicates that the medication is proven to be effective (or ineffective); level B indicates that it is probably effective (or ineffective); level C indicates that it may be effective (or ineffective); and level U indicates that there is insufficient or conflicting evidence to support or disprove the use of the medication (Antonaci et al. 2016) (Table 9.1).

9.2 Need of Alternative Therapies

Eliminating the triggers prevents migraines, but doing so requires the sufferer to be aware of the circumstances that lead to migraine attacks. A huge variety of possible triggers for an intense migraine episode are there, including hunger, stress, trigger foods, and lack of sleep. Not a single treatment exists for migraines. When migraine is moderate, avoiding triggers can benefit, but if the pain is severe or regular, seeking medical assistance is required. Evidence-based and well-established

Table 9.1 Pharmacotherapeutic options for migraine (Antonaci et al. 2016)

Category	Example	Mode of action
Non-specific drugs		
Analgesics and non-steroidal anti-inflammatory drugs (NSAIDS)	Acetylsalicylic acid, Ibuprofen, Naproxen, Diclofenac, Paracetamol, Metamizol, Phenazone, Tolfenamic acid, Acetylsalicylic acid + acetaminophen + caffeine	Inhibition of the enzyme cyclooxygenase (COX)
Specific drugs for migraine		
Ergot byproducts	Dihydroergotamine and Ergotamine	Agonistic effect on 5-hydroxytryptamine (serotonin) receptors
Triptans	Sumatriptan, Zolmitriptan, Naratriptan, Rizatriptan, Eletriptan, Frovatriptan, Almotriptan	Neurotransmitter release inhibition, primarily through 5-HT1B/1D receptors (trigeminovascular afferents and trigeminal nucleus caudalis), at both peripheral and central trigeminal nociceptive terminals
Novel acute and preventive treatments		
Geptans	BI44370TA, BMS-927711, MK-3207, NCT01613248, Olcagepant, Telcagepant	Calcitonin gene-related peptide antagonists
Ditans	Lasmiditan	Serotonin 5HT1F agonists
Glutamate receptor antagonists	AMPA, NMDA and kainate	Glutamate inhibition, which is produced by trigeminal ganglion neurons that express 5-HT1B/1D/1F receptors

NMDA *N*-methyl-D-aspartate, *AMPA* α amino-3-hydroxy-5-methyl-4-isoazolepropionic acid

pharmaceutical alternatives are available for the acute treatment of migraine attacks. But in terms of pharmaco-prophylaxis, situations aren't all that robust (Charles 2018). Significant drawbacks associated with a number of the choices include adverse effects and inadequate safety and efficacy data. When thinking about treating children and adolescents, these factors are especially crucial (Oskoui et al. 2019a, b). Thus, new and creative therapeutic approaches are required due to the dearth of adequate nonpharmacological therapy choices (Brighina et al. 2019). More research into the causes and symptoms of migraine has led to the development of more promising alternative therapeutic techniques (Al-Quliti and Assaedi 2016). The field of neuromodulation, which involves altering the activity of neurons in the brain using electrical or chemical means, is the most rapidly expanding field of study (Al-Quliti and Assaedi 2016).

The goals of this chapter are to offer a streamlined description of the investigation on migraine in terms of pharmacological management and explore potential drug-free alternatives for the active treatment and prophylaxis. Further, the chapter explains what neuromodulation is, how it works, how it may be used to treat migraine, and the several types of neuromodulation treatments available, each with its own advantages and disadvantages. Methods of neurostimulation such as remote electronic neuromodulation, cranial stimulation and many more have been addressed. When making clinical judgments on neuromodulation, it is important to consider both invasive and non-invasive methods, therefore it is important to evaluate the clinical relevance and scientific validity of the available evidence. Neurostimulation in migraine has been the subject of many investigations, and this article explores the limits of these studies and possible future therapies.

9.3 Neuromodulation

The use of electrical, magnetic, or chemical stimulation to activate, regress, alter, or modify processes in the central or peripheral nervous system is known as neuromodulation (Krames et al. 2009). It entails the targeted application of a chemical, electrical, or magnetic stimulus to specific brain regions within the body in order to modulate nerve activity. The process of neuromodulation is characterised by its adaptability, reversibility, and lack of harm. It's becoming clear that using this technology instead of more conventional drug therapy, can cure migraines more safely and efficiently (Martel et al. 2015). Due to their demonstrated efficacy and safety, non-invasive neuromodulation therapies such as single-pulse transcranial magnetic stimulation and non-invasive vagus nerve stimulation (nVNS) are gaining popularity among patients and clinicians (Schwedt and Vargas 2015). As of right now, it has been shown that this technology has a lot to give, especially to vulnerable patient populations including expecting mothers, people who have a hard time taking medication, or people who find it ineffective. Non-pharmacological neuromodulation methods could be more affordable in certain circumstances (Tiwari and Agrawal 2022).

9.3.1 Mode of Action

In order to affect or activate the central or peripheral pain pathways, neuromodulation uses electrical or magnetic pulsations. In order to reduce the pain ratings, this treatment aims to regulate the pain systems. When pain circuits are stimulated electrically or magnetically, central neurotransmitters can be altered (Oshinsky et al. 2014). These changes may be used to disrupt the processes responsible for the genesis in the treatment of acute migraine attacks. These modifications are intended to lessen the central sensitization that causes chronic headaches in order to prevent them. Conventional approaches involve stimulating the nervous system through the skin in a central or peripheral manner. To control the headache mechanisms connected to pain, an electric current or a changeable magnetic field is used. Prolonged treatment may have long-term preventive effects, even if both delivery modalities are suitable for addressing acute symptoms due to their fast responses (Puledda and Shields 2018). Different neuromodulation approaches are depicted in Fig. 9.2, altogether with their corresponding sites of action or targets. With a device that uses electrical or magnetic stimulations to change or modulate brain cell activity, neuromodulation is carried out. These devices all function in various ways. While some are used preventively, others stop attacks. The concept of modifying neural route activity is the same in each of them. These devices also go by the name of stimulators. Devices for neuromodulation may be electrical, magnetic, or temperature-changing. Additionally, they may need to be surgically positioned, be portable, invasive, or non-invasive (Tiwari and Agrawal 2022).

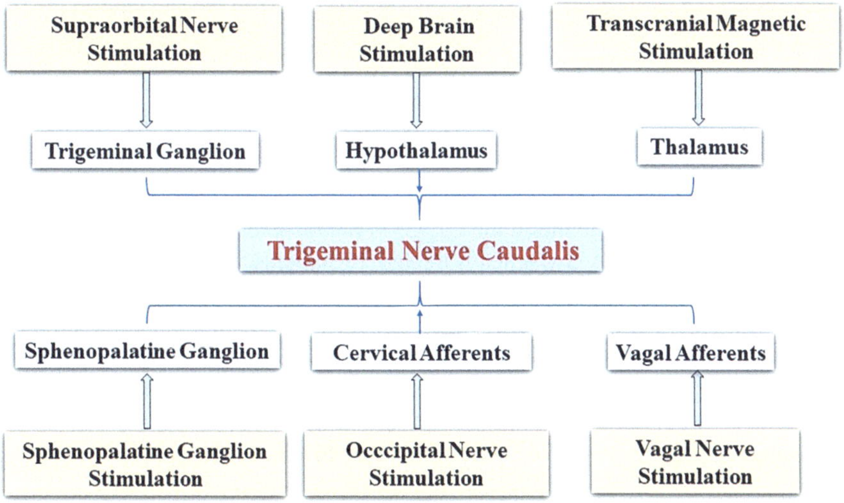

Fig. 9.2 Techniques of neuromodulation and their locations of action

9.4 Invasive Neuromodulation Techniques

Since most invasive neuromodulation therapies need surgical operations (electrodes, batteries, etc.), they have primarily been used on the most severely impaired migraine patients, or those who have chronic migraine.

9.4.1 Occipital Nerve Stimulation (ONS)

ONS was initially proposed to treat occipital neuralgia using the same equipment that was utilized to treat neuropathic pain by spinal cord stimulation. Additionally, primary headaches, such as migraines, were treated with it. The mechanisms by which ONS prevents migraines are unknown. The trigeminocervical complex's (TCC) convergent input from trigeminal and cervical afferents provides the background for the ONS's proposal to treat migraines and other primary headaches (Bartsch and Goadsby 2003). However, the impact latency seen in numerous patients points to a more intricate mechanism of action, which is confirmed by functional imaging research. According to Matharu et al. (2004), a PET scan revealed activation in the rostro-dorsal pons, which is hypothesised to include migrainous generators, as well as the anteroposterior cingulate cortex and other pain matrix-related areas. Based on limited data, the European Headache Federation stated that the use of ONS looks appropriate for individuals who suffer from chronic migraines (Martelletti et al. 2013). Since the mean incidence of all ONS-related problems in the treatment of chronic migraines was predicted to be 66%, the use of ONS in this setting is currently limited (Yang et al. 2016). Because of the high levels of stimulation and the fact that ONS is typically used day and night, there is also a risk of early battery depletion. Lastly, a lead migration related to neck motion may exacerbate ONS and necessitate surgical revision when it does (Lanteri-Minet and Donnet 2020).

9.4.2 Cervical Spinal Cord Stimulation (CSCS)

According to the fundamental theory of CSCS for primary headache management, the trigeminocervical complex will be stimulated more by electrical pulses injected directly into the dorsal columns at the C2–C3 vertebral level than by stimulating the greater occipital nerve. A Swiss team treated 17 patients with refractory chronic migraine using CSCS; in a retrospective cohort, 12 of the patients showed a 50% response (De Agostino et al. 2015). A few years ago, a proof-of-concept experiment was conducted to examine the safety and effectiveness of CSCS in treating refractory chronic migraine; however, the results are still awaited. The European Headache Federation states that CSCS should be carefully avoided in patients with migraine until adequate proof is available (Martelletti et al. 2013).

9.4.3 Sphenopalatine Ganglion Stimulation (SPGS)

The business company Autonomic Technologies has created a neuromodulation device that may be implanted chronically and is specifically intended for acute SPGS. It can be used as needed. The neurostimulator device is screwed to the zygomatic process with a plate, and the lead is positioned in contact with the sphenopalatine ganglion. It is implanted in the pterygopalatine fossae, which runs the length of the maxillary bone's posterior wall. Instead of requiring a battery, the neurostimulator is fuelled and triggered by radiofrequency energy through a remote controller. Although a randomised controlled study for the acute treatment of migraines has just concluded and the findings are being waited upon, the primary focus of this device's clinical development was the acute treatment of cluster headaches (Fontaine et al. 2018). As discussed, limitations of invasive neuromodulation approaches include safety, acceptability, and a dearth of evidence that can be applied universally. Hence, alternatives are required.

9.5 Non-invasive Neuromodulation Techniques

The non-invasive technique in this context refers to the stimulation that can be applied electrically or in another way without requiring any form of skin penetration. It is also known as external stimulation. Neuromodulation is another name for this process because it affects central pain processing structures rather than stimulating peripheral nerve systems, which is probably not the main mechanism for pain/headache relief. In context with neurostimulation, referral is about mechanical, electrical, thermal or magnetic stimulation of nerve fibres (Fig. 9.3). Non-invasive neurostimulation has been more well-known recently as a low-risk, successful treatment for migraines in some individuals (Evers 2021).

9.5.1 Magnetic Neurostimulation

9.5.1.1 Single-Pulse Transcranial Magnetic Stimulation (sTMS)

A magnetic beat is applied to the occipital area of the scalp in TMS, a non-invasive technique for neural stimulation (Barker et al. 1985). Based on experimental models, the mode of action is believed to be through suppression of cortical-spreading depression and thalamocortical signalling (Andreou et al. 2016). It has been discovered that sTMS works well for managing episodic migraine headaches acutely. Four pulses twice a day is the dosage for preventative therapy; three consecutive pulses as needed, repeated every 15 min if relief is not achieved, is the dosage for acute therapy. The major trials evaluating the use of sTMS in treating migraines reported side effects such as paraesthesia, light-headedness, scalp discomfort, dizziness, tinnitus, and the worsening of migraine These side effects were generally well-tolerated and did not significantly differ between sTMS and sham devices (Blech and Starling 2020). Seizures and implants and conductive materials in the upper torso and head

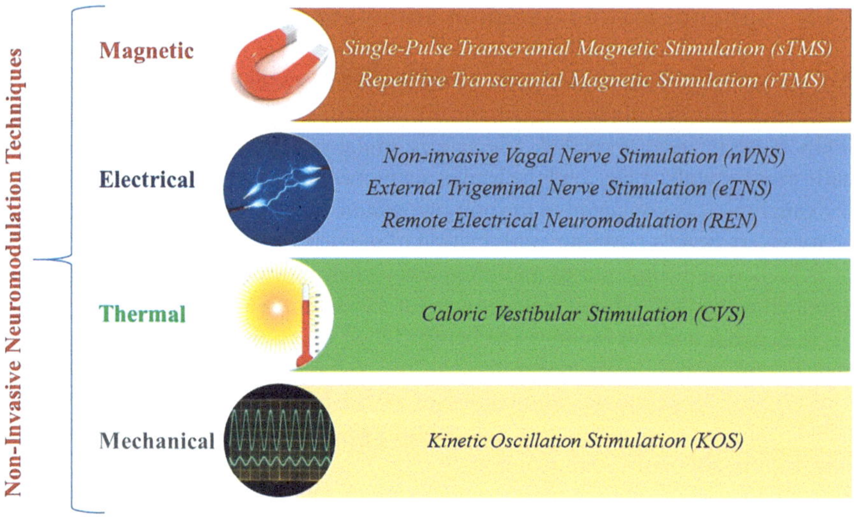

Fig. 9.3 Neuromodulation modalities in treatment of migraine

(such as vagal nerve stimulators and pacemakers) are contraindications for use (Blech and Starling 2020).

9.5.1.2 Repetitive Transcranial Magnetic Stimulation (rTMS)

Numerous researches have investigated the potential of rTMS to prevent episodic and chronic migraines (Grazzi et al. 2021). Misra et al. (2013) showed the preventive value of rTMS for patients with persistent migraines in a randomized, placebo-controlled research that had few side effects. Additional double-blind trials and randomized control groups produced positive outcomes with no side effects (Conforto et al. 2014; Dalla Libera et al. 2013). However, one study brought attention to minor side effects of rTMS, including low drowsiness that was observed in one patient (Misra et al. 2013). Regretfully, the results cannot be broadly applied due to the variations in the methods, location, and stimulation parameters throughout the investigations. Additionally, the brief follow-up periods have diminished the research' dependability, despite the fact that overall, the findings are favourable (Lan et al. 2017).

9.5.2 Electrical Neurostimulation

9.5.2.1 Non-invasive Vagal Nerve Stimulation (nVNS)

The anterolateral region of the neck, which covers the vagus nerve's descent, is stimulated electrically by a tiny, portable device. This method is called non-invasive vagal nerve stimulation. Most of the body's parasympathetic innervation originates from this nerve. Research has demonstrated that activation of the vagus nerve can alter neural regions such as the locus coeruleus, nucleus tractus solitarious, and trigeminal spinal tract that are believed to be involved in the pathogenesis of migraine (Blech and Starling 2020). The FDA initially approved "gammaCore" (produced by electroCore, Inc.) for use in treating cluster headaches. For prevention, the usual dosage for vagal nerve stimulation with "gammaCore" is two stimulations (separately lasting 2 min) three times a day. When using acute therapy, two stimulations are usually administered as soon as the headache appears, with more stimulations, if necessary, at the 20-min and 2-h marks. Based on the outcomes of the nVNS as Acute Therapy For Migraine (PRESTO) trial, the FDA approved the "gammaCore" device (Tassorelli et al. 2018). In this trial, using the "gammaCore" device, migraine patients were randomized to either sham treatment or non-invasive vagus nerve stimulation as an acute therapy. Statistically significant differences were found favouring non-invasive vagus nerve stimulation in many secondary endpoints, that comprises pain relief at 30 and 60 min, but not meeting its primary endpoint (pain freedom at 120 min). According to Kinfe et al. (2015), VNS is generally well-tolerated and may have advantages beyond improving migraine symptoms, such as improving depression and sleep quality. Some of the negative effects that have been reported include dizziness, tinnitus, lip/facial droop, stiff neck, frequent urination, shoulder pain, and mild disorientation (Silberstein et al. 2016).

9.5.2.2 External Trigeminal Nerve Stimulation (eTNS)

With the use of a device that facilitates external trigeminal nerve stimulation (eTNS), the forehead is stimulated, thereby activating the ophthalmic nerve (V1) supratrochlear and supraorbital branches. Its exact mode of action is unknown, but it most likely involves modulating trigeminal nerve pain. Meanwhile, FDG-PET studies have demonstrated enhanced metabolism in specific cortical regions, such as the orbitofrontal and rostral anterior cingulate cortices, which are thought to be involved in the pathways leading to headache pain (Magis et al. 2017). In the PREMICE (PREvention of MIgraine using CEfaly) study, a double-blind RCT, efficacy was shown in 67 subjects' use of eTNS (using the "Cefaly" device) for migraine prevention in 2013. The treatment arm's average number of migraine days upgraded from 6.84 to 4.88 with a significance value of 0.023. A monthly migraine incident, 50% response rate, headache days, and acute migraine medication use were also improved. As a result, the FDA approved the "Cefaly" gadget in 2014 for the prevention of migraines. A 60-min session for acute therapy using the "Cefaly" and one 20-min treatment per day for preventive use are recommended. Similar to other neuromodulation tools, eTNS is generally widely accepted. According to Magis et al. (2013), there have been reports of forehead paraesthesias (particularly when used outside of an acute migraine attack), headaches, tiredness, an infrequent allergic skin reaction, sleeplessness, and reversible forehead skin irritation. Metallic head implants, pacemaker devices, intracardiac lines and recent face trauma are among the contraindications for use (Blech and Starling 2020).

9.5.2.3 Remote Electrical Neuromodulation (REN)

A more current neuromodulation goal for primary headache diseases including migraine is remote electrical neuromodulation (REN). This treatment for the upper arm peripheral nerves is thought to elicit conditioned pain modulation (CPM), which can block pain in distant regions by activating C and Aδ noxious fibres above depolarization thresholds (Yarnitsky et al. 2019). Yarnitsky et al. (2019) conducted a randomized, double-blind, placebo-controlled multi-centre trial involving REN with 296 participants. The results showed that 66.7% of patients experienced pain relief and the net therapeutic gain was 27.9% with a significance value $p < 0.0001$. Furthermore, the pain relief was also observed to be sustained 48 h after treatment. Consequently, in May 2019, the FDA approved the device "Nerivio Migra" under the de novo categorization for the acute treatment of migraines in individuals 18 years of age and above (USFDA 2019). Wearable on the upper arm, the device uses a smartphone application to control 45 min of stimulation per treatment. Using the phone app, the patient can alter the intensity level and they are advised to set it to the highest level of stimulation that does not exceed their perceived pain threshold. Although documented side effects in the trials comprised of itching, arm/hand numbness, warmth sensation, discomfort, muscle spasms and tingling, adverse events with REN are not prevalent (Yarnitsky et al. 2019). Uncontrolled epilepsy, congestive heart failure, and active implanted medical devices (hearing aid implants, heart pacemakers etc.) are among the conditions that preclude using this medication (Blech and Starling 2020).

9.5.3 Thermal Neurostimulation

All of the stimulation methods that have been covered in this chapter thus far have their roots in either direct electrical stimulation or indirect electrical stimulation through the use of magnetic fields. Other methods of stimulating the brain for treating migraines include temperature or mechanic-based stimulations.

9.5.3.1 Caloric Vestibular Stimulation (CVS)

Caloric vestibular stimulation (CVS) is a technique that involves using a device that resembles an earphone to stimulate the ear with frequent and periodic temperature changes (Evers 2021). Clinical applications of caloric vestibular stimulation (CVS) include the diagnosis of vestibular dysfunction syndromes and evaluation of brainstem function in a range of neurologic disorders. Caloric vestibular stimulation has been suggested as a potential method for neuromodulation of the brainstem because migraine routes seem to involve dysfunctional brainstem areas (such as the reticular formation, locus coeruleus and periaqueductal grey matter) (Dieterich and Brandt 2008). A multi-center, randomized, parallel-arm, placebo-controlled clinical trial assessed the efficacy of CVS in preventing episodic migraines in 2017. The FDA classified the Scion NeuroStim, LLC-developed ThermoNeuroModulation (TNM) device as a non-significant risk for investigation in episodic migraine. Episodic migraine patients without aura, ranging in age from 18 to 65 years, were identified.

After 3 months of treatment, the treatment arm significantly reduced migraine days compared to the placebo ($p = 0.12$), and improved monthly pain scores and acute medication use reduction according to Wilkinson et al. (2017). Of the 100 patients who were randomised, 81 finished the study. Consequently under de novo classification in 2018, the device received FDA clearance for the preventative treatment of episodic migraine in patients above 12 years of age. It is necessary to place over-the-ear headphones into the external ear canal in order to use this device. It is a non-invasive, DC-powered device. For device control, it has a touch-screen display. According to Black et al. (2016), it functions by delivering regulated heat waves to the ear canal, specifically intended to target the vestibular organs. Although it is usually well accepted, clinical trials have reported side effects such as impaired vision, nausea, increased headaches, ear pain, tinnitus, and itching and irritation. Active ear infections, ruptured tympanic membranes, and cochlear implantation are among the contraindications for use (Blech and Starling 2020).

9.5.4 Mechanical Neurostimulation

9.5.4.1 Kinetic Oscillation Stimulation (KOS)

Kinetic oscillation stimulation (KOS) is a technique that puts high-frequency, oscillating pressure on to the nasal mucosa with the use of an intranasal plastic catheter. The two short trials in the same group (Juto and Hallin 2015; Li et al. 2016) showed that this stimulation method effectively reduced the number of acute migraine attacks, one of which was under control.

9.6 Limitations

While the outcomes of non-invasive and surgical neuromodulation techniques for migraines have shown promise, numerous issues are still unresolved. The FDA must approve a number of these devices under trials, before healthcare professionals can utilise them as prophylactic or preventative treatments. Neuromodulatory operations are frequently recommended only in refractory disorders due to their considerable upfront cost. There is also continuing research into the precise targets of neuromodulation and therapeutic mechanisms of action. Additionally, there is evidence to suggest that the devices work better for chronic migraine than for episodic migraine. However, given that the majority of the research focused on chronic migraine and that very few examined refractory episodic migraine, there may be bias in this. The effectiveness of these gadgets in treating a "normal" episodic migraine is simply unknown. There have been no comparative studies conducted with regard to other subtypes of migraine. Despite the fact that these devices are not as effective as oral drug prophylaxis, many patients accept them because of the adverse effects of oral drugs. However, a few of patients exhibit complete non-acceptance to any kind of device. Furthermore, there is a widespread belief that invasive stimulation methods like occipital nerve stimulation or sphenopalatine ganglion stimulation may be more effective than non-invasive devices. Limitations include effects like user error, special treatment reactions from customers, complications from surgeries, and surgical errors.

9.7 Prospects on Future Treatments

Studies on the acute and preventive effectiveness of neuromodulatory techniques are currently focusing on their primary and secondary outcomes. It will be advantageous to include criteria such as the minimum duration of stimulation adjustment therapy periods and requirements for pharmacological refractoriness. Furthermore, several studies have indicated that individuals with concurrent medication overuse had worse outcomes than those without overuse, raising questions about whether these patients should be excluded from neurostimulation research. If at all feasible, results from individuals who also have a headache from concurrent drug overuse should be independently reviewed in order to answer this question. Furthermore, given the high rate of complications, particularly in ONS, the manufacturers are called upon to further improve both implantation techniques, including neurosurgical training, and their devices (e.g., by improving the durability of electrodes, leads and battery life, or by further developing rechargeable microstimulators) (McGreevy et al. 2012; Jurgens and Leone 2013). A first retrospective study for hypothalamic deep-brain stimulation in chronic cluster headaches shows that neuromodulatory techniques can be cost effective, despite the relatively high initial costs for the device and the implantation process (McGreevy et al. 2012; Jurgens and Leone 2013). Future research plans must consider the constricting availability of healthcare resources. The cost-effectiveness of both invasive and non-invasive neuromodulatory procedures must be shown. Thanks to technical breakthroughs, smaller and more wirelessly capable devices have now been made. This enables the insertion of small leads and does away with the need for a larger implanted power supply. These advancements may find use in the management of migraines. Modern technology and an expanding amount of knowledge indicate that neuromodulation has a promising future.

9.8 Conclusion

Common medical conditions like migraine and headache carry a significant financial and social cost. Since medications such as beta-blockers, triptans, NSAIDs, tricyclic antidepressants, and antiepileptics only help specific patients, technology is crucial to the development of new therapeutic approaches. Medical device approvals by the Food and Drug Administration are primarily granted based on safety, rather than therapeutic efficacy. A small number of these devices were certified by the FDA and made available for clinical use since large-scale, rigorous, and randomised control trials were not feasible at the time. Potential therapeutic activity may be revealed by future post-market investigations. What non-invasive neuromodulation technologies have in common are safety, tolerability, limited contraindications, and the capacity to treat main headache conditions both immediately and preventively. These advantages may be especially helpful for people who need a more efficient treatment plan or an alternative to oral medication, as well as for individuals who are managing multiple medical conditions. A safe, effective, and non-invasive treatment option for migraine is neuromodulation. More research is required in particular groups, such as children, and on a range of headache conditions, including chronic migraine, medication misuse headache, and post-traumatic headache, in order to strengthen their evidence base and therapeutic application.

References

Akerman S, Holland PR, Goadsby PJ (2011) Diencephalic and brainstem mechanisms in migraine. Nat Rev Neurosci 12(10):570–584. https://doi.org/10.1038/nrn3057

Al-Quliti KW, Assaedi ES (2016) New advances in prevention of migraine: review of current practice and recent advances. Neurosci J 21(3):207–214. https://doi.org/10.17712/nsj.2016.3.20150506

Andreou AP, Holland PR, Akerman S, Summ O, Fredrick J, Goadsby PJ (2016) Transcranial magnetic stimulation and potential cortical and trigeminothalamic mechanisms in migraine. Brain 139(7):2002–2014. https://doi.org/10.1093/brain/aww118

Antonaci F, Ghiotto N, Wu S, Pucci E, Costa A (2016) Recent advances in migraine therapy. Springerplus 5(1):637. https://doi.org/10.1186/s40064-016-2211-8

Barker AT, Jalinous R, Freeston IL (1985) Non-invasive magnetic stimulation of human motor cortex. Lancet 325(8437):1106–1107. https://doi.org/10.1016/s0140-6736(85)92413-4

Bartsch T, Goadsby PJ (2003) Increased responses in trigeminocervical nociceptive neurons to cervical input after stimulation of the dura mater. Brain 126(8):1801–1813. https://doi.org/10.1093/brain/awg190

Black RD, Rogers LL, Ade KK, Nicoletto HA, Adkins HD, Laskowitz DT (2016) Non-invasive neuromodulation using time-varying caloric vestibular stimulation. IEEE J Transl Eng Health Med 4:1–10. https://doi.org/10.1109/jtehm.2016.2615899

Blech B, Starling AJ (2020) Noninvasive neuromodulation in migraine. Curr Pain Headache Rep 24:1–7. https://doi.org/10.1007/s11916-020-00914-3

Bonfert M, Straube A, Schroeder AS, Reilich P, Ebinger F, Heinen F (2013) Primary headache in children and adolescents: update on pharmacotherapy of migraine and tension-type headache. Neuropediatrics 44(01):003–019. https://doi.org/10.1055/s-0032-1330856

Brighina F, Raieli V, Messina LM, Santangelo G, Puma D, Drago F, Rocchitelli L, Vanadia F, Giglia G, Mangano S (2019) Non-invasive brain stimulation in pediatric migraine: a perspective from evidence in adult migraine. Front Neurol 10:364. https://doi.org/10.3389/fneur.2019.00364

Bussone G (2004) Pathophysiology of migraine. Neurol Sci 25:s239–s241. https://doi.org/10.1007/s10072-004-0295-3

Charles A (2018) The pathophysiology of migraine: implications for clinical management. Lancet Neurol 17(2):174–182. https://doi.org/10.1016/s1474-4422(17)30435-0

Conforto AB, Amaro E Jr, Gonçalves AL, Mercante JP, Guendler VZ, Ferreira JR, Kirschner CC, Peres MF (2014) Randomized, proof-of-principle clinical trial of active transcranial magnetic stimulation in chronic migraine. Cephalalgia 34(6):464–472. https://doi.org/10.1177/0333102413515340

Dalla Libera D, Colombo B, Nuara A, Spagnolo F, Straffi L, Chieffo R, Coppi E, Di Maggio G, Ferrari L, Houdayer E, Velikova S (2013) Repetitive transcranial magnetic stimulation (rTMS) as preventive treatment of chronic migraine: a safe approach with a promising effect. Neurology 80:S59.005

De Agostino R, Federspiel B, Cesnulis E, Sandor PS (2015) High-cervical spinal cord stimulation for medically intractable chronic migraine. Neuromodulation 18(4):289–296. https://doi.org/10.1111/ner.12236

Dieterich M, Brandt T (2008) Functional brain imaging of peripheral and central vestibular disorders. Brain 131(10):2538–2552. https://doi.org/10.1093/brain/awn042

Evers S (2021) Non-invasive neurostimulation methods for acute and preventive migraine treatment—a narrative review. J Clin Med 10(15):3302. https://doi.org/10.3390/jcm10153302

Fontaine D, Santucci S, Lanteri-Minet M (2018) Managing cluster headache with sphenopalatine ganglion stimulation: a review. J Pain Res 11:375–381. https://doi.org/10.2147/JPR.S129641

Grazzi L, Toppo C, D'Amico D, Leonardi M, Martelletti P, Raggi A, Guastafierro E (2021) Non-pharmacological approaches to headaches: non-invasive neuromodulation, nutraceuticals, and behavioral approaches. Int J Environ Res Public Health 18(4):1503. https://doi.org/10.3390/ijerph18041503

Headache Classification Committee of the International Headache Society (IHS) (2013) The international classification of headache disorders (beta version). Cephalalgia 33(9):629–808. https://doi.org/10.1177/0333102413485658

Jurgens TP, Leone M (2013) Pearls and pitfalls: neurostimulation in headache. Cephalalgia 33(8):512–525

Juto JE, Hallin RG (2015) Kinetic oscillation stimulation as treatment of acute migraine: a randomized, controlled pilot study. Headache 55(1):117–127. https://doi.org/10.1111/head.12485

Kinfe TM, Pintea B, Muhammad S, Zaremba S, Roeske S, Simon BJ, Vatter H (2015) Cervical non-invasive vagus nerve stimulation (nVNS) for preventive and acute treatment of episodic and chronic migraine and migraine-associated sleep disturbance: preliminary findings from a prospective observational cohort study. J Headache Pain 16(1):1–11. https://doi.org/10.1186/s10194-015-0582-9

Krames ES, Peckham PH, Rezai A, Aboelsaad F (2009) What is neuromodulation? In: Neuromodulation. Academic, pp 3–8. https://doi.org/10.1016/B978-0-12-374248-3.00002-1

Lambert GA, Truong L, Zagami AS (2011) Effect of cortical spreading depression on basal and evoked traffic in the trigeminovascular sensory system. Cephalalgia 31(14):1439–1451. https://doi.org/10.1177/0333102411422383

Lan L, Zhang X, Li X, Rong X, Peng Y (2017) The efficacy of transcranial magnetic stimulation on migraine: a meta-analysis of randomized controlled trails. J Headache Pain 18:1–7. https://doi.org/10.1186/s10194-017-0792-4

Lanteri-Minet M, Donnet A (2020) Neuromodulation in migraine. In: Neuromodulation in headache and facial pain management: principles, rationale and clinical data. Springer, Cham, pp 147–156

Li TQ, Wang Y, Hallin R, Juto JE (2016) Resting-state fMRI study of acute migraine treatment with kinetic oscillation stimulation in nasal cavity. Neuroimage Clin 12:451–459. https://doi.org/10.1016/j.nicl.2016.08.014

Lipton RB, Stewart WF, Diamond S, Diamond ML, Reed M (2001) Prevalence and burden of migraine in the United States: data from the American Migraine Study II. Headache 41(7):646–657. https://doi.org/10.1046/j.1526-4610.2001.041007646.x

Magis D, Sava S, d'Elia TS, Baschi R, Schoenen J (2013) Safety and patients' satisfaction of transcutaneous supraorbital neurostimulation (tSNS) with the Cefaly® device in headache treatment: a survey of 2,313 headache sufferers in the general population. J Headache Pain 14:1–8. https://doi.org/10.1186/1129-2377-14-95

Magis D, D'Ostilio K, Thibaut A, De Pasqua V, Gerard P, Hustinx R, Laureys S, Schoenen J (2017) Cerebral metabolism before and after external trigeminal nerve stimulation in episodic migraine. Cephalalgia 37(9):881–891. https://doi.org/10.1177/0333102416656118

Martel MO, Finan PH, Dolman AJ, Subramanian S, Edwards RR, Wasan AD, Jamison RN (2015) Self-reports of medication side effects and pain-related activity interference in patients with chronic pain: a longitudinal cohort study. Pain 156(6):1092. https://doi.org/10.1097/j.pain.0000000000000154

Martelletti P, Jensen RH, Antal A, Arcioni R, Brighina F, de Tommaso M, Franzini A, Fontaine D, Heiland M, Jürgens TP, Leone M (2013) Neuromodulation of chronic headaches: position statement from the European Headache Federation. J Headache Pain 14:1–16. https://doi.org/10.1186/1129-2377-14-86

Matharu MS, Bartsch T, Ward N, Frackowiak RS, Weiner R, Goadsby PJ (2004) Central neuromodulation in chronic migraine patients with suboccipital stimulators: a PET study. Brain 127(1):220–230. https://doi.org/10.1093/brain/awh022

McGreevy K, Hameed H, Erdek MA (2012) Updated perspectives on occipital nerve stimulator lead migration: case report and literature review. Clin J Pain 28:814–818

Misra UK, Kalita J, Bhoi SK (2013) High-rate repetitive transcranial magnetic stimulation in migraine prophylaxis: a randomized, placebo-controlled study. J Neurol 260:2793–2801. https://doi.org/10.1007/s00415-013-7072-2

Mungoven TJ, Henderson LA, Meylakh N (2021) Chronic migraine pathophysiology and treatment: a review of current perspectives. Front Pain Res 2:705276. https://doi.org/10.3389/fpain.2021.705276

Oshinsky ML, Murphy AL, Hekierski H Jr, Cooper M, Simon BJ (2014) Noninvasive vagus nerve stimulation as treatment for trigeminal allodynia. Pain 155(5):1037–1042. https://doi.org/10.1016/j.pain.2014.02.009

Oskoui M, Pringsheim T, Billinghurst L, Potrebic S, Gersz EM, Gloss D, Holler-Managan Y, Leininger E, Licking N, Mack K, Powers SW (2019a) Practice guideline update summary: pharmacologic treatment for pediatric migraine prevention: report of the Guideline Development, Dissemination, and Implementation Subcommittee of the American Academy of Neurology and the American Headache Society. Neurology 93(11):500–509. https://doi.org/10.1212/WNL.0000000000008105

Oskoui M, Pringsheim T, Holler-Managan Y, Potrebic S, Billinghurst L, Gloss D, Hershey AD, Licking N, Sowell M, Victorio MC, Gersz EM (2019b) Practice guideline update summary: acute treatment of migraine in children and adolescents: report of the Guideline Development, Dissemination, and Implementation Subcommittee of the American Academy of Neurology and the American Headache Society. Neurology 93(11):487–499. https://doi.org/10.1212/WNL.0000000000008095

Puledda F, Shields K (2018) Non-pharmacological approaches for migraine. Neurotherapeutics 15(2):336–345. https://doi.org/10.1007/s13311-018-0623-6

Puledda F, Messina R, Goadsby PJ (2017) An update on migraine: current understanding and future directions. J Neurol 264:2031–2039. https://doi.org/10.1007/s00415-017-8434-y

Schwedt TJ, Vargas B (2015) Neurostimulation for treatment of migraine and cluster headache. Pain Med 16(9):1827–1834. https://doi.org/10.1111/pme.12792

Silberstein SD, Calhoun AH, Lipton RB, Grosberg BM, Cady RK, Dorlas S, Simmons KA, Mullin C, Liebler EJ, Goadsby PJ, Saper JR (2016) Chronic migraine headache prevention with noninvasive vagus nerve stimulation: the EVENT study. Neurology 87(5):529–538. https://doi.org/10.1212/wnl.0000000000002918

Tassorelli C, Grazzi L, De Tommaso M, Pierangeli G, Martelletti P, Rainero I, Dorlas S, Geppetti P, Ambrosini A, Sarchielli P, Liebler E (2018) Noninvasive vagus nerve stimulation as acute therapy for migraine: the randomized PRESTO study. Neurology 91(4):e364–e373. https://doi.org/10.1212/wnl.0000000000005857

Tiwari V, Agrawal S (2022) Migraine and neuromodulation: a literature review. Cureus 14(11):e31223. https://doi.org/10.7759/cureus.31223

U.S. Food and Drug Administration (2019). https://www.accessdata.fda.gov/cdrh_docs/pdf18/DEN180059.pdf. Accessed 25 Dec 2023

Wilkinson D, Ade KK, Rogers LL, Attix DK, Kuchibhatla M, Slade MD, Smith LL, Poynter KP, Laskowitz DT, Freeman MC, Hoffer ME (2017) Preventing episodic migraine with caloric vestibular stimulation: a randomized controlled trial. Headache 57(7):1065–1087. https://doi.org/10.1111/head.13120

Yang Y, Song M, Fan Y, Ma K (2016) Occipital nerve stimulation for migraine: a systematic review. Pain Pract 16(4):509–517. https://doi.org/10.1111/papr.12303

Yarnitsky D, Dodick DW, Grosberg BM, Burstein R, Ironi A, Harris D, Lin T, Silberstein SD (2019) Remote electrical neuromodulation (ren) relieves acute migraine: a randomized, double-blind, placebo-controlled, multicenter trial. Headache 59(8):1240–1252. https://doi.org/10.1111/head.13551

Novel Treatment Approaches for the Management of Migraine Pain

10

Harish Selvaraj, Suhashini Shanmuganathan, Jean Sophy Roy, Gayathri Gopal, and Senthil Visaga Ambi

10.1 Introduction

Migraine, characterized as a chronic neurovascular headache disorder, are typically marked by moderate to severe headache episodes that are followed by phonophobia, vomiting, and nausea. In addition, about one third of migraineurs experience brief neurological symptoms known as migraine auras. In women under 50, where prevalence is highest, migraine is the most disabling condition and have a significant socioeconomic burden. Increased sub-cortical excitability, trigeminovascular activation and release of the neuropeptide calcitonin gene-related peptide (CGRP), a member of the larger calcitonin family, have all been continuously shown to play a crucial and causative role in migraine pathogenesis, even though the precise pathophysiological mechanisms are still unknown (Olesen and Ashina 2011; Al-Hassany et al. 2023). The onset of migraine attacks, including the premonitory and aura phase, is largely dependent on the participation of neuronal mechanisms (e.g., cortical spreading depression and sensitization of perivascular sensory nerve terminals) and structures (e.g., the brainstem and hypothalamus). It appears that in rats, mechanosensitive meningeal nociceptive neurons cannot be activated or sensitized by CGRP alone. However, since its discovery, monoclonal antibodies and gepants—two recently approved (preventive) treatments—that either target CGRP or its

H. Selvaraj · S. Shanmuganathan · J. S. Roy
Biopharmaceutical Research Lab, Anusandhan Kendra-1, School of Chemical and Biotechnology, SASTRA Deemed-to-be-University, Thanjavur, Tamil Nadu, India

G. Gopal · S. V. Ambi (✉)
Biopharmaceutical Research Lab, Anusandhan Kendra-1, School of Chemical and Biotechnology, SASTRA Deemed-to-be-University, Thanjavur, Tamil Nadu, India

Department of Bioengineering, School of Chemical and Biotechnology, SASTRA Deemed-to-be-University, Thanjavur, Tamil Nadu, India
e-mail: senthilvisagaambi@scbt.sastra.edu

S. Bhatt et al. (eds.), *Management of Migraine Pain*,
https://doi.org/10.1007/978-981-97-4529-6_10

receptor—have made CGRP their main pharmacological target (Al-Hassany et al. 2023). Despite being effective and tolerable in a considerable proportion of patients with episodic and chronic migraines, a notable proportion of migraineurs are categorized as "non-responders" who show little to no response to these CGRP-receptor targeted treatments. This underscores the necessity of investigating the possible function of substitute substrates, which could potentially function as targets for therapeutic interventions in a range of migraineurs. Additionally, given the intricate pathophysiology of migraine and the wide range of patient presentations, it is crucial to investigate these targets further as different migraine patients may be affected by distinct signaling pathways involving neuropeptides and neurotransmitters (Bertels and Pradhan 2019; Nisar et al. 2023). In this vein, this chapter would discuss the potential drug targets of migraine management, highlighting the significance of each target in alleviating migraine onset and the associated mechanisms. Our primary objective was to present an overview of the available clinical data for these targets, considering that they could be useful targets for pharmacological treatments in the future and a useful addition to the migraine therapy inventory already in use.

10.1.1 Painful Pathway—Migraine

Migraine, a neurological disorder characterized by recurrent headaches and are a condition that transcends mere discomfort, probing deep into the intricate pathways of the brain (Puledda et al. 2023). Migraines have emerged as a global health concern, attacking individuals across diverse demographics. According to the Global Burden of Disease Study, migraine rank as the second highest cause of disability worldwide, surpassing conditions such as diabetes, epilepsy and Alzheimer's disease (Steiner et al. 2013). The prevalence of migraines is not confined to a specific age group, gender or geographic location, presenting a significant burden on healthcare systems. The unpredictable nature of migraine attacks, coupled with intense symptoms such as nausea and sensitivity to light and can lead to reduced productivity and a diminished quality of life. Moreover, these effects extend to the emotional and psychological well-being of the individuals. The hallmark characteristic of migraine is a throbbing, pulsating pain localized to one side of the head which persists for a few hours to days (Ong et al. 2018). The challenge in diagnosis and management of this disorder lies in the variability in the symptoms, triggers and the temporal nature of migraine. There are two common subtypes of migraine—Migraine with Aura (MA) and Migraine without Aura (MO). Migraine without Aura is the most dominant, affecting around 70% of the individuals, while approximately 30% suffer with Migraine with Aura. MA is largely accompanied by aura symptoms such as visual disturbances, muscle weakness and motor disturbances (Foroozan and Cutrer 2009).

10.1.2 Pathogenesis of Migraine

The pathogenesis of migraine is a multifaceted process and involves a combination of genetic, environmental and neurobiological factors. The painful process of a migraine attack involves a cascade of events, from the induction of signals to the perception of intense pain in the head (Sprenger and Goadsby 2009). It is an intricate disorder, initiating with a combination of genetic predisposition and environmental triggers. Environmental triggers, ranging from stress and sleep deprivation to hormonal fluctuations, act as the catalysts for the initiation of a migraine attack (Dodick 2018). Cortical Spreading Depression (CSD) is a major phenomenon central to the migraine attack. It is an intense wave of depolarization that sweeps across the cerebral cortex. Understanding the complex mechanisms of a migraine can help in developing therapeutic approaches (Edvinsson et al. 2012).

10.1.2.1 Possible Mediators Involved in Migraine Pathophysiology

Migraine pathophysiology involves a complex interplay of various mediators that contribute to the initiation, progression and resolution of migraine attacks. These mediators encompass neurotransmitters, neuropeptides, inflammatory molecules and vasoactive substances. Understanding these mediators is essential for developing targeted therapies (Eller and Goadsby 2020).

10.1.2.2 Calcitonin Gene-Related Peptide (CGRP)

Released from both central and peripheral terminals of trigeminal nerve fibres, CGRP plays an integral role in the vasodilation observed during migraines. This neuropeptide acts as a potent vasodilator, relaxing and widening blood vessels, remarkably in the meninges surrounding the brain, leading to increased blood flow and the characteristic throbbing headache (Edvinsson et al. 2012). Moreover, CGRP is implicated in neurogenic inflammation, a process where inflammatory mediators are released in reaction to nerve activation. The release of CGRP during migraines not only fosters inflammation of the blood vessels but also contributes to the sensitization of pain pathways, lowering the threshold for pain perception. CGRP has been allied with cortical spreading depression (CSD), a phenomenon underlying the visual disturbances and neurological symptoms characteristic of the aura (De Tommaso et al. 2021). CGRP's actions are interceded through specific receptors, CGRP receptor type 1 (CGRP-R1) and CGRP receptor type 2 (CGRP-R2), which are abundantly present in the trigeminal system and blood vessels, emphasizing the targeted nature of its effects (Lionetto et al. 2012).

10.1.2.3 Serotonin (5-HT)

Serotonin, also known as 5-hydroxytryptamine (5-HT) acts as a neurotransmitter and vasoactive substance and contributes to the modulation of pain perception and vascular tone (Haanes and Edvinsson 2019). Serotonin acts as a neuromodulator in the central nervous system, influencing the transmission of pain signals. Alterations in serotonin levels have been affiliated to the onset and progression of migraine attacks (Monteith and Goadsby 2011). A decrease in serotonin release is thought to lower the pain threshold, making individuals more susceptible to migraine triggers. Serotonin interacts with these neuropeptides, influencing vasodilation, neurogenic inflammation, and the sensitization of pain pathways. This interaction is a key component of the trigeminovascular theory (Dodick 2018). Serotonin has vasoconstrictive effects on blood vessels, including those in the brain. During a migraine attack, there is evidence of alterations in cerebral blood flow, and serotonin is implicated in these changes. The interplay between serotonin and blood vessels contributes to the complex regulation of vascular tone in the context of migraines (Suwannasual et al. 2018). Variations in genes related to serotonin receptors and transporters have been identified in individuals with a predisposition to migraines. These genetic factors may influence the responsiveness to serotonin-based treatments and the overall susceptibility to migraines (Stam et al. 2005).

10.1.2.4 Brain Structures Involved in Migraine Pathophysiology

The intricate pathophysiology of migraines involves a network of anatomical structures within the brain. The cerebral cortex, interlinked with higher cognitive functions, experiences Cortical Spreading Depression (CSD), a phenomenon triggering the aura phase of migraines. Blood vessels, both extracranial and intracranial, contribute to the pulsating pain through vasodilation and vasoconstriction. The trigeminal nerve is a major cranial nerve which becomes activated, releasing neuropeptides that lead to neurogenic inflammation and the transmission of pain signals to the brainstem (Levy et al. 2007). Pain processing centres in the brainstem and the sensory relay station, the thalamus, play crucial roles in the perception of pain during migraines. The hypothalamus, involved in circadian rhythm regulation, exhibits disruptions, contributing to the rhythmic nature of migraines. Serotonin receptor sites, the pituitary gland, and meningeal structures such as the dura mater all play roles in hormonal regulation and vascular changes associated with migraines (Diener et al. 2015). Furthermore, the amygdala, a component of the limbic system central to emotional processing, may influence migraine susceptibility through its connection to stress and emotional triggers (Diener and Limmroth 2001).

10.1.3 Mechanism of Action

The role of neurogenic inflammation in migraines is closely tied to the trigeminovascular system, a complex network involving the trigeminal nerve and blood vessels. It involves activation of the trigeminovascular system resulting from the primary dysfunction in the brainstem, responsible for vascular tone regulation. The initiation of a migraine attack often involves the activation of the trigeminal nerve due to various triggers, including hormonal changes, stress, certain foods and environmental stimuli (Seidel et al. 2022). This activation results in cranial vasodilation mediated by vasoactive neuropeptides, particularly substance P and calcitonin gene-related peptide (CGRP). These chemicals play a critical role in the transmission of pain signals. Substance P plays a multifaceted role in the inflammatory process by promoting vasodilation, increasing vascular permeability and recruiting immune cells to the site of inflammation (Diener 2003). It also acts as a mediator of pain transmission, enhancing the perception of pain signals. CGRP, on the other hand, is a key player in vasodilation. When released, CGRP causes the dilation of blood vessels in the meninges, the protective layers surrounding the brain. This vasodilation contributes to the headache phase of migraines and is a hallmark of neurogenic inflammation. The combination of substance P and CGRP release results in a localized inflammatory response in the meninges, a phenomenon known as neurogenic inflammation (Zobdeh et al. 2021). Additionally, the release of pro-inflammatory substances can sensitize the trigeminal nerve, making it more responsive to subsequent stimuli—a phenomenon known as peripheral sensitization. The onset of pain pathways and the release of inflammatory mediators account for the throbbing pain characteristic of migraines. The pain signals are transmitted to the brain, leading to the perception of headache. Immune cells, such as mast cells, are also implicated in neurogenic inflammation. Mast cells release histamine and other inflammatory substances, contributing to vascular changes and the overall inflammatory cascade. The intricate interplay between neurotransmitters, immune cells and blood vessels creates a neuroinflammatory environment that sustains and amplifies the migraine attack (Green et al. 2019).

10.1.4 Phases of Pain

10.1.4.1 Premonitory Phase

The premonitory phase of a migraine, also known as the prodrome phase, occurs before the onset of the actual headache. This phase is experienced by some individuals who are affected by migraines, though not everyone with migraines will have this premonitory phase. The symptoms during this phase can serve as warning signs that a migraine attack is imminent. These symptoms typically appear hours to days before the headache itself (Russo et al. 2017). Common traits of the premonitory phase may include mood changes, fatigue, food cravings, neck stiffness, increased thirst and urination, sensitivity to light and sound and visual disturbances such as blurring or seeing flashes of light. Various studies have revealed hypothalamic activation and alteration in homeostasis including skipped meals and improper sleep cycles as the major triggers for the onset of these symptoms. PET and functional imaging studies have proved altered activities of thalamus and thalamo-cortical regions in migraine patients when compared to controls during migraine attack (Andreou and Edvinsson 2019).

10.1.4.2 Aura Phase

One of the key components of migraine is the prevalence of the aura. A migraine aura is a set of neurological symptoms that can exhibit as visual, sensory or motor disturbances. Cortical Depression Spreading (CSD) is considered as one of the leading physiological causes of the migraine aura. CSD refers to a wave of altered neuronal and glial activity that travels across the cerebral cortex, affecting brain function (Fraser et al. 2019). CSD is marked by an influx of Na, Ca and H_2O and an efflux of K, proton, ATP and glutamate caused by modifications in the transmembrane ion channels. Experimental models of cortical spreading depression (CSD) can help identify potential migraine therapeutic targets and modulators. Experimental models of CSD can help identify potential migraine therapeutic targets and modulators (Olsen 1995).

10.1.4.3 Headache Phase

The headache phase is the most recognizable and often the most debilitating stage of a migraine attack. It typically follows the aura phase and is characterized by intense, throbbing pain on one side of the head accompanied by nausea, vomiting and sensitivity to light and sound. It can last anywhere from a few hours to several days. The severity and duration of the pain can vary among individuals. During the headache phase, blood vessels in the brain become dilated and inflamed, causing intense pain. The activation of the trigeminovascular system is the major trigger for the inception of the headache phase (Levy 2010). It is believed to be related to abnormal brain activity and imbalances in certain chemicals in the brain. One possible mechanism responsible for this is the activation of astrocytes led by the opening of the neuronal pannexin 1 channel. This activation may lead to the production of cytokines and prostanoids, which sensitize meningeal nociceptors. In addition to the physical symptoms, individuals experiencing a migraine headache may also suffer from cognitive impairments such as difficulty concentrating or thinking clearly. This can make it challenging for them to carry out daily tasks or participate in social activities (Kung et al. 2011).

10.2 Potential Drug Targets

The fundamental physiology of the migraine is linked to the trigeminovascular system's abnormal activation. The trigeminovascular neurons innervate several pain-sensitive structures in the head and communicate with the trigeminal cervical complex (TCC) in conjunction with afferents from the upper cervical cord. Trigeminal system activation causes neurogenic inflammation, heightened pain sensitivity and a persistent headache. It also releases a number of neuropeptides which are crucial for pain signalling and modulation, such as substance P and calcitonin gene-related peptide (CGRP) (Mungoven et al. 2021; Nisar et al. 2023). New discoveries in the field of migraine pathophysiology have broadened the scope of potential therapy targets and produced a plethora of innovative therapeutic alternatives. We highlight multiple therapeutic options that are being explored and subjected to clinical trials. These capable targets comprise of potassium channels, δ-opioid receptors (DOR), pituitary adenylate cyclase-activating polypeptide (PACAP), adenosine, transient receptor potential ion channels (TRP) and acid-sensing ion channels (ASIC) (Nisar et al. 2023).

10.2.1 Pituitary Adenylate Cyclase-Activating Polypeptide (PACAP)

Through mechanisms apart from those involving CGRP, nitric oxide and potassium, PACAP is associated with migraines (Ceriani et al. 2019). A rapid decline in nocifensive behaviour, allodynia and hyperalgesia following the peripheral administration of PACAP was observed. In an animal model, CGRP and PACAP mAbs were unable to prevent PACAP and CGRP-induced light aversion, respectively. An in vivo study suggested that PACAP MAbs were unable to act on the signalling molecules initiated by the CGRP pathway or potassium channels. During the ictal phase, there was a negative correlation observed between the duration of the migraine and the elevated serum level of PACAP38. Patients with chronic migraines (CM) had higher levels of PACAP than either episodic migraine (EM) patients or healthy controls. Various mAbs that target the PACAP pathway were pondered in clinical settings. While AMG301 targets the PAC1 receptor, ALD1910 binds specifically to PACAP38 and PACAP27. ALD1910 (NCT04197349) finished its phase I, double-blind, randomized, placebo-controlled trial in 2019, but no results have been released so far. In 343 randomized patients on monthly migraine days or days when they took specific migraine medications, AMG301 was not effective over the placebo in a phase IIA, randomized, double-blind, placebo-controlled trial. The lack of effectiveness could be brought about by variations in AMG301's affinity for PAC1 receptors, variations in the concentration at which it is achieved at the target, or possibly insufficient selective inhibition of the PAC1 receptor to lessen the frequency of migraine attacks. Different PACAP38 mAbs are being assessed in two ongoing clinical trials in an attempt to forbid migraines that are resistant to treatment. Further understanding of PACAP mAbs for migraine prevention will be provided by these studies (Ceriani et al. 2019; Do et al. 2019; Nisar et al. 2023).

10.2.2 Adenosine Receptor

Adenosine is a desirable target for migraines, as it is linked to energy, metabolism and the sleep-wake cycle. Adenosine reuptake inhibitors like dipyridamole cause headaches in both healthy people and migraine sufferers. Due to its role in the breakdown product of adenosine triphosphate (ATP) and its binding to ATP-gated P2X3 receptors, adenosine is also associated with the pathophysiology of migraines. Adenosine as a target is further supported by two recent systematic reviews of preclinical and clinical research on its role in migraine signalling pathways (Olesen and Ashina 2011). In models of peripheral pain, the A1 receptor agonist suppresses nociceptive signalling. Selective stimulation of the A1 receptor reduces neural activation without causing vasoconstriction, as demonstrated in animal investigations of trigeminovascular nociception. When combined with a reduction in non-evoked pain behaviour and evoked mechanical hyperalgesia following surgical paw incision, A1 receptor response raises the possibility of using the A1 receptor as a target for pain management. Additionally, the alpha-2 adenosine (A_{2A}) receptor that cause nociceptive behaviour would serve as another target for migraine. Electrical stimulation of the trigeminal ganglion (TG) upregulates the A_{2A} receptor whose antagonist inhibits the vasodilation, suggesting that the A1 and A_{2A} homeostasis might be crucial to reduce dural vasodilation. Studies also provide information on the activation of the A1 receptor and inhibition of the A_{2A} receptor may serve as a possible target for migraine treatment (Nisar et al. 2023).

10.2.3 δ-Opioid Receptors

The four basic types of opiod receptor includes mu receptor (MOR), kappa receptor (KOR), delta receptor (DOR), nociception receptor (NOR) and zeta receptor (ZOR) which have a vital role in modulating different kinds of pain. But the majority of medications that target the μ-opioid receptor (MOR) are not suited for migraine discomfort. Numerous investigations on both humans and animals substantiate the alpha 1 (A1) receptor's function as a target for migraine treatment. Delta Opioid receptors and CGRP are expressed simultaneously in 40% of the neurons wherein the DOR agonist is expected to have an anti-nociceptive effect. This effect might be due to the prevention of CGRP release from the neurons and associated signalling. However, clinical trials to substantiate that DOR agonist is useful in migraine is very limited. Still, preclinical studies on DOR agonist demonstrated that it inhibited the cortical spreading depression (CSD) which is a typical symptom of migraine and it also did not disturb the seizure potential, suggesting that the DOR agonist can be used for the acute migraine and prophylaxis of migraine management (Nisar et al. 2023).

10.2.4 Potassium Channels

K_{ATP} channels belonging to the family of inward rectifying transmembrane potassium channels (Kir) and big-conductance calcium activated K^+ channels are the potassium channels associated with migraine, that are found in the trigeminal ganglion (TG) and trigeminal nucleus caudalis (TNC). Levcromakalim, a K_{ATP} channel agonist, is frequently employed in Chronic Migraine rat models. Levcromakalim treatment on a regular basis significantly increased the expression of c-Fos in brain regions linked to migraine, including the spinal trigeminal nucleus and medial prefrontal cortex. The cerebral and extracerebral arteries are vasodilatorily affected by substances that cause headaches and migraines. The Middle Meningeal Artery (MMA) and the superficial temporal artery were dilated with levcromakalim, but not the Middle Cerebral Artery (MCA). This is likely due to the restricted blood-brain barrier crossing capacity. Given that levcromakalim can trigger migraine attacks, it makes sense to investigate the K_{ATP} channel blockers as a possible treatment for migraines. Glibeenclamide, a sulfonylurea anti-diabetic medication, was found to inhibit spontaneous cephalic hypersensitivity and CGRP production by non-specifically blocking K_{ATP} channel opening. A potential future anti-migraine therapy could involve selectively blocking Kir6.1 or SUR2B K_{ATP} channel subunits in big cerebral and meningeal arteries, since they do not have any effect on the central nervous system, heart or hormones (insulin release) (Olesen and Ashina 2011; Do et al. 2019; Nisar et al. 2023).

10.2.5 Transient Receptor Potential Ion Channels

TRP channels constitute a sizable family of cation channels. They are triggered by variations in the influx of ions such as Ca^{2+} and Na^{+}, which cause modifications in the voltage across cell membranes and initiate signaling pathways via secondary messenger systems. The subchannels of TRP include TRPV1, TRPV2, TRPV3, TRPV4, TRPA1 and TRPM8, known as thermo-TRP channels and are sensitive to temperature changes. Preclinical research demonstrates that the antagonistic mechanism of TRPV1 acts through CGRP. Umbellulone, a substance that naturally induces migraines and familial pain syndrome is linked to TRPA1, which is significantly co-expressed with TRPV1. In vivo studies with icilin (a TRPM8 activator) experienced cutaneous allodynia. This effect is lessened when TRPM8 antagonist (AMG1161) is administered beforehand. Moreover, TRPM8 mediates the behavioural response of cold stimuli on CGRP. Due to the decreased sensitivity to the cold stimuli in people with migraine protective alleles, TRPM8 have a crucial spot on the onset of migraines. Although TRP channels are a desirable target for medication to treat inflammatory pain disorders, changes in thermoregulation might be a major drawback. TRPV1 and TRPA1 are implicated in sensitivity to environmental stimuli like heat, pH changes and natural migraine triggers, and they can act as a bridge in relating CGRP and headaches (Al-Hassany et al. 2023; Nisar et al. 2023).

10.2.6 Acid Sensing Ion Channels

Acid Sensing Ion Channels (ASIC) or voltage-sensitive cation channels are mostly found in the nervous system. ASIC inhibitors have demonstrated possible benefits in treating migraines. They suppress trigeminal activation, CSD and allodynia, as seen in preclinical studies. Amiloride is also used as an antihypertensive and diuretic; however, it blocks ASIC specifically and does not affect epithelial sodium channels, which reduces its impact on blood pressure (Do et al. 2019). In animal studies, amiloride also reduced CSD and trigeminal activation. Amiloride prevented the development of extra and sustained-cephalic allodynia, so corroborating the notion that migraine-related behaviours are caused by ASIC activity. ASIC3 inhibitors cause a three-fold reduction in the liberation of CGRP from trigeminal neurons, indicating that ASICs also impact other neuropeptides. Though preclinical research on CSD and allodynia reveals encouraging findings, detailed clinical studies are needed to confirm the effectiveness of ASIC inhibitors in migraine patients (Nisar et al. 2023).

10.2.7 Nitric Oxide

The pathophysiology of migraines has been strongly linked to Nitric Oxide (NO) generation and signalling. Nitroglycerin, an NO donor and established human migraine trigger, has been widely employed in both animal models of migraine and human migraine provocation research. A strong vasodilator, nitroglycerin causes a headache quickly in both migraineurs and healthy subjects. Nevertheless, migraineurs experience a postponed migraine attack several hours following this vasodilatory impact, indicating that vascular changes and migraine induction are not necessarily related.

Soluble guanylyl cyclase (sGC), the intracellular receptor for NO, results in the elevation in the levels of the second messenger, cyclic guanosine monophosphate (cGMP). This, in turn, causes the opening of big conductance calcium-activated potassium (BKCa) and adenosine triphosphate-sensitive potassium (KATP) channels, which are found to be key players in the progression of migraine attacks (Olesen and Ashina 2011). GW274150, a highly specific inhibitor of iNOS, did not demonstrate efficacy in acute or preventive treatment. A different study found that NXN-188, an nNOS inhibitor and serotonin (5-HT1B/1D) receptor agonist, prevented rat dura mater, TG and TNC from releasing immunoreactive CGRP (iCGRP) when stimulated. Nevertheless, in the same study, NXN-413, a specific nNOS inhibitor, blocked the release of iCGRP from dura mater but not from the TG or TNC. For this reason, more trustworthy clinical trials—possibly with larger sample sizes—are required to find new medication candidates that target this biological pathway (Bertels and Pradhan 2019).

10.2.8 Phosphodiesterase-3

In cerebral vascular cells, CGRP stimulates adenylate cyclase transmembrane receptors and results in the elevated levels of second messenger cAMP. Headaches are caused by intracellular cAMP buildup, which is broken down by phosphodiesterase-3 (PDE3). PDE3 and cyclic nucleotide-gated ion channels are expressed in the dura mater, TG, middle cerebral artery and basilar artery of rats' trigeminovascular system. Consequently, PDE3 modulation of cAMP levels and cyclic nucleotide-gated ion channel activation may contribute to the pathophysiology of migraine-like episodes. Indeed, PDE3 is expressed in cerebral arteries, vascular smooth muscle cells and the TG. As cilostazol inhibits PDE3, intracellular cAMP builds up and KATP channels open more easily. This results in cerebral artery dilatation and vascular smooth muscle relaxation without compromising local blood flow. As a result, cilostazol initiates a mechanism downstream of both CGRP and PACAP. It also has a high success rate in inducing migraine-like events in both healthy individuals and 86% of patients with migraine without aura. It may play the following role in migraine headaches: regulating nociceptive input in areas of the trigeminovascular system where pain is processed by neurons. These findings suggest that PDE3 may have a part in the pain pathways that migraine sufferers encounter (Olesen et al. 2009; Bertels and Pradhan 2019).

10.2.9 Phosphodiesterase-5

cGMP is broken down by an intracellular enzyme called phosphodiesterase-5 (PDE5). PDE5 inhibition raises intracellular cGMP levels, which relax smooth muscle cells and stimulate neurons. On the other hand, CGRP works by raising cAMP levels inside cells. PDE5 inhibitors, including tadalafil and sildenafil, were first developed for cardiovascular conditions, but their efficacy was subsequently shown towards headache induction. The intracellular signalling pathways for cAMP and cGMP are affected by CGRP and sildenafil, respectively. However following the administration of CGRP and sildenafil, patients develop migraine attacks with matching clinical symptoms, indicating that these two intracellular pathways most likely converge in a downstream common denominator accountable for the molecular cascade that initiates migraines. As a result, there is growing preclinical evidence that the cAMP and cGMP pathways have a reciprocal effect. Significantly, the sildenafil injection caused more migraine attacks in patients with migraine than CGRP (89% vs.67%), possibly indicating a stronger ability to generate migraines. MMA dilates both intradural and extradural following CGRP or PDE5 inhibitor infusion; however, there was no apparent temporal relationship between arterial dilatation and pain, indicating that this might be due to the initiation of perivascular dural afferents. Thus, the important biological mechanisms of migraine attack onset may

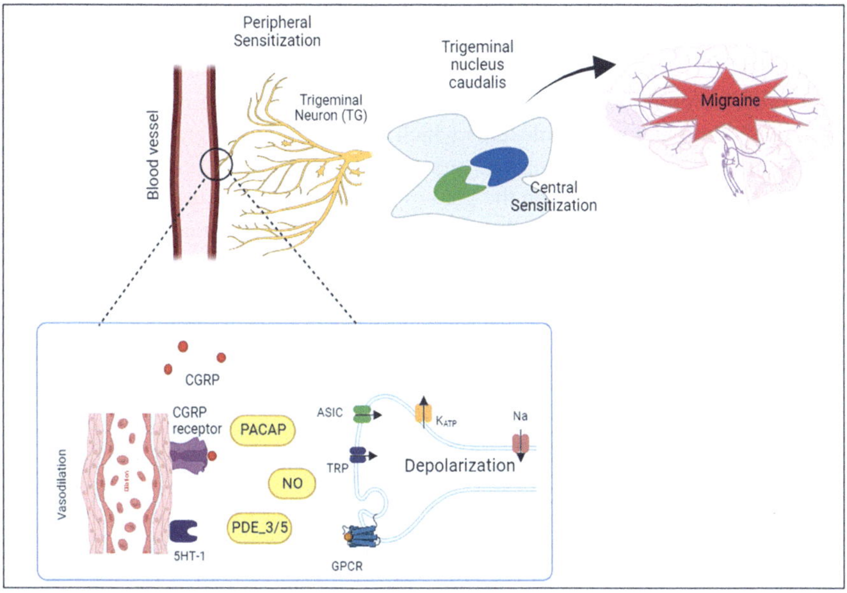

CGRP- calcitonin gene-related peptide ; **HT**- hydroxytryptamine; **ASIC**- Acid Sensing Ion Channels; $\mathbf{K_{ATP}}$- Potassium Channel; **Na** – Sodium Channel; **GPCR** – G protein coupled receptor; **NO** - Nitric Oxide; **PACAP** - Pituitary adenylate cyclase-activating polypeptide **PDE** – Phosphodiesterase; **TRP**- Transient Receptor Potential Ion Channels

Fig. 10.1 Potential Drug targets for migraine management

be the exasperation of perivascular sensory nerve terminals (Olesen et al. 2009; Sehgal et al. 2014; Bertels and Pradhan 2019; Al-Hassany et al. 2023) (Fig. 10.1).

10.3 Acute Treatment of Migraine

Simple analgesics like acetaminophen or nonsteroidal anti-inflammatory medicines (NSAIDs) are used in abortive (symptomatic) migraine therapy. Other medications include triptans, antiemetics, CGRP antagonists, lasmiditan and dihydroergotamine. The choice of a particular agent is contingent upon the unique characteristics of the patient, such as the nature and severity of their symptoms, coexisting medical problems, and their past treatment response (Ramadan and Buchanan 2006). Early in the course of the headache, abortive therapies are typically more effective; large single dosages typically work better than repeated modest ones. Oral medications may be less effective for certain patients due to poor absorption resulting from vomiting and stomach stasis brought on by migraines. The fundamentals of treating acute migraines include using the right medication at the right dose in the onset of the headache, restricting the number of days a patient can receive acute medication to 2 or 9 days per month, adding an antiemetic or preferring parenteral medication in patients who undergo acute pain or nausea at first, and evaluating the side effect profile before choosing a medication. Certain patients may react well to one medication while not responding effectively to another. The drug's effectiveness should be evaluated after at least two attacks before switching to a substitute (Todd and Schwedt 2022). The ideal strategy is to strategize therapy such that, depending on the intensity of each episode, the patient can return to full function in 2 h. Research on the progression of novel medications on the management of acute migraines is underway. Identification of molecules which are specific to migraines, having less side effects—particularly vascular adverse effects—can be conveniently delivered, and are easily accessible, will be the goal of future acute migraine treatment medications. Hepatotoxicity has forced the suspension of CGRP antagonist studies, the results of which were eagerly anticipated. Research on novel serotonin receptor agonists (particularly those that target the 1F receptor), AMPA/Kainate receptor antagonists, orexins, nitric oxide, prostanoids, COX-1/COX-2 inhibitors, and novel combinations of sumatriptan, dihydroergotamine, and diclofenac potassium is still ongoing (Tedavisi 2013; Todd and Schwedt 2022).

10.3.1 Ergot Alkaloids

Ergot preparations are still used to treat acute migraine attacks, despite the introduction of migraine-specific, more selective, and medications with less side effects. Ergot alkaloids have greater adverse effects and a less specific agonist activity on serotonin receptors than triptans. Dihydroergotamine, one of the ergot alkaloids that is not available in our country, has advantages in the management of migraine due to its extended half-life, low risk of medication overuse headache, and ability to be administered intramuscularly, subcutaneously, intravenously and nasally. Ergotamine tartarate can produce drug overuse headaches and has systemic vascular adverse effects, hence it should not be used in the management of migraines. Their usage should be restricted to young kids with extremely rare incidents who become unresponsive to other acute therapy medications (Tedavisi 2013).

10.3.2 Triptans

Triptans, which are serotonin 1b/1d agonists, are useful in the acute management of migraine. The purpose of triptans is to specifically treat migraines. Triptan responsiveness, however, would not be regarded as a migraine diagnosis as other main and secondary headaches may also get better with triptan treatment. Trigeminal nerve activation is a crucial part of the chain reaction that initiates and prolongs a migraine attack. The trigeminovascular system is related to distinct 5-HT receptor subtypes, such as 5-HT1B, 5-HT1D and 5-HT1F. Triptans are believed to work by three main processes, which include presynaptic dorsal root activation (5-HT 1D), peripheral neuronal inhibition (5-HT 1B) and intracranial artery vasoconstriction (5-HT 1D). The precise mode of action of triptans is unknown. Moreover, they might influence the activity of 5-HT 1F receptors and descending inhibitor pain pathways. In 60% of the patients who become unresponsive to NSAI medications, triptans were effective. In cases of really severe attacks or in people who do not react to non-migraine-specific medications, they can be utilized as first-line treatments. The potent agonists of the 5-HT1B and 5-HT1D receptors, as well as some of the potent 5-HT1F agonists, are sumatriptan and the other triptans that are presently available on the market: almotriptan, eletriptan, frovatriptan, naratriptan, rizatriptan and zolmitriptan. Later research revealed that sumatriptan inhibits the emission of proinflammatory and vasoactive chemicals (e.g. CGRP), which are linked to migraine and neurogenic inflammation which might be related to migraine. Sumatriptan significantly altered the management of acute migraines since it frequently resulted in both significant pain relief and the amelioration or complete absence of accompanying phonophobia, photophobia and nausea symptoms. But when clinical evidence mounted, sumatriptan's drawbacks were revealed. More than 50% of victims were not completely pain-free from oral medication, and over 30% of victims who initially reacted experienced headache recurrence. Most second-generation triptans fulfilled their promise of improved pharmacokinetic characteristics. Additionally, they expanded the selection of therapies specifically designed to treat migraines, resulting in a more individualized course of treatment. Regretfully, improved preclinical pharmacodynamic and kinetic profiles did not translate into a significant and meaningful clinical improvement over sumatriptan, in terms of maximising efficacy, lowering adverse events or getting rid of cardiovascular complications (Ramadan and Buchanan 2006; Tedavisi 2013; Todd and Schwedt 2022).

10.3.3 5-HT$_{1d}$ Agonists and 5-HT$_{IF}$ Agonists

5HT1D receptor agonists have no vasoactive characteristics and are strong inhibitors of dura plasma protein extravasation (PPE), an animal model that has some relevance to migraine. Furthermore, these receptors are exhibited by peptidergic nociceptors in an activation-dependent manner. In animal models of migraine, two non-vasoactive 5-HT1D selective agonists have demonstrated some potential. In a human-randomised, controlled study (RCT), PNU-142633 did not show statistical significance; participants complained of chest pain and other cardiovascular symptoms. The complaints of chest problems on PNU-142633 require additional clarification about the non-vasoactive hypothesis of 5-HT1D agonist. Since 5-HT1D receptors are present on cardiac valves, one intriguing theory is that 5-HT1D agonists are cardio but not vaso-active, and that the symptoms in the chest are valvular in origin (Ramadan and Buchanan 2006). The 5-HT1F receptor is implicated in migraine, according to multiple lines of evidence. 5-HT1F agonists are not vasoconstrictive and are useful in preclinical models of migraine, according to animal studies. Furthermore, it is argued that the PPE model activity and affinity are correlated at the 5-HT1F receptor but not at the 5-HT1D or 5-HT1B receptors. Additionally, intravital imaging shows that LY334370, a selective 5-HT1F agonist, decreases single cell firing in the TNC without affecting the cerebral artery lumen. On the other hand, LY334370 development was terminated due to animal toxicity (Ramadan and Buchanan 2006).

10.3.4 Dopamine Agonist

Acute migraine treatment frequently makes use of a group of dopamine antagonists. Acute migraine victims have been studied for the potential therapeutic benefit of droperidol, an in vivo antagonist of dopamine effects. Intramuscular droperidol significantly reduces acute migraine pain and related symptoms, according to a large RCT with over 300 acute migraine patients. Adverse effects are concerning because over 30% of patients had ataxia, and sleeplessness or anxiety was also prevalent (Ramadan and Buchanan 2006).

10.3.5 Glutamate Modulators

On the trigeminal system, metabotropic glutamate (mGlu) receptors, ionotropic N-methyl D-aspartate (NMDA) and non-NMDA (α-amino-3-hydroxy-5-methyl-4-isoxazolepropionic acid [AMPA], kainic acid [KA]) glutamate receptor subtypes are widely distributed. Moreover, extracellular glutamate is raised by painful stimulation of the trigeminal nerve, which in turn activates TNC neurons. Furthermore, 5-HT1B, 5-HT1D and 5-HT1F receptor-positive neurons in the TG were express glutamate receptors. This suggests that 5-HT1 autoreceptors may prevent presynaptic glutamate release. Finally, a number of preclinical pharmacological investigations have linked trigemino-vascular trafficking to the glutamatergic system (Ramadan and Buchanan 2006).

10.3.6 Adenosine Receptors

Human TG contains the A1 receptor protein, which is how adenosine controls nociception through its A1 receptor. Moreover, trigeminovascular activation can be inhibited by prejunctional selective A1 agonists, both in the trigeminal nucleus and by the suppression of CGRP release in the cranial circulation. Finally, glutamate may influence pain perception by inhibiting prejunctional A1 receptors by the release of adenosine. In animal models of migraine, the strong and selective A1 agonist GR79236 has been demonstrated to impart anti-nociceptive effects. GR79236 decreased the cat's nociceptive transmission from the TG nerve and the venous levels of CGRP. GR79236 reduced trigeminally elicited dural vasodilation and the activity of neurons in the TNC that are activated by nociceptive input in rats. GR79236 treatment, however, had no effect on capsaicin-induced CGP release in the pig. In an open-label investigation involving 12 adult healthy female volunteers, GR79236 effectively suppressed the nociceptive blink reflex at levels that are inhibitory of trigeminal neurons. Human studies for conventional migraine treatment are planned, however there are uncertainties that A1 agonist target may have systemic pharmacodynamic characteristics that would prevent further research (Ramadan and Buchanan 2006).

10.3.7 CGRP Antagonists

The action of CGRP in the paroxysmal headache cascade is propped up by multiple lines of evidence. Initially, CGRP is found on the trigeminal nerve's sensory terminals and is released when the nerve is stimulated. Additionally, as part of the trigemino-vascular reflex, which inhibits excessive cerebral vascular contraction, CGRP mediates dilation of the cerebral vasculature and increases cerebral blood flow. Moreover, nociceptors on cerebral arteries can be activated by CGRP-induced vasodilation. Intravenous human CGRP injection causes migraine-like headaches in individuals who are susceptible to migraine. Patients with migraine (n = 126) were randomised to receive either a placebo or BIBN4096BS intravenously at doses of 0.25 mg, 0.5 mg, 1 mg, 2.5 mg, 5 mg or 10 mg. When using the optimum dose (2.5 mg) of BIBN4096BS, a significantly higher proportion of patients (66%) reported headache alleviation (improvement from moderate or severe pain to mild or none at 2 h) compared to those who received a placebo (26.8%). The most frequent adverse event associated with BIBN4096BS was paraesthesia. BIBN4096BS's mode of action is unaffected by variations in cerebral blood flow (Ramadan and Buchanan 2006). Ubrogepant has demonstrated efficacy in treating acute migraine in numerous clinical trials. Within 4 h of a single migraine attack, adult patients with episodic migraine were randomly assigned in a 1:1:1 ratio to receive one tablet of ubrogepant 100 mg, ubrogepant 50 mg or a placebo and a second dose or the patient's own rescue medication was allowed 2–48 h after the first dose. Based on efficacy data spanning 1327 participants, the percentage of patients experiencing pain relief at 2 h post-treatment was higher in those administered 100 mg and 50 mg of ubrogepant as opposed to placebo. Evidence from clinical trials supports the effectiveness of rimegepant in treating acute migraines. Patients who received rimegepant were more likely than those who received a placebo to be pain-free after 2 h in a clinical trial involving 1466 migraine patients. In a study involving 1405 adult migraineurs, individuals given zavegepant 10 mg had a higher likelihood of being pain-free after 2 h compared to those given a placebo (Todd and Schwedt 2022).

10.3.8 Nitric Oxide Synthase Inhibitors

Endothelial cells, platelets, neutrophils, and the brain contain NO, a lipophilic gas that is membrane-permeable and possesses vasodilatory properties. Experiments conducted in both preclinical and clinical settings validate the usefulness of NO inhibition as a target for migraine treatments. Nitroglycerin (NTG) administered intravenously releases NO, produces migraine in over 60% of migraineurs and stimulates trigeminal neurons in test subjects. The rat dura mater experiences delayed PPE due to NTG injection. Cortical spreading depression (CSD) in cats causes the emission of NO, which results in the emission of CGRP, vasodilation and nociceptor activation, thereby coupling neuronal activity with vascular tone. In the clinical setting, NG-methyl-L-arginine hydrochloride (546C88 [LNMAH]), an inhibitor of nitric oxide synthase (NOS) was proved to be successful. A non-selective NOS inhibitor named 546C88 was given intravenously to migraineurs on the course of acute attack in a limited RCT. The 2-h headache response rate was 67% on 546C88, compared to 14% on placebo. It is thought that the effects of 546C88 on humans are unaffected by vascular alterations because cerebral blood flow velocities did not vary (Ramadan and Buchanan 2006; Todd and Schwedt 2022).

10.4 Preventative Medication

Migraine can be viewed as an abnormal physiological condition characterized by a heightened susceptibility to intermittent neuronal excitability (Welch 2003). The stressors such as emotional stress, a sudden shift in female sex hormone balance and head injury have a significant impact on the abnormal physiology. These stressors further decrease the threshold for neuronal firing, leading to closely time-locked events of CSD. These events closely coincide with vasomotor changes, inflammation and activation of nociceptors, resulting in head pain, migraine neurological aura and other related symptoms (Ramadan and Buchanan 2006). Additionally, migraine is a highly debilitating condition that significantly affects an individual's financial circumstances, familial bonds and daily commitments at work and school (Amiri et al. 2022). These facts highlight that a subset of migraineurs are in desperate need of preventative medication (Al-Quliti and Assaedi 2016). According to findings from the American Migraine Prevalence and Prevention Study, it was discovered that a mere 13% of patients were receiving prophylactic treatment, despite the fact that approximately 38% of migraine patients required such treatment (Tronvik et al. 2006). Preventative therapy aims to decrease the frequency, duration or intensity of attacks. It also offers additional advantages such as improving the effectiveness of acute treatments, enhancing a patient's functional ability and reducing disability (Silberstein 2009). The possible ways in which migraine prophylaxis works involve raising the activation threshold for migraines by stabilizing the active nervous system, increasing pain relief, reducing sensitivity in the peripheral and central nervous systems, inhibiting inflammation caused by nerves and adjusting the levels of activity in the sympathetic, parasympathetic and serotonergic systems (Tronvik et al. 2006). When selecting a drug for prophylaxis, its efficiency, potential side effects and any comorbid diseases associated with the migraine are to be noted. The commonly used drugs for prophylactic treatment include beta-adrenergic blockers, antidepressants, calcium channel antagonists, botulinum neurotoxins, nonsteroidal anti-inflammatory drugs, serotonin antagonists and anticonvulsant drugs. If preventive medication is required, it is recommended to select an agent from the first-line categories. The selection must be made on the basis of the drug's relative efficacy as proven in double-blind placebo-controlled trials, its side effect profile, and the patient's preference. Other factors to consider include the presence of coexistent and comorbid conditions (Silberstein 2009).

10.4.1 β-Adrenergic Blockers

Of all the agents used for prophylactic migraine treatment, beta-adrenoceptor blockers have been studied extensively and are the most commonly prescribed. Among the beta-adrenoceptor blockers, propranolol and metoprolol have been extensively studied for the prevention of migraines and are widely acknowledged as effective (Holroyd et al. 1991). Although different doses of propranolol and metoprolol have been tested in various trials, their meta-analysis indicates that a daily dose of 160 mg of propranolol and 200 mg of metoprolol can be considered effective prophylactic doses. On the other hand, there is ongoing debate regarding the preventive effectiveness of b-adrenoceptor blockers for migraines. It is uncertain whether this property applies to all blockers in this drug class or only specific members with distinct properties (Limmroth and Michel 2001). Clinical studies also provide evidence for the effectiveness of timolol (20–30 mg/day) and bisoprolol (5 mg/day) in the prevention of migraines. Atenolol and nadolol have a moderate effect in reducing the frequency of migraine attacks (van de Ven et al. 1997; Schellenberg et al. 2008). Beta-blockers with intrinsic sympathomimetic activity, such as acebutolol, alprenolol, oxprenolol and pindolol were shown ineffective for preventing migraines (Silberstein 2009). Due to the lack of proven animal models, the region where prophylactic -adrenoceptor blockers exhibit their effects has yet to be determined. Some β-blockers such as propranolol, metoprolol, oxprenolol, and alprenolol exhibit high lipophilicity, enabling them to penetrate the central nervous system effectively. In contrast, atenolol, nadolol and practolol demonstrate either slight or no lipophilicity (Cruickshank and Prichard 1994). Forssman et al. (1983), Ryan et al. (1983), Freitag and Diamond 1984; Johannsson et al. 1987) demonstrated that atenolol and nadolol were found beneficial in mitigating migraine attacks, it is not mandatory that these medications require high lipophilicity or the ability to reach the central nervous system in order to be successful in migraine prophylaxis. Several β-adrenoceptor blockers, such as propranolol and pindolol, were demonstrated to have a better affinity for 5-HT receptors, including 5-HT1A, 5-HT1B/D and 5-HT2 receptors that were either candidates for acute migraine therapy or other preventative drugs. (Silberstein 2009) shows that propranolol suppresses the NO synthesis by inhibiting inducible NO synthase and also blocks the kainate-induced currents and works in tandem with N-methyl-d-aspartate blockers to lower neuronal activity and stabilize the membrane. Beta-blockers should not be used in patients with asthma, chronic obstructive pulmonary disease, congestive heart failure, atrioventricular conduction abnormalities, peripheral vascular disease, brittle diabetes or Raynaud's disease. The behavioural adverse effects (AEs) of all β-blockers include weariness, lethargy, melancholy, nightmares, sleep disturbances, sleepiness and hallucinations (Silberstein 2009).

10.4.2 Antidepressants

10.4.2.1 Tricyclic Antidepressants

Amitriptyline was found effective for migraine prevention (Couch et al. 1976; James and Hassanein 2015). Amitriptyline is a mixed serotonin-norepinephrine reuptake pump inhibitor, it is believed to enhance descending noxious inhibition, or the natural pain-control processes that travel from the brainstem to the spinal cord as well as the trigeminal nucleus caudalis. The blockade of Alpha2-adrenoceptors has been demonstrated to inhibit the antinociceptive effect of amitriptyline and hence it is believed that the efficacy of amitriptyline is partially mediated by its action as an α_2-agonist. It is crucial to note, however, that amitriptyline has a variety of additional impacts on channels and receptors (Gray et al. 1999). Antihistaminic and antimuscarinic effects are found as the results of the action of amitriptyline on the sodium channel blocker. The effectiveness of this treatment in preventing migraines is not dependent on its antidepressant properties. Moreover, its ability to prevent migraines manifests earlier in the course of treatment than its expected efficacy in treating depression. The effective dosage range of tricyclic antidepressants is broad, necessitating an individualized determination of the appropriate dose. Lower doses, often around 25 mg per day, was found effective for the management of migraines when compared to traditional antidepressant doses. If a satisfactory response is not achieved, the dosage should be gradually increased to a maximum of 100 mg per day. In elderly patients, it is recommended to start with lower doses, typically around 10 mg per day (Moraczewski et al. 2023). Adverse events (AEs) are frequently observed in individuals who use Tricyclic Antidepressants (TCAs).

Antimuscarinic AEs encompass symptoms such as a dry mouth, a metallic taste, epigastric distress, constipation, dizziness, mental confusion, tachycardia, palpitations, blurred vision and urinary retention. Additionally, other AEs may occur, including weight gain (although this is rarely seen with protriptyline), orthostatic hypotension, reflex tachycardia and palpitations (Silberstein 2009). Special notice must be given when dealing with elderly patients due to the potential risk of cardiac conduction disorders, as well as confusion and delirium (Tronvik et al. 2006).

10.4.2.2 Selective Serotonin and Serotonin-Noradrenaline Reuptake Inhibitors

Although these drugs exhibits less effectiveness for migraine prevention compared to tricyclic antidepressants, they are considered as a viable treatment option due to their better tolerability in terms of side effects (Punay and Couch 2003). Studies demonstrate that the fluoxetine and venlafaxine are efficient in inhibiting migraine prophylaxis at the doses 10–40 mg/day for the fluoxetine and 150 mg/day for the latter with doses initiating from 37.5 mg/mL (Ozyalcin et al. 2005; Adly et al. 1991; Bulut et al. 2004).

10.4.3 Calcium Channel Antagonist

Calcium channel antagonist was tested in migraine based on the belief that vasoconstriction triggers the onset of migraines and by maintaining a cerebral vasodilatory tone, it is possible to prevent it. However, it is now evident that calcium channel antagonist also affects various other mechanisms in the migraine pathway (Ramadan and Buchanan 2006). Verapamil has been found to hinder the activity of neuronal NOS (nitric oxide synthase), which ultimately leads to the reduction of pain perception. Additionally, it has the capability to impede hyperalgesia, meaning it can alleviate heightened sensitivity to pain and also can enhance the effectiveness of pain relief provided by opioids and acetaminophen (Omote et al. 1995). The suggested dosage for this medication is 5–10 mg per day. Potential side effects may include weight gain, drowsiness, dry mouth, low blood pressure, worsening of depression and uncommon extrapyramidal reactions (Tronvik et al. 2006). Therefore, calcium channel antagonists are a beneficial choice for individuals with hypertension or those who are unable to utilize β-blockers due to the presence of side effects.

10.4.4 Anticonvulsants

Anticonvulsants are being increasingly recommended for the management of migraines due to the existence of well-conducted placebo-controlled trials. The efficiency of valproate and topiramate in treating migraines has been proven to be high within this group for patients who cannot handle high dosages of a single medicine, a cocktail of topiramate and sodium valproate at smaller doses may be a viable therapy choice (Intolerant and Monotherapy 2011).

10.4.5 Topiramate

Topiramate is effective in treating both episodic and chronic migraines. It is used as a migraine prophylaxis in numerous countries. Topiramate is commonly known as a "dirty drug" due to its ability to block multiple channels, including voltage-dependent sodium channels and high-voltage-activated L-type calcium channels (Cutrer et al. 2001). Topiramate exerts its action by blocking Ca^{++} and Na^{+} channels, inhibiting glutamate activity, inhibiting carbonic anhydrase and enhancing GABA production (Tronvik et al. 2006; Calabresi et al. 2007; Galletti et al. 2009). Furthermore, research has demonstrated that topiramate can effectively decrease the emission of CGRP from trigeminal neurons when exposed to depolarizing stimuli (Sprenger et al. 2018). In a minimized migraine trial, it was shown that low-dose (50 mg per day) topiramate treatment do not affect CGRP plasma levels. This finding suggests that the effectiveness of topiramate in treating migraines is independent on the CGRP pathway in humans. Paraesthesias are frequently observed in the victims. Additional effects include fatigue, loss of appetite, nausea, diarrhoea, abdominal pain and weight loss (Diener et al. 2007).

10.4.6 Valproate

Valproic acid or sodium valproate was found as the effective medication for the management of prophylactic migraine. It is typically administered in doses ranging from 500 to 2000 mg per day, either as a single dose or divided into multiple doses, and is available in a long-acting preparation (Rothrock 1997; Shaygannejad et al. 2006). Valproate exerts its effects by enhancing the transmission of GABA, which is a neurotransmitter that inhibits neural activity. Additionally, it inhibits the function of low-threshold T-type calcium channels, prevents the activation of voltage-dependent sodium channels, and reduces the extravasation of plasma proteins (Calabresi et al. 2007; Galletti et al. 2009). Additionally, valproate was found to suppress protein kinase C, which is a regulator of glutamatergic neurotransmission. Another study revealed that valproate inhibits the NF-κB pathway, potentially offering an additional mechanism of action in migraine. This finding was observed in the in vivo model of migraine, where nitro-glycerine was used as a trigger (Sprenger et al. 2018). The common side effects encompass nausea and vomiting. During the later stages, individuals may also experience hand tremors and hair loss. However, sedation and cognitive function disturbances are rare occurrences. Severe side effects, although seen infrequently, may include hepatitis and pancreatitis. Other contraindications to consider are conditions such as thrombocytopenia, pancytopenia and bleeding disorders that may affect the safety or effectiveness of the treatment.

10.4.7 Angiotensin Modulators

Angiotensin converting enzyme (ACE) inhibitors, like lisinopril, have the ability to modify vasoreactivity, adjust sympathetic tone and facilitate the breakdown of pro-inflammatory substances including substance P, enkephalin and bradykinin. It was also found that lisinopril was well tolerated and with adverse events being mild or moderate in nature (Schrader et al. 2001). Tronvik et al. (2006) investigate the effectiveness of candesartan, an angiotensin II receptor blocker, in preventing migraines. The study was randomized, double-blind, and placebo-controlled, with a crossover design. Over a period of 12 weeks, participants taking placebo experienced an average of 18.5 headache days, while those taking candesartan experienced an average of 13.6 headache days. The results were statistically significant ($P = 0.001$) in the intention-to-treat analysis, with a sample size of 57 participants. Among the participants, 18 (31.6%) showed a reduction of half in headache days compared to placebo, and 23 (40.4%) showed a reduction of at least 50% in migraine days. It was found that candesartan, an angiotensin II receptor blocker, has shown effectiveness in its treatment. Furthermore, its tolerability profile was found to be comparable to that of the placebo.

10.4.8 Botulinum Toxins

Clostridium botulinum produces seven BoNT serotypes (A, B, C1, D, E, F, and G). All of these are known to inhibit acetylcholine release with different targets and mechanisms of action. BoNT is commercially available as onabotulinumtoxinA, abobotulinumtoxinA and rimabotulinumtoxinB (BoNTB). Though the mode of action of BoNT is still unknown, and has been used in case of episodic migraines (Herd et al. 2019).

10.5 Non-pharmacological Approaches for Migraine Management

Prevailing treatment options for migraines are associated with inadequate efficacy, poor tolerance and increased risk of complications with long-term use for certain subgroups affected by this neurological disorder. This may lead to medication overuse with acute migraine, low treatment compliance and subsequently migraine chronification. Hence there is a compelling demand for non-invasive and non-pharmacological forms of treatment. To pay attention to the issue, multiple advancements in recent years such as Neuromodulation, Behavioural therapy and Nutraceuticals, along with more efficient management of migraine triggers, have been followed (Puledda and Shields 2018; Haghdoost and Togha 2022).

10.5.1 Management of Triggers Associated with Migraine Onset

Modifications in day-to-day habits to prevent migraine are preferable since they usually involve minimal side effects and can be made part of one's daily routine, hence reducing frequent annoyance for those undergoing treatment. This may result in increased patient compliance, and any inexorable pharmacological intervention in future might result in a higher success rate since association of both non-pharmacological and pharmacological approaches delivers higher efficacy. The first step in the prevention of migraines involves identifying and thoroughly documenting the triggers associated with it, to later circumvent it. The factors responsible may be categorized into the following classifications (Haghdoost and Togha 2022).

10.5.1.1 Environmental Factors

Stress is reported to be a significant migraine trigger for both men and women (Kesserwani 2021). Migraine onset frequency and stress are both components of a vicious cycle, where both complement each other, resulting in each component amplifying the other. Prevalence and severity of stress are crucial determinants in the chronification of migraine headaches (Haghdoost and Togha 2022). Stress may also affect sleep quality or worsen already existing sleep abnormalities, prompting further neurological disarray. Several theories such as dopaminergic system dysregulation and accumulation of toxic waste substances due to long-term sleep disruption in chronic migraineurs have been proposed to explain sleep disturbance as migraine triggers (Arguello et al. 2017). Clinical studies on changes in weather as a possible trigger for migraines have so far yielded inconclusive results. This is speculated as the complexity of the pathophysiology of the condition and potentially unsuitable study design, since a considerable amount of those who deal with migraines regularly indicate weather changes as a reliable predictor of migraine onset.

The inconclusive trial outcomes also demonstrate that the weather would not be the primary deciding factor and co-existing factors also play a role in migraine onset and severity in a non-linear manner (Hoffmann et al. 2015). Clinical data in Peru established a statistically significant association between long-term exposure to high-altitude conditions and the frequency of migraine onset. They also established a relationship between the altitude and the intensity of the migraine attack (Linde et al. 2017). Sensory overload was identified as a potential trigger for migraines but further studies have yielded conflicting results. Bright lights are a frequently reported migraine trigger; the disturbance in the circadian rhythm caused by them is speculated to be possibly responsible for migraine onset. Since lights of all wavelengths do not trigger migraines, novel therapeutic strategies involving integrative lighting to regulate the circadian rhythm are now explored (Lin et al. 2023). Odours and sounds are also well-studied migraine triggers, with one study identifying the frequency range and acoustic properties of sounds that may trigger migraine episodes (Ishikawa et al. 2019).

10.5.1.2 Pre-existing Medical Conditions and Treatment

Migraine incidence is associated with several physiological and mental disorders such as depression, anxiety disorders, cardiovascular disorders and gastrointestinal disorders, but the manner of association between them is yet to be elucidated due to contradictory outcomes from studies undertaken (Haghdoost and Togha 2022). Medication overuse headaches (MOH) refer to headaches resulting from overuse or extended use of pharmaceuticals meant for short-term pain management. Such medication use significantly contributes to the chronification of migraines, but the prognosis is shown to improve after successful long-term withdrawal from MOH-causing medication (Pryse-Phillips et al. 1998). Other drugs such as H_2 receptor blockers, oral contraceptives, reserpine, atenolol, danazol etc. are known to be potential triggers of migraines.

10.5.1.3 Other Factors

Diet habits are believed to be correlated to migraines although the biological mechanisms behind the hypothesis are not yet well elucidated. Fasting (long periods without food) and certain food items (alcoholic beverages, chocolate, caffeine, tyramine, monosodium glutamate, etc.) are the other factors that are often reported and to be addressed in the migraine patients. The same food items may not necessarily serve as a trigger for all those affected. Identifying triggers unique to the patient may significantly contribute to more efficient avoidance of migraine onset (Gazerani 2020; Haghdoost and Togha 2022). Strenuous physical activity in combination with dehydration is a potential migraine trigger; whereas regular physical activity in moderation with adequate hydration and rest exhibits a protective effect against migraine onset. A sedentary lifestyle is linked to migraines with higher incidence and severity (Haghdoost and Togha 2022).

In women, migraine episodes may be heavily influenced by hormonal factors. Decreased oestrogen levels in the plasma is associated with migraine episodes in women, especially after periods of high oestrogen levels in serum. A reduced incidence of migraines is recorded in the second and third trimester of pregnancy whereas an increased incidence is recorded postpartum. Migraines triggered by menstruation are reported to be more severe and disabling than migraines during the rest of the cycle (Sacco et al. 2012).

10.5.2 Preventive and Curative Measures for Migraine Management in a Clinical Setting

On occasion, the management of migraine triggers and pharmacological methods may be insufficient to avoid migraine onset or to attenuate the intensity of migraine. In such cases, non-pharmacological and non-invasive techniques might be a potential alternative. They have the advantage of avoiding typical side effects whilst possibly providing insight into the pathophysiology of migraines. Topics explored in this avenue include nutraceuticals, acupuncture, behavioural therapy, and neuromodulation (Puledda and Shields 2018).

10.5.2.1 Nutraceuticals

Nutraceuticals refer to supplements that serve as pharmaceutical alternatives in addition to providing nutritional benefits. In patients affected by migraines, nutraceuticals are slowly gaining popularity due to their perceived benefits. Riboflavin has been widely explored in this regard, the rationale being that migraines may be caused due to mitochondrial dysfunction in the brain. Since riboflavin is involved in the synthesis of coenzymes essential for unhindered energy production in the mitochondrion, it was hypothesized to have a curative effect on migraines. Clinical trials conducted to test that hypothesis have yielded inconclusive and conflicting results. Even then, there appears to be sufficient convincing evidence that riboflavin might be effective enough to be recommended as a prophylactic treatment.

Coenzyme Q10 has a function similar to that of riboflavin in the mitochondrion; in the operation of the electron transport chain. Since clinical trials yielded inconclusive results for Coenzyme Q10 as well, more investigation is required to justify the use of Coenzyme Q10 as a nutraceutical for migraine treatment. In contrast, limited evidence from clinical trials conducted investigating magnesium for this purpose suggests that it might be effective. Other nutraceuticals explored for the management of migraines include coenzyme Q10, *Petasites hybridus* root extract, *Tanacetum parthenium,* and magnesium (Puledda and Shields 2018; Haghdoost and Togha 2022).

10.5.2.2 Acupuncture and Behavioural Therapy

Trials conducted testing acupuncture as a choice have yielded unsatisfactory results, indicating that it is not suited for treatment. Since it has had partial success, it is speculated that the biological effect of acupuncture may be independent of the arrangement of needles on the patient.

Behavioural Therapy involves the utilization of techniques that manage mental state to treat migraine symptoms without targeting the biological pain mechanisms directly. For example: relaxation techniques to cope better with migraine episodes, biofeedback training to monitor and self-regulate certain parameters (blood pressure, heart rate, etc.) associated with migraine onset and severity and psychotherapeutic techniques such as Cognitive Behavioural Therapy to manage stress on a short-term basis. Behavioural Therapy has proven to be effective in clinical trials as well (Puledda and Shields 2018; Haghdoost and Togha 2022).

10.5.2.3 Neuromodulation

Neuromodulation concerning migraine treatment refers to stimulation of the central or peripheral nervous system via an electric current or magnetic field, targeting the trigeminovascular or cortical network to alleviate a migraine episode. It is a relatively novel field, showing outstanding success rates with a low risk for undesirable effects. Long-term studies on a larger scale are necessary to confirm previously observed success rates. There are two modalities by which Neuromodulation is carried out: Invasive and Non-Invasive.

Non-invasive neuromodulation (NIN) typically involves the external stimulation of a nerve with the goal of pain relief. NIN may be used in a preventive and also curative capacity for acute episodes. Clinical trials for springTMS (single-pulse transcranial magnetic stimulation to modify the excitability of cortical neurons and certain associated circuits to treat acute migraine episodes), gammaCore (transcutaneous vagus nerve stimulator that is used to deliver electric current to the cervical branch of the vagus nerve to prevent migraine onset in those with chronic migraines), Cefaly (transcutaneous supraorbital stimulation in the occipital area for prevention and acute pain treatment in chronic migraine patients) and tDCS (anodal transcranial direct current stimulation where the scalp is exposed to anodal activating current for the prevention of migraine onset) are under investigation for migraine pain. Other methods under investigation include PMES (Percutaneous Mastoid Stimulation via induction of the fastigial nucleus in the cerebellum) and non-painful brachial electric stimulation (electrical stimulation in the cutaneous layer of the arm). These trials have provided preliminary evidence that non-invasive neuromodulation has the potential for migraine management. Invasive neuromodulation involves the placement of probes in proximity to the target nerve to supply electrical signals. Methods under investigation include occipital nerve stimulation (electrical pulses supplied to the greater occipital nerve, to target neurones in the trigeminocervical complex), Sphenopalatine Ganglion Stimulation and High Cervical Spinal cord stimulation. They are regarded a last resort for migraine treatment due to risks associated with surgical implantation. A few preliminary studies have yielded inadequate results and so large-scale clinical trials are required to confirm the efficacy of these devices (Bajaj and Dulebohn 2018; Puledda and Shields 2018; Tiwari and Agrawal 2022).

10.6 Conclusion

According to statistics, this is the most incapacitating illness affecting individuals of ages less than 50. Many pharmaceutical companies are targeting the chronic neurovascular headache condition due to its debilitating symptoms and persistent socioeconomic cost. Even if there are now more medications available, it might not be enough. Numerous drugs are currently used for both acute therapy and migraine prevention. Treatment for acute migraines aims to minimize the use of resources, such as diagnostic imaging, clinic visits and ambulatory infusion centre visits, while providing rapid relief from pain and related symptoms without recurrence, and aims to enhance function. The pathophysiology of migraines remains incompletely understood. Research continuously shows that the emission of the neuropeptide CGRP has a vital role in the beginning of migraines. As a result, the CGRP receptor is the focus of numerous contemporary and novel pharmaceutical treatments. However, a significant portion of migraineurs do not react well to these drugs. These patients, who are categorized as non-responders, highlight the necessity for ongoing research on other medication targets, for the management of migraines. This chapter highlights the CGRP targets, giving major importance to the non-CGRP targets with the association of migraine onset, where the targets included metabotropic receptors like PACAP, intracellular molecules like NO, PDE3/5 and ion channels like TPR, potassium channels and ASIC channels. The novel therapeutic option for both acute and prophylactic treatment has been discussed with appropriate clinical trial data. Non-pharmacological approaches have also been identified as the potential anti-migraine treatment option. However, studies are required to substantiate the significance of such treatments. Ultimately, in order to help the sizeable number of "non-responders," it is imperative that alternate pharmacological therapy for CGRP receptor antagonists be identified and developed. Although effective treatments for these unique targets have not yet been produced by current research, it might still be able to unfold the mechanisms underlying potential future treatments.

References

Adly C, Straumanis J, Chesson A (1991) Fluoxetine prophylaxis of migraine, pp 3–6

Al-Hassany L, Boucherie DM, Creeney H et al (2023) Future targets for migraine treatment beyond CGRP. J Headache Pain 24:1–18. https://doi.org/10.1186/s10194-023-01567-4

Al-Quliti KW, Assaedi ES (2016) New advances in prevention of migraine: review of current practice and recent advances. Neurosciences 21:207–214. https://doi.org/10.17712/nsj.2016.3.20150506

Amiri P, Kazeminasab S, Nejadghaderi SA et al (2022) Migraine: a review on its history, global epidemiology, risk factors, and comorbidities. Front Neurol 12:1–15. https://doi.org/10.3389/fneur.2021.800605

Andreou AP, Edvinsson L (2019) Mechanisms of migraine as a chronic evolutive condition. J Headache Pain 20:117. https://doi.org/10.1186/s10194-019-1066-0

Arguello D, Kali S, Thomas PD (2017) 乳鼠心肌提取 HHS public access. Physiol Behav 176:139–148. https://doi.org/10.1111/head.13358.Sleep

Bajaj J, Dulebohn SC (2018) Migraine, surgical interventions. StatPearls

Bertels Z, Pradhan AAA (2019) Emerging treatment targets for migraine and other headaches. Headache 59:50–65. https://doi.org/10.1111/head.13585

Bulut S, Berilgen MS, Baran A et al (2004) Venlafaxine versus amitriptyline in the prophylactic treatment of migraine: randomized, double-blind, crossover study. Clin Neurol Neurosurg 107:44–48. https://doi.org/10.1016/j.clineuro.2004.03.004

Calabresi P, Galletti F, Rossi C et al (2007) Antiepileptic drugs in migraine: from clinical aspects to cellular mechanisms. Trends Pharmacol Sci 28:188–195. https://doi.org/10.1016/j.tips.2007.02.005

Ceriani CEJ, Wilhour DA, Silberstein SD (2019) Novel medications for the treatment of migraine. Headache 59:1597–1608. https://doi.org/10.1111/head.13661

Couch JR, Ziegler DK, Hassanein R (1976) Amitriptyline in the prophylaxis of migraine. Effectiveness and relationship of antimigraine and antidepressant effects. Neurology 26:121–127. https://doi.org/10.1212/wnl.26.2.121

Cruickshank JM, Prichard BNC (1994) Beta-blockers in clinical practice, 2nd edn. Churchill Livingstone, Edinburgh

Cutrer FM, Silberstein SD, Mathew NT, Rozen TD (2001) Antiepileptic drugs in migraine, cluster headache, and mood disorders. Headache 41:1–3. https://doi.org/10.1046/j.1526-4610.2001.01154.x

De Tommaso M, Vecchio E, Quitadamo SG et al (2021) Pain-related brain connectivity changes in migraine: a narrative review and proof of concept about possible novel treatments interference. Brain Sci 11:1–21. https://doi.org/10.3390/brainsci11020234

Diener HC (2003) RPR100893, a substance-P antagonist, is not effective in the treatment of migraine attacks. Cephalalgia 23:183–185. https://doi.org/10.1046/j.1468-2982.2003.00496.x

Diener HC, Limmroth V (2001) Advances in pharmacological treatment of migraine. Expert Opin Investig Drugs 10:1831–1845. https://doi.org/10.1517/13543784.10.10.1831
Diener HC, Bussone G, Van Oene JC et al (2007) Topiramate reduces headache days in chronic migraine: a randomized, double-blind, placebo-controlled study. Cephalalgia 27:814–823. https://doi.org/10.1111/j.1468-2982.2007.01326.x
Diener HC, Charles A, Goadsby PJ, Holle D (2015) New therapeutic approaches for the prevention and treatment of migraine. Lancet Neurol 14:1010–1022. https://doi.org/10.1016/S1474-4422(15)00198-2
Do TP, Guo S, Ashina M (2019) Therapeutic novelties in migraine: new drugs, new hope? J Headache Pain 20:1–13. https://doi.org/10.1186/s10194-019-0974-3
Dodick DW (2018) A phase-by-phase review of migraine pathophysiology. Headache 58:4–16. https://doi.org/10.1111/head.13300
Edvinsson L, Villalón CM, Maassenvandenbrink A (2012) Basic mechanisms of migraine and its acute treatment. Pharmacol Ther 136:319–333. https://doi.org/10.1016/j.pharmthera.2012.08.011
Eller M, Goadsby PJ (2020) Migraine: a brain state amenable to therapy. Med J Aust 212:32–39. https://doi.org/10.5694/mja2.50435
Foroozan R, Cutrer FM (2009) Transient neurologic dysfunction in migraine. Neurol Clin 27:361–378. https://doi.org/10.1016/j.ncl.2008.11.002
Forssman B, Lindblad CJ, Zbomikova V (1983) Atenolol for migraine prophylaxis. Headache 23:188–190. https://doi.org/10.1111/j.1526-4610.1983.hed2304188.x
Fraser CL, Hepschke JL, Jenkins B, Prasad S (2019) Migraine aura: pathophysiology, mimics, and treatment options. Semin Neurol 39:739–748. https://doi.org/10.1055/s-0039-1700525
Freitag FG, Diamond S (1984) Nadolol and placebo comparison study in the prophylactic treatment of migraine. J Am Osteopath Assoc 84:343–347
Galletti F, Maria L, Corbelli I et al (2009) Progress in neurobiology pathophysiological basis of migraine prophylaxis. Prog Neurobiol 89:176–192. https://doi.org/10.1016/j.pneurobio.2009.07.005
Gazerani P (2020) Migraine and diet. Nutrients 12:1–11. https://doi.org/10.3390/nu12061658
Gray AM, Pache DM, Sewell RDE (1999) Do alpha 2 -adrenoceptors play an integral role in the antinociceptive mechanism of action of antidepressant compounds? pp 161–168
Green DP, Limjunyawong N, Gour N et al (2019) A mast-cell-specific receptor mediates neurogenic inflammation and pain. Neuron 101:412–420.e3. https://doi.org/10.1016/j.neuron.2019.01.012
Haanes KA, Edvinsson L (2019) Pathophysiological mechanisms in migraine and the identification of new therapeutic targets. CNS Drugs 33:525–537. https://doi.org/10.1007/s40263-019-00630-6
Haghdoost F, Togha M (2022) Migraine management: non-pharmacological points for patients and health care professionals. Open Med 17:1869–1882. https://doi.org/10.1515/med-2022-0598
Herd CP, Tomlinson CL, Rick C et al (2019) Cochrane systematic review and meta-analysis of botulinum toxin for the prevention of migraine. BMJ Open 9:1–8. https://doi.org/10.1136/bmjopen-2018-027953
Hoffmann J, Schirra T, Lo H et al (2015) The influence of weather on migraine—are migraine attacks predictable? Ann Clin Transl Neurol 2:22–28. https://doi.org/10.1002/acn3.139
Holroyd KA, Penzien DB, Cordingley GE (1991) Propranolol in the management of recurrent migraine: a meta-analytic review. Headache 31:333–340. https://doi.org/10.1111/j.1526-4610.1991.hed3105333.x
Intolerant R, Monotherapy F (2011) Brief communication low-dose Topiramate plus sodium Divalproate for positive, pp 129–132. https://doi.org/10.1111/j.1526-4610.2011.02035.x
Ishikawa T, Tatsumoto M, Maki K et al (2019) Identification of everyday sounds perceived as noise by migraine patients. Intern Med 58:1565–1572. https://doi.org/10.2169/internalmedicine.2206-18
James R, Hassanein RS (2015) Amitriptyline in
Johannsson V, Nilsson LR, Widelius T et al (1987) Atenolol in migraine prophylaxis a double-blind cross-over multicentre study. Headache 27:372–374. https://doi.org/10.1111/j.1526-4610.1987.hed2707372.x

Kesserwani H (2021) Migraine triggers: an overview of the pharmacology, biochemistry, atmospherics, and their effects on neural networks. Cureus 13:1–9. https://doi.org/10.7759/cureus.14243

Kung TA, Guyuron B, Cederna PS (2011) Migraine surgery: a plastic surgery solution for refractory migraine headache. Plast Reconstr Surg 127:181–189. https://doi.org/10.1097/PRS.0b013e3181f95a01

Levy D (2010) Headache currents migraine pain and nociceptor activation—where do we stand? Headache 50:909–916

Levy D, Burstein R, Kainz V et al (2007) Mast cell degranulation activates a pain pathway underlying migraine headache. Pain 130:166–176. https://doi.org/10.1016/j.pain.2007.03.012

Limmroth V, Michel MC (2001) The prevention of migraine: a critical review with special emphasis on β-adrenoceptor blockers. Br J Clin Pharmacol Suppl 52:237–243. https://doi.org/10.1046/j.0306-5251.2001.01459.x

Lin TH, Yang CC, Lee SY et al (2023) The effect of bright light therapy in migraine patients with sleep disturbance: a prospective, observational cohort study protocol. Front Aging Neurosci 14:1–6. https://doi.org/10.3389/fnagi.2022.1041076

Linde M, Edvinsson L, Manandhar K et al (2017) Migraine associated with altitude: results from a population-based study in Nepal. Eur J Neurol 24:1055–1061. https://doi.org/10.1111/ene.13334

Lionetto L, Negro A, Palmisani S et al (2012) Emerging treatment for chronic migraine and refractory chronic migraine. Expert Opin Emerg Drugs 17:393–406

Monteith TS, Goadsby PJ (2011) Acute migraine therapy: new drugs and new approaches. Curr Treat Options Neurol 13:1–14. https://doi.org/10.1007/s11940-010-0105-6

Moraczewski J, Awosika AO, Aedma KK (2023) Tricyclic antidepressants. StatPearls

Mungoven TJ, Henderson LA, Meylakh N (2021) Chronic migraine pathophysiology and treatment: a review of current perspectives. Front Pain Res 2:1–15. https://doi.org/10.3389/fpain.2021.705276

Nisar A, Ahmed Z, Yuan H (2023) Novel therapeutic targets for migraine. Biomedicines 11:1–12. https://doi.org/10.3390/biomedicines11020569

Olesen J, Ashina M (2011) Emerging migraine treatments and drug targets. Trends Pharmacol Sci 32:352–359. https://doi.org/10.1016/j.tips.2011.02.016

Olesen J, Tfelt-Hansen P, Ashina M (2009) Finding new drug targets for the treatment of migraine attacks. Cephalalgia 29:909–920. https://doi.org/10.1111/j.1468-2982.2008.01837.x

Olsen TS (1995) Pathophysiology of the migraine aura: the spreading depression theory. Brain 118:307. https://doi.org/10.1093/brain/118.1.307

Omote K, Iwasaki H, Kawamata M, Saoh O, Namiki A (1995) Effects of verapamil on spinal anesthesia with local anesthetics. Anesth Analg 80:444–448. https://doi.org/10.1097/00000539-199503000-00002

Ong JJY, Wei DYT, Goadsby PJ (2018) Recent advances in pharmacotherapy for migraine prevention: from pathophysiology to new drugs. Drugs 78:411–437. https://doi.org/10.1007/s40265-018-0865-y

Ozyalcin SN, Talu GK, Kiziltan E, Yucel B (2005) The efficacy and safety of venlafaxine in the prophylaxis of migraine. Headache 45(2):144–152

Pryse-Phillips WEM, Dodick DW, Edmeads JG et al (1998) Guidelines for the nonpharmacologic management of migraine in clinical practice. CMAJ 159:47–54

Puledda F, Shields K (2018) Non-pharmacological approaches for migraine. Neurotherapeutics 15:336–345. https://doi.org/10.1007/s13311-018-0623-6

Puledda F, Silva EM, Suwanlaong K, Goadsby PJ (2023) Migraine: from pathophysiology to treatment. J Neurol 270:3654–3666. https://doi.org/10.1007/s00415-023-11706-1

Punay NC, Couch JR (2003) Antidepressants in the treatment of migraine headache. Curr Pain Headache Rep 7(1):51–54

Ramadan NM, Buchanan TM (2006) New and future migraine therapy. Pharmacol Ther 112:199–212. https://doi.org/10.1016/j.pharmthera.2005.04.010

Rothrock J (1997) Clinical studies of valroate for migraine prophylaxis. Cephalalgia 17:81–83

Russo A, Silvestro M, Tedeschi G, Tessitore A (2017) Physiopathology of migraine: what have we learned from functional imaging? Curr Neurol Neurosci Rep 17:95. https://doi.org/10.1007/s11910-017-0803-5
Ryan RE, Ryan RE, Sudilovsky A (1983) Nadolol: its use in the prophylactic treatment of migraine. Headache 23:26–31. https://doi.org/10.1111/j.1526-4610.1983.hed2301026.x
Sacco S, Ricci S, Degan D, Carolei A (2012) Migraine in women: the role of hormones and their impact on vascular diseases. J Headache Pain 13:177–189. https://doi.org/10.1007/s10194-012-0424-y
Schellenberg R, Lichtenthal A, Wöhling H et al (2008) Nebivolol and metoprolol for treating migraine: an advance on β-blocker treatment? Headache 48:118–125. https://doi.org/10.1111/j.1526-4610.2007.00785.x
Schrader H, Stovner LJ, Helde G et al (2001) Prophylactic treatment of migraine with angiotensin converting enzyme inhibitor (lisinopril): randomised, placebo controlled, crossover study. Br Med J 322:19–22. https://doi.org/10.1136/bmj.322.7277.19
Sehgal SA, Hassan M, Rashid S (2014) Pharmacoinformatics elucidation of potential drug targets against migraine to target ion channel protein KCNK18. Drug Des Devel Ther 8:571–581. https://doi.org/10.2147/DDDT.S63096
Seidel MF, Hügle T, Morlion B et al (2022) Neurogenic inflammation as a novel treatment target for chronic pain syndromes. Exp Neurol 356:114108. https://doi.org/10.1016/j.expneurol.2022.114108
Shaygannejad V, Janghorbani M, Ghorbani A, Ashtary F, Zakizade N, Nasr V (2006) Comparison of the effect of topiramate and sodium valporate in migraine prevention: a randomized blinded. Headache 46:642–648. https://doi.org/10.1111/j.1526-4610.2006.00413.x
Silberstein SD (2009) Preventive migraine treatment. Neurol Clin 27:429–443. https://doi.org/10.1016/j.ncl.2008.11.007
Sprenger T, Goadsby PJ (2009) Migraine pathogenesis and state of pharmacological treatment options. BMC Med 7:1–5. https://doi.org/10.1186/1741-7015-7-71
Sprenger T, Viana M, Tassorelli C (2018) Current prophylactic medications for migraine and their potential mechanisms of action. Neurotherapeutics 15:313–323. https://doi.org/10.1007/s13311-018-0621-8
Stam AH, Haan J, Frants RR et al (2005) Migraine: new treatment options from molecular biology. Expert Rev Neurother 5:653–661. https://doi.org/10.1586/14737175.5.5.653
Steiner TJ, Stovner LJ, Birbeck GL (2013) Migraine: the seventh disabler. Cephalalgia 33:289–290. https://doi.org/10.1177/0333102412473843
Suwannasual U, Lucero JA, McDonald JD, Lund AK (2018) Exposure to traffic-generated air pollutants mediates alterations in brain microvascular integrity in wildtype mice on a high-fat diet. Environ Res 160:449–461. https://doi.org/10.1016/j.envres.2017.10.029
Tedavisi MA (2013) Acute treatment of migraine. Arch Neuropsychiatry 50:26–29. https://doi.org/10.4274/Npa.y7299
Tiwari V, Agrawal S (2022) Migraine and neuromodulation: a literature review. Cureus 14:e31223. https://doi.org/10.7759/cureus.31223
Todd J Schwedt IG (2022) Acute treatment of migraine in adults. UpToDate, Wolters Kluwer, pp 1–42
Tronvik E, Stovner LJ, Helde G, Sand T, Bovim G (2006) Prophylactic treatment of migraine with an angiotensin II receptor blocker a randomized controlled trial. Med Clin (Barc) 126:467–472. https://doi.org/10.1157/13086327
van de Ven LLM, Franke CL, Koehler PJ (1997) Prophylactic treatment of migraine with bisoprolol: a placebo controlled study. Cephalalgia 17:596–599
Welch K (2003) Contemporary concepts of migraine pathogenesis. Neurology 61:S2–S8. https://doi.org/10.1212/WNL.61.8_suppl_4.S2
Zobdeh F, Ben Kraiem A, Attwood MM et al (2021) Pharmacological treatment of migraine: drug classes, mechanisms of action, clinical trials and new treatments. Br J Pharmacol 178:4588–4607. https://doi.org/10.1111/bph.15657

11 Currently Available Interventions for the Management of Migraine Pain

Rakesh Kumar Singh, Shivam Kumar Pandey,
Shreya Singh, Apoorva Chittoda,
and Sakshi Arun Chaudhari

11.1 Introduction

Migraine is recognized as the sixth most disabling global burden of disease among neurological disorders. The intricate biology of migraine involves a complex interplay of genetic predisposition, behavioral factors, and environmental influences that alter the brain's processing of sensory stimuli, rendering the individuals more susceptible to certain triggers. It is characterized by recurrent, unilateral, and moderate-to-severe headache episodes, accompanied by nausea and heightened sensitivity to light and sound, and pose a significant global health challenge (Goadsby et al. 2017).

The term "migraine" has its origins in the Greek word "hemikrania," later translated to Latin as "hemigranea" and further evolved into the French term "migraine." This common neurological disorder is a major cause of disability and work absenteeism. The complex events of the migraine attack lasts from hours to days, often occur without aura in 75% of cases (Rose 1995).

The International Headache Society's classification committee categorizes migraines into different subtypes (Steiner 2004). The migraine without aura involves recurrent headache attacks lasting 4 to 72 h, typically unilateral, pulsating, and aggravated by physical activity. The associated symptoms include nausea and sensitivity to light and sound. The migraine with aura presents minute-long, reversible attacks with unilateral symptoms, affecting motor, brainstem, language and speech, retinal, visual, sensory, and headache functions. The chronic migraine persists for

R. K. Singh (✉) · S. K. Pandey · S. Singh · A. Chittoda
Department of Pharmacology and Toxicology, National Institute of Pharmaceutical Education and Research (NIPER), Raebareli, Lucknow, Uttar Pradesh, India
e-mail: rakesh.singh@niperraebareli.edu.in

S. A. Chaudhari
Department of Regulatory Toxicology, National Institute of Pharmaceutical Education and Research (NIPER), Raebareli, Lucknow, Uttar Pradesh, India

S. Bhatt et al. (eds.), *Management of Migraine Pain*,
https://doi.org/10.1007/978-981-97-4529-6_11

more than 3 months, with migraine-like episodes at least 8 days per month. The probable migraine refers to symptomatic attacks not fitting other headache types.

A significant percentage of the world's population suffers from this common and debilitating neurological disorder (Steiner 2004). The global prevalence of migraines surpasses one billion people, showcasing the severity and widespread impact of this neurological disorder. The prevalence rate varies across different age groups, regions, and gender demographics (GBD 2016 Headache Collaborators 2018). Migraine prominently affects individuals aged 25–55, with notable occurrences in school-aged children (up to 10%) and adolescence (up to 28%) (Steiner et al. 2020). The gender disparities are evident, with a higher prevalence among women (18–25%) compared to men (6–8%) (Stovner et al. 2014).

However, the accurate assessment of migraine prevalence seems to be challenging. Migraine exhibits a diverse array of symptoms beyond headaches, such as sensitivity to sound and light, nausea, and in some cases aura, which adds to the complexity of diagnosis and reporting (El-Metwally et al. 2020). Underreporting and misdiagnosis due to the variability and complexity of migraine symptoms contributes to difficulties in determining precise prevalence figures. The associated healthcare expenses, including consultations, medications, and potential hospitalizations, combined with reduced productivity and absenteeism at work or school pose substantial economic costs (Mayans and Walling 2018). Raising awareness, fostering accurate diagnosis, and developing effective treatment strategies are paramount to mitigating the far-reaching effects of migraines on individuals and society. The continued research efforts are aimed at understanding the underlying mechanisms, triggers, and variations in migraines, seeking to enhance management approaches and therapeutic interventions, ultimately aiming to alleviate the considerable burden imposed by migraines on a global scale (Abu-Arafeh et al. 2010).

For individuals, families, and society, migraine is a very burdensome condition. A person's ability to function in a variety of roles and contexts such as professional, academic, social, familial, and personal ones can be impacted by migraines (Leonardi et al. 2005; Buse et al. 2016; Blumenfeld et al. 2011). The fact that migraines are most common in the second, third, fourth, and fifth decades of life increases the burden on society. According to the Global Burden of Disease (GBD) study, the highest rate of migraine-related disability (20.3 million years lived with disability, YLD) was seen in women within the 15 and 49 years age group (GBD 2016 Headache Collaborators 2018). Those with migraines may be taking care of young children or aging parents during this period, which increases the likelihood of a negative impact on quality of life and disability. It is also a time when their work productivity is frequently at its highest.

Migraine imposes a far-reaching impact, significantly altering the lives of individuals and casting a substantial societal burden. On a personal level, the throbbing, intense pain coupled with accompanying symptoms such as nausea, vomiting, sensitivity to sound, lights, or even odors disrupts daily routines, hindering work performance, social interactions, and personal relationships (Hareendran et al. 2018). The unpredictability of migraine attacks instills fear and anxiety, leading to emotional distress and mental health challenges. The chronic migraine pain often leads to a decreased quality of life, contributing to feelings of frustration, isolation, and depression (Messali et al. 2016).

Functionality takes a severe hit during a migraine episode, with individuals often incapacitated, unable to perform routine tasks or fulfill work obligations. This results in frequent absenteeism from work or school, leading to reduced productivity and income instability (Lipton et al. 2018). The financial strain is compounded by medical expenses, including doctor visits, medications, and sometimes emergency room treatments, adding to the overall economic burden. The reliance on medications to manage migraines can introduce further complications. The overuse of painkillers might exacerbate headaches, leading to more frequent and severe attacks, a condition known as medication-overuse headaches. Additionally, some medications might have adverse side effects, impacting overall health and potentially requiring further medical intervention (Lipton et al. 2016). Society too, feels the weight of migraines. The healthcare systems face increased demands, from consultations and treatments to diagnostic tests and specialized care (Lipton et al. 2016). The educational institutions witness compromised academic performance, absenteeism, and limited participation in extracurricular activities among students battling migraines (Buse et al. 2013). Moreover, the societal fabric is strained as migraines can impede social interactions and strain relationships. The cancelled societal plans, missed gatherings, and withdrawal from social activities due to unpredictable attacks contribute to feelings of isolation and can strain personal connections, affecting the overall social well-being of individuals (Minen et al. 2016).

Comprehending the extensive and intricate impact of migraines on individuals and society underscores the importance of holistic support systems, effective treatment strategies, and heightened awareness initiatives. Addressing these multifaceted challenges is pivotal to mitigate the pervasive burden of migraines on both the individual and the larger societal framework (Stewart et al. 2010). The significant impact of migraines on an individual's quality of life emphasizes the importance of effective treatment. The effective migraine treatment is essential not only to relieve pain during attacks but also to prevent and minimize the overall impact of migraines on a person's life, improving functionality and well-being (Burch 2019). The aim of this chapter is to discuss the currently available interventions for the management of migraine pain.

11.2 Pathophysiology of the Migraine

The previous vascular theory of migraine, which holds that vasodilation causes headaches and vasoconstriction causes aura, is no longer valid (Amin et al. 2013). These days, theories propose that a number of primary neuronal impairments result in a cascade of extracranial and intracranial alterations that give rise to migraine (Burstein et al. 2015). There are several stages of a migraine attack: the premonitory, aura, headache, and postdrome. A higher vulnerability during these stages of development may cause them to advance or worsen. Every stage has a unique physiological mechanism involving a particular region of the brain, and other

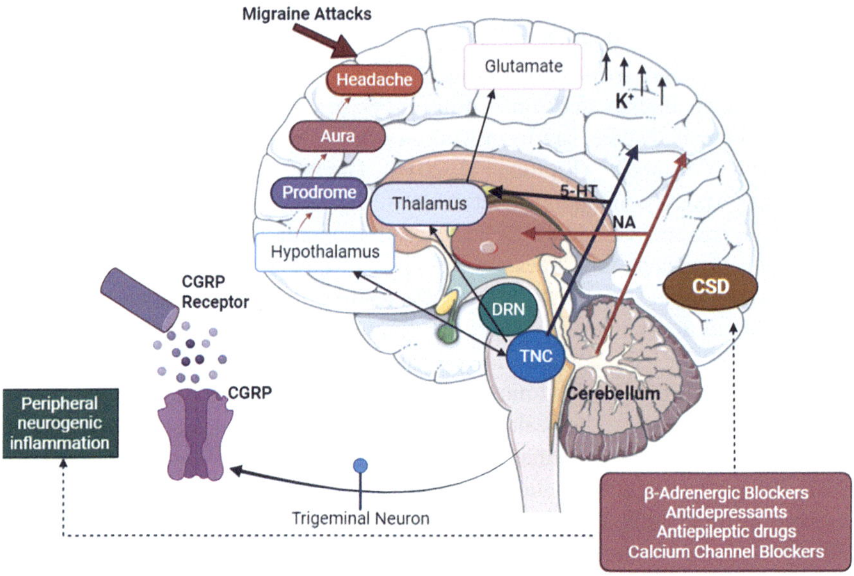

Fig. 11.1 The pathogenesis of migraine involves peripheral and central sensitization of the trigeminal system, with CGRP release contributing to neurogenic inflammation. Cortical spreading depression (CSD) disrupts normal regulatory mechanisms, affecting trigeminal neuron responses. Antidepressants may influence CSD by preserving 5-HT and NA neurotransmission, while propranolol reduces excitability via beta-adrenergic blockage

pathophysiological factors are included. There are four distinct phases to migraine attacks and other pathophysiological factors included (Fig. 11.1).

11.2.1 Prodrome

The prodromal phase involves premonitory symptoms associated with hypothalamus activation, particularly dopamine release. These symptoms may precede headaches by 24–48 h, with about 77% of patients experiencing them. Females are more susceptible (81%) than males (64%) (Karsan and Goadsby 2020; Laurell et al. 2016).

11.2.2 Aura

Approximately 25% of migraine cases exhibit aura, involving changes in blood circulation, neurovascular integration, and cortical function (Hansen and Charles 2019; van Dongen and Haan 2019). Aura may precede or occur concurrently with the headache, typically lasting less than an hour. Positive symptoms, such as visual disturbances and sensory phenomena, result from neuronal activity, while negative symptoms indicate a temporary loss of function. Aura must be completely reversible.

11.2.3 Headache

The headache phase involves further modifications to blood flow and function in the brainstem, thalamus, hypothalamus, and cortex. Headaches are usually unilateral, pulsating, and intensify over the first few hours. Symptoms such as photophobia, allodynia, nausea, vomiting, phonophobia, lachrymation, rhinorrhea, and osmophobia correlate with headache intensity. The duration can vary from a few hours to several days.

11.2.4 Postdrome

The postdromal phase is characterized by persistent blood alterations and symptoms even after the headache subsides. Pain sensitivity in the same location as the previous headache, tiredness, light-headedness, difficulty focusing, and euphoria are common postdromal symptoms.

11.2.5 Trigeminovascular System Activation and Neurogenic Inflammation

The trigeminovascular system comprises pseudounipolar sensory neurons originating from the upper cervical dorsal nerve and the trigeminal ganglion. These neurons innervate large venous sinuses, cerebral blood vessels, dura mater, and pial vessels. Convergence at the trigeminal nucleus caudalis (TNC) explains the distribution of headaches in the head and upper neck. The TNC projects to the ventroposterior medial nucleus of the thalamus and the sensory cortex, with additional fibers reaching subcortical regions, including the cerebellum, midbrain, and reticular formation. The limbic system receives nociceptive information, contributing to the emotional reactions associated with migraines (Akerman et al. 2011).

11.2.6 Cortical Spreading Depression (CSD)

Cortical spreading depression (CSD) is a propagating wave of depolarization involving neurons and glial cells across the cerebral cortex. This wave is hypothesized to induce inflammation in structures susceptible to pain, triggering the release of pro-inflammatory mediators. CSD is implicated in the aura phase, altering the blood-brain barrier permeability and delaying trigeminal nociceptor activation, the primary cause of migraine headaches. This phenomenon may explain aura-free migraines in specific brain regions where depolarization waves are undetectable (Zhang et al. 2011).

11.2.7 CGRP's Function in Migraine Headaches

Calcitonin gene-related peptide (CGRP), a neuropeptide found in trigeminal ganglia nerves, plays a crucial role in migraine headaches. CGRP can significantly dilate cerebral blood vessels, mediating the vasodilatory mechanism of neurological inflammation and transmitting trigeminovascular nociceptive signals from intracranial vessels to the central nervous system (Zhang et al. 2011). In summary, the pathophysiology of migraine involves a complex interplay of neuronal and vascular factors, with distinct stages contributing to the manifestation of symptoms. The evolving understanding of migraine mechanisms has led to targeted therapeutic approaches, such as CGRP receptor antagonists, aiming to address specific components of the pathophysiological cascade.

11.3 Current Pharmacological Interventions for Migraines

Acute migraine attacks can be debilitating, significantly affecting the quality of life for individuals experiencing them. This section delves into the pharmacological interventions designed to alleviate the immediate symptoms of migraines, providing relief, and improving patient outcomes. The quest for effective migraine management has led to the evolution of pharmacological interventions that span a spectrum of mechanisms and targets. From traditional treatments like nonsteroidal anti-inflammatory drugs (NSAIDs) and triptans to more recent innovations like calcitonin gene-related peptide (CGRP) inhibitors, the pharmacological armamentarium has expanded significantly (Olesen and Ashina 2011). This evolution reflects not only advances in our understanding of migraine pathophysiology but also a commitment to providing tailored and effective options for individuals with varying clinical profiles. A fundamental aspect of migraine management is the recognition of the heterogeneity among individuals experiencing migraine. Factors such as comorbidities, contraindications, and patient preferences play pivotal roles in shaping treatment plans. This chapter aims to explore the nuances of currently available pharmacological interventions, emphasizing the importance of patient-centered care that considers the unique characteristics of each individual.

11.3.1 Acute Migraine Medications

The treatment of acute migraine attacks relies on a variety of medications designed to provide rapid relief from symptoms.

11.3.1.1 Triptans

Triptans constitute a category of medications extensively employed in the acute management of migraine. They exert their effects primarily through serotonin receptor agonism, specifically targeting 5-HT1B and 5-HT1D receptors. This activation leads to vasoconstriction and inhibition of neuropeptide release, mitigating the pain associated with migraine (Saxena and Ferrari 1989; Waeber 2003). Despite their efficacy, triptans are not devoid of limitations, including contraindications in patients with cardiovascular diseases. Sumatriptan, a groundbreaking development in migraine pharmacotherapy, holds the distinction of being the pioneering member of the triptan class of drugs (Derry et al. 2012). The synthesis of sumatriptan represented a strategic departure from conventional approaches, seeking to harness the therapeutic benefits observed with ergot-derived compounds while mitigating the vasoconstrictive side effects. The culmination of this endeavor resulted in the formulation of a novel drug that specifically targeted 5-HT1B and 5-HT1D receptors. An exceptional aspect of sumatriptan's introduction to the therapeutic landscape was its availability in injectable subcutaneous form, marking the initiation of the global dissemination of the triptan class (Derry et al. 2012). This mode of administration provided a rapid and effective means of delivering the drug, offering relief to individuals experiencing acute migraine attacks. The advent of injectable subcutaneous sumatriptan paved the way for subsequent triptans to follow suit, collectively establishing the triptan class as a significant advancement in migraine management. As of the present, seven distinct triptans are marketed, each with unique characteristics and applications. Sumatriptan, along with its counterparts such as rizatriptan, zolmitriptan, naratriptan, almotriptan, and frovatriptan, exhibits binding affinity to both 5-HT1B and 5-HT1D receptors (Pleş et al. 2023). This shared receptor interaction underlines the commonality in the pharmacological mechanism among these triptans, emphasizing their collective role in modulating serotonin receptors to alleviate migraine symptoms. Triptans, generally considered safe and well tolerated in patients lacking concomitant vascular disease, represent a cornerstone in the pharmacological management of migraine (de Boer et al. 2023). However, their utilization is contraindicated in specific clinical scenarios, including cardio or cerebrovascular disease, uncontrolled hypertension, ischemic bowel, hemiplegic migraine, migraine with brainstem aura, pregnancy, or in individuals at risk for pregnancy. Additionally, the administration of triptans is discouraged within 24 h of using another triptan or ergotamine/dihydroergotamine (Roberto et al. 2015).

Despite their overall safety profile, triptans can elicit sensations such as chest pressure and tingling in the limbs in approximately more than 7% of cases. It is noteworthy, however, that these sensations, while relatively common, have not demonstrated an associated risk of cardiovascular ischemia. Importantly, triptans do not exert their effects on the 5-HT2A receptors, mitigating concerns related to an elevated risk of serotonin syndrome, whether used in monotherapy or in combination with selective serotonin reuptake inhibitors (SSRIs) and serotonin-noradrenaline reuptake inhibitors (SNRIs) (Dobrek and Głowacka 2023). This nuanced evaluation of the safety considerations surrounding triptan use underscores the delicate balance between their established efficacy in migraine management and the necessity for a judicious approach in certain clinical contexts. The delineation of contraindications and the clarification of potential side effects contribute to a more comprehensive understanding of the clinical landscape, facilitating informed decision making by clinicians and ensuring the optimal and safe use of triptans in the treatment of migraines. Certain triptans, including sumatriptan, eletriptan, almotriptan, and naratriptan, incorporate a sulfonyl group or a sulfonamide moiety. Despite this structural similarity to sulfur-containing compounds, patients with sulfur allergies typically tolerate these triptans well. However, in cases where previous allergic reactions to sulfonamides have been severe, opting for a sulfur-free triptan, such as rizatriptan, frovatriptan, or zolmitriptan, is a prudent consideration. While generally well-tolerated, prolonged and frequent use of triptans can lead to Medication Overuse Headache (MOH) in susceptible individuals (Diener et al. 2016). To mitigate this risk, it is recommended to restrict the use of triptans to less than 10 days per month (Wammes-van der Heijden et al. 2006). This limitation aims to balance the effective relief provided by triptans with the need to avoid the development of MOH, emphasizing the importance of judicious and strategic use in migraine management protocols. These guidelines serve as a valuable framework for healthcare practitioners, offering insights into optimal prescribing practices and considerations that contribute to the safe and effective utilization of triptans in the multifaceted landscape of migraine treatment.

11.3.1.2 Nonsteroidal Anti-inflammatory Drugs (NSAIDs)

NSAIDs, such as ibuprofen and naproxen, are commonly employed as first-line agents for acute migraine relief (Mayans and Walling 2018). Their mechanism involves the inhibition of cyclooxygenase, reducing prostaglandin synthesis and subsequent inflammation (Suthisisang et al. 2010). NSAIDs are particularly useful in mild to moderate migraine attacks and are often preferred in individuals with contraindications to triptans (Tajti et al. 2015). However, long-term use should be approached cautiously due to potential adverse effects on the gastrointestinal and renal systems. In the acute treatment of migraine, NSAIDs find extensive usage. Their fundamental mode of action centers on inhibiting neurogenic inflammation and reversing central sensitization, linked to migraine. Multiple double-blind and randomized controlled studies have affirmed the effectiveness of NSAIDs, including ibuprofen, naproxen sodium, acetylsalicylic acid (ASA), and diclofenac potassium in treating migraine (Peters 2019; Straube et al. 2011). Ibuprofen, a commonly used NSAID, functions as a nonselective COX inhibitor and is rapidly absorbed from the gastrointestinal tract, ensuring a swift onset of action. Liquid-filled capsules with enhanced solubility demonstrate a faster onset and a more effective response in alleviating headaches at 1 h when compared to the traditional tablet formulation (Cristofoletti and Dressman 2016). With a short half-life of 2 h, Ibuprofen may require repeated dosing, and there are indications that higher doses do not yield increased efficacy beyond the common 400-mg dose (Varrassi et al. 2020). Diclofenac potassium is typically prescribed at 50 mg, with a recommended maximum daily dose of 150 mg. Demonstrating consistent effectiveness, Naproxen is utilized for treating moderate-to-severe migraine attacks, boasting a prolonged half-life of 14 h. However, its onset of action is comparatively slower than Ibuprofen. Naproxen sodium, distinguished for cardiovascular safety among NSAIDs, is preferred, with typical doses ranging from 500 to 550 mg and a maximum daily dosage of 1375 mg (Aminoshariae et al. 2016; Ong and De Felice 2018). Some clinicians advocate for Naproxen over Naproxen sodium in sodium-restricted patients. Aspirin

(ASA) is taken at doses of 975–1000 mg, with a daily maximum of 4000 mg. Because of active salicylate metabolites, ASA has a prolonged half-life of 6 h. Effervescent ASA acts more quickly, showing efficacy that is similar to sumatriptan 50 mg and greater than that of naproxen sodium 500 mg (Ong and De Felice 2018). When paired with metoclopramide 10 mg, ASA improves relief from migraine-related nausea and vomiting. Patients with peptic ulcer disease, inflammatory bowel disease, a history of gastric bypass surgery, ASA allergy, renal impairment, or who are taking anticoagulants should not take NSAIDs (Szeto et al. 2020). To reduce the risk of medication-overuse headache (MOH), NSAID use should be limited to 14 days or less per month.

In contrast, Acetaminophen, commonly known as paracetamol, was developed in the nineteenth century. While its precise mechanism of action remains incompletely understood, paracetamol is thought to exert its effects through the central inhibition of prostaglandin synthesis (Graham et al. 2013). Unlike NSAIDs, paracetamol does not directly inhibit COX functions outside the central nervous system. Its central action does not involve direct blockage of an active site but rather entails the reduction of COX activity. Widely used, paracetamol has proven efficacy in managing migraine, often employed for mild to moderately severe migraine attacks. Additionally, acetaminophen can be utilized in combination with other medications. The amalgamation of aspirin/paracetamol/caffeine has shown superior efficacy compared to any single component administered alone at equivalent doses.

11.3.1.3 Antiemetics

Nausea and vomiting frequently accompany migraine, necessitating the inclusion of antiemetic agents in acute treatment strategies. Medications like metoclopramide and prochlorperazine target dopamine receptors, alleviating nausea and improving the overall tolerability of migraine medications (Ong and De Felice 2018). Their dual action as antiemetic and analgesic agents make them valuable additions to comprehensive migraine management plans. Medications falling within this category prove beneficial not only in the acute treatment of migraine but also as supplementary treatment for the nausea commonly experienced during migraine attacks. The suggested mechanism of action for these agents is thought to stem from their properties as dopamine antagonists, counteracting the reported hypersensitivity to dopamine observed during migraine attacks. Metoclopramide, frequently administered in the emergency department (ED) to address nausea and vomiting, exhibits several advantageous effects (Láinez et al. 2013). It mitigates gastric stasis, augments the absorption of concurrently administered medications, and has demonstrated effectiveness in treating migraines, even during pregnancy. As mentioned earlier, the standard dose of metoclopramide is 10 mg, deemed safe for administration up to four times a day (Ong and De Felice 2018; Láinez et al. 2013). Intravenous metoclopramide is the recommended first-line treatment for migraines in the emergency department (ED), according to a meta-analysis of 13 randomized controlled trials (Ungrungseesopon and Wongtanasarasin 2022). The analysis also indicates that metoclopramide has a primary effect on migraines beyond its well-established antiemetic properties. Although both chlorpromazine and prochlorperazine have antiemetic effects resulting from histaminergic receptor antagonism (H1), they are recognized as dopamine receptor antagonists, targeting D1, D2, D3, and D4 receptors. Prochlorperazine can be taken orally at a dose of 10 mg, up to four times per day, or via a suppository at a dose of 10 mg, with a daily limit of 50 mg. However, there is a higher risk of extrapyramidal side effects (Becker 2015). All the drugs are safe when combined with other frequently prescribed medications to treat migraine attacks. However, there are potential side effects of a dopamine blockade, which include fatigue, weight gain, sedation, xerostomia, blurred vision, QT interval prolongation, cardiac arrhythmias, and postural hypotension. Additionally, a dopamine blockade can result in extrapyramidal symptoms like akathisia, tremors, dystonia, drug-induced pseudo-parkinsonism, and tardive dyskinesia. Finally, an increase in prolactin release may cause sexual dysfunction, galactorrhea, and infertility.

11.3.1.4 CGRP Inhibitors: Role and Emerging Therapies

Calcitonin Gene-Related Peptide (CGRP) inhibitors have emerged as groundbreaking therapeutic agents in the management of migraine (Russo and Hay 2023). The migraine, a complex neurological disorder, involves the activation of trigeminovascular pathways and the release of neuropeptides, particularly CGRP, leading to vasodilation and inflammation (Muddam et al. 2023). CGRP, a neuropeptide widely distributed in the central and peripheral nervous systems, plays a pivotal role in migraine pathophysiology. Released from trigeminal nerve endings, CGRP induces vasodilation, neurogenic inflammation, and the sensitization of pain pathways. Understanding these processes has led to the development of CGRP inhibitors as targeted interventions. Moreover, CGRP exerts its effects through binding to CGRP receptors, predominantly located in vascular smooth muscle cells and neurons. The activation of these receptors results in vasodilation and the release of pro-inflammatory mediators, contributing to the neurogenic inflammation characteristic of migraines. CGRP inhibitors primarily consist of monoclonal antibodies designed to specifically target and neutralize CGRP or its receptors. By binding to CGRP and preventing its interaction with receptors, these antibodies disrupt the cascade of events leading to migraine attacks (Schoenen et al. 2020). In addition to monoclonal antibodies, small molecules acting as CGRP receptor antagonists have been developed. These compounds bind to CGRP receptors, inhibiting their activation and downstream signaling. The diversity in CGRP inhibitor formulations offers clinicians options tailored to patient preferences and therapeutic needs. Clinical trials evaluating CGRP inhibitors, including erenumab, fremanezumab, and galcanezumab, consistently demonstrate a significant reduction in migraine frequency and severity (Murray et al. 2022). These findings extend across various patient populations, encompassing both episodic and chronic migraine sufferers. Beyond migraine reduction, CGRP inhibitors contribute to improvements in the overall quality of life. Patients report enhanced functionality, reduced disability, and increased satisfaction with their migraine management. The efficacy of CGRP inhibitors extends to patients with a history of inadequate responses to other preventive treatments (Murray et al. 2022). In the pursuit of optimizing treatment adherence and convenience, extended-release formulations of CGRP inhibitors are under investigation.

Prolonged pharmacokinetics and less frequent dosing schedules offer potential advantages, addressing practical challenges associated with regular injections. Combination therapies involving CGRP inhibitors and other migraine preventive agents are being explored (Mechtler et al. 2021). The rationale behind these approaches lies in targeting multiple pathways implicated in migraine pathophysiology simultaneously. Trials assessing the safety and efficacy of such combinations are underway. CGRP inhibitors represent a transformative chapter in migraine therapeutics, offering remarkable efficacy and improvements in patient quality of life. As research advances, the exploration of novel targets, extended-release formulations, and combination therapies hold promise for more. Newer medications targeting the CGRP pathway have been developed for migraine treatment. These include both oral and injectable formulations, such as Ubrogepant (oral CGRP receptor antagonist), Rimegepant (oral CGRP receptor antagonist), and Erenumab (CGRP monoclonal antibody) (Chiang and Schwedt 2020). A variety of CGRP antagonists have been developed to address both the acute and preventive treatment of migraine headaches. Monoclonal antibodies, targeting either the CGRP receptor or ligand, are administered periodically to prevent migraine.

Additionally, small-molecule CGRP antagonists, known as "gepants," are available in oral or intranasal formulations, with specific agents designed for either acute treatment or migraine prevention (Dos Santos and da Silva 2022). In December 2019, Ubrogepant obtained approval from the US Food and Drug Administration (FDA) for the treatment of acute migraine in adults (Hutchinson et al. 2021). Administered as a sole oral dose of 75 mg, Rimegepant has demonstrated effectiveness. Notably, it has shown efficacy in individuals who had an inadequate response to triptans previously. Furthermore, Rimegepant has received approval from the US Food and Drug Administration (FDA) for the prevention of migraine when administered at a dosage of 75 mg once every other day (Croop et al. 2021).

The safety of using a CGRP antagonist within the 2–4-h window of a triptan or ergotamine agent is not firmly established. There is a need for additional long-term data on CGRP antagonists to delineate their safety profile and tolerability more precisely. CGRP antagonists are frequently employed for patients who exhibit an inadequate response to or have contraindications (e.g., coronary artery disease) for triptans and/or antiemetics. Among the options for acute migraine management is Ubrogepant, which is administered as a single oral tablet at either 50 or 100 mg (Dodick et al. 2019). The dose may be repeated after 2 h, with a maximum daily dosage of 200 mg. Recently developed monoclonal antibodies targeting CGRP or its receptors have shown promising results in clinical trials. These drugs represent a paradigm shift in migraine treatment, offering a more targeted and potentially safer approach. However, their long-term safety profile and cost-effectiveness remain subjects of ongoing research. Administered as a single intranasal spray at 10 mg, Zavegepant offers a swift absorption and onset of action (Martirosov et al. 2023). Nasal delivery of a CGRP-receptor antagonist presents a rapid effect, making it a favorable choice for patients experiencing nausea and/or vomiting and who are unable to tolerate oral alternatives. The US Food and Drug Administration (FDA) approved Zavegepant for the treatment of acute migraine in 2023 (Larik et al. 2023).

As our understanding of migraine pathophysiology expands, new and more targeted interventions, such as CGRP inhibitors, offer hope for enhanced efficacy and improved patient outcomes. However, addressing the challenges associated with acute migraine management requires a holistic approach, considering both the individual needs of patients and the broader landscape of healthcare delivery. The ongoing pursuit of innovative solutions holds the promise of further advancing the field and alleviating the burden of migraine on individuals and society.

11.3.2 Preventive Migraine Medications

11.3.2.1 Beta-Blockers

Beta-blockers, traditionally employed for cardiovascular conditions, have found a niche in migraine prevention (Fumal and Schoenen 2008). Primarily acting by antagonizing beta-adrenergic receptors, they modulate neurotransmitter release, vascular tone, and central pain processing. Propranolol, metoprolol, and timolol are commonly prescribed members of this class. Propranolol, a non-selective beta-blocker, stands as a prototypical agent extensively studied for migraine prophylaxis (Dakhale et al. 2019). Clinical trials have consistently demonstrated its efficacy in reducing migraine frequency, severity, and duration. The benefits extend to both episodic and chronic migraine sufferers. Numerous clinical trials and real-world studies have attested to the efficacy of beta-blockers in reducing migraine frequency and severity. Propranolol has demonstrated significant benefit, leading to its approval for migraine prevention (Dakhale et al. 2019). Metoprolol and timolol also exhibit efficacy, with considerations for individual patient characteristics influencing the choice of agent. Beta-blockers are generally well-tolerated, but caution is advised in patients with cardiovascular comorbidities (Danesh and Gottschalk 2019). Dosing, titration, and potential side effects, such as fatigue and sexual dysfunction, require careful consideration. Special attention is warranted in specific populations, such as pregnant women, where the benefits must be weighed against potential risks.

The choice among these agents often hinges on individual patient characteristics, including comorbidities and tolerability. Initiating beta-blocker therapy involves a process of titration guided by clinical response and potential side effects. Blood pressure, heart rate, and symptoms such as fatigue or dizziness merit close monitoring during this association phase. Regular follow-ups ensure ongoing efficacy and the absence of adverse effects (Danesh and Gottschalk 2019). Certain contraindications, such as heart failure or severe asthma, limit the use of beta-blockers. Special populations, including pregnant women, necessitate a thoughtful risk-benefit analysis. In such cases, alternative migraine prophylactic agents may be considered. Beta-blockers stand as valuable additions to the arsenal of migraine prophylactic agents. Their diverse mechanisms of action, encompassing both central and peripheral effects, underline their multifaceted impact on migraine pathophysiology. While propranolol has garnered significant attention, the broader class of beta-blockers offers flexibility in tailoring treatment to individual patient profiles. The ongoing scientific exploration of their nuanced actions and considerations in clinical application positions beta-blockers as pivotal players in the evolving landscape of migraine management.

11.3.2.2 Antidepressants

Tricyclic antidepressants, including amitriptyline and nortriptyline, have demonstrated efficacy in migraine prevention. While their precise mechanism in migraine remains unclear, modulation of serotonin and norepinephrine reuptake, as well as anti-inflammatory effects, are proposed mechanisms (Burch 2021). Additionally, SSRIs and SNRIs, which are commonly used for mood disorders, have shown promise in certain migraine populations. While the exact mechanisms are intricate, the modulation of serotonin levels and central pain pathways is postulated. Fluoxetine and venlafaxine are among the antidepressants explored in this context. Antidepressants offer versatility in treating comorbid depression or anxiety, common in individuals with migraine. Amitriptyline, with its long-established use, is often considered a first-line choice (Argoff 2011). Individual patient profiles, potential side effects, and the gradual titration of doses are integral aspects of prescribing antidepressants for migraine prevention.

11.3.2.3 Antiepileptic Drugs

Antiepileptic drugs, known for their neuro-modulatory effects, have emerged as effective preventive agents for migraine (Diener et al. 2015). Valproic acid and topiramate have garnered attention for their ability to modulate neurotransmitters, stabilize neuronal membranes, and reduce cortical excitability. Clinical trials and meta-analyses support the use of valproic acid and topiramate in migraine prevention (Diener et al. 2015). Their dosages, however, require careful consideration, balancing efficacy with potential adverse effects. Individual patient characteristics, including comorbidities and reproductive considerations, play a pivotal role in drug selection. Beyond valproic acid and topiramate, ongoing research explores novel antiepileptic drugs, such as gabapentin and lamotrigine, for their potential in migraine prevention (Johannessen Landmark 2008; Löscher and Schmidt 2006). The intricate interplay between antiepileptic mechanisms and migraine pathophysiology unfolds as researchers delve into the molecular underpinnings.

11.3.2.4 Calcium Channel Blockers (CCBs)

CCBs primarily recognized for their cardiovascular applications have emerged as a class of medications with potential efficacy in migraine management (Zobdeh et al. 2021). This chapter delves into the intricate pharmacology of CCBs, their proposed mechanisms of action in mitigating migraines, and the current state of scientific understanding regarding their role in migraine prophylaxis. Calcium channels play a crucial role in neurotransmitter release, neuronal excitability, and vascular smooth muscle contraction. In migraine pathophysiology, abnormal calcium signaling contributes to cortical spreading depression, trigeminovascular activation, and the subsequent release of vasoactive neuropeptides, including calcitonin gene-related peptide (CGRP) (Close et al. 2019). CCBs are classified into dihydropyridines (e.g., amlodipine), non-dihydropyridines (e.g., verapamil), and mixed-action CCBs (e.g., nimodipine). Each subclass exhibits specific pharmacological properties, including variations in selectivity for certain types of calcium channels. Dihydropyridines primarily target vascular smooth muscle calcium channels, leading to vasodilation. Non-dihydropyridines, such as verapamil, additionally impact cardiac calcium channels and exert central effects, potentially influencing migraine pathophysiology beyond vascular changes (Wang et al. 2021). Non-dihydropyridines, particularly verapamil, exhibit central effects by blocking calcium channels in neurons. This central action may modulate neurotransmitter release, including glutamate, implicated in cortical spreading depression, a phenomenon linked to migraine aura. Clinical studies evaluating the efficacy of dihydropyridine CCBs in migraine prophylaxis have shown mixed results (Joshi et al. 2021). While some trials suggest a reduction in migraine frequency, others report limited efficacy. Variability in patient populations, dosages, and study designs contributes to the heterogeneity in findings. Verapamil, a non-dihydropyridine CCB, has demonstrated efficacy in certain migraine populations, particularly those with chronic migraine and cluster headaches (Urits et al. 2020). Its combined vasodilatory and central effects make it a unique candidate for individuals with specific migraine phenotypes. The effectiveness of CCBs in migraine prevention is intricately linked to dosage, titration, and ongoing monitoring. Individual response variations, potential side effects, and considerations for coexisting medical conditions necessitate a personalized approach to dosing. CCBs are generally well-tolerated, but potential adverse effects include hypotension, peripheral edema, and constipation. Contraindications, such as heart block or severe heart failure, must be carefully considered, especially in the context of migraine patients with comorbid cardiovascular conditions.

Comparative studies between beta-blockers, antidepressants, antiepileptic, and CCB drugs offer insights into their relative efficacy, tolerability, and side effect profiles. Such analyses aid clinicians in tailoring preventive strategies to individual patient needs, considering not only the migraine profile but also comorbidities and lifestyle factors.

11.4 Currently Available Non-pharmacological Interventions

11.4.1 Lifestyle Modifications

11.4.1.1 Trigger Identification and Management

Anything that causes an intense migraine headache to develop upon exposure or withdrawal is considered a migraine trigger. A migraine attack can be caused by a variety of precipitating causes or triggers (Amery and Vandenbergh 1987). Determining the trigger factors may lead to a better understanding of the power, frequency, or duration of assaults. Many individuals have had a variety of trigger variables looked at and found. Two types of stimuli may be distinguished from each other: (a) behavioral stimuli, including stress, emotions, hunger, and sleep disturbances; (b) environmental stimuli, like strong light, perfumes, and atmospheric conditions; (c) dietary products, such as chocolate, wine, coffee, ice cream, aged cheese, and so forth; (d) substances, such as tyramine, aspartame, and nitrates; and (e) changes in hormone levels, among other things (Peroutka 2014). Measurable risk variables before the attack may make it possible to predict future migraine episodes with ease and accuracy. This option opens the door to the prospect of a new paradigm for preventative therapy, one that focuses on preventing attacks rather than treating symptoms after they happen (Giffin et al. 2003). Patients who get more than four or five headaches each month should determine and adjust their trigger factors. Moreover, stress is indeed a major player in the world of migraines, affecting roughly half to two-thirds of migraine sufferers. That significant temporal correlation between daily stress and migraine episodes speaks volumes about the crucial role stress plays in triggering these debilitating headaches. The exact mechanisms by which stress triggers migraines are complex and not fully understood. However, researchers believe it involves a chain reaction of events in the brain. Stress activates the fight-or-flight response, releasing stress hormones like cortisol and adrenaline. These hormones can cause changes in blood flow and neurotransmitter levels in the brain, potentially making it more susceptible to migraines. Changes in sleep patterns and disrupted circadian rhythms due to stress can also contribute to migraines (Holm et al. 1997). Stress may precede the migraine attack by a few days. Other potent triggers like fatigue, auditory, visual, olfactory etc. must be tracked and avoided (Martin and Behbehani 2001). Most of the trigger factors can be considered incident triggers and can even reside on the causal pathways to headache attacks (e.g., stress) (Houle et al. 2017). On the other hand, certain trigger candidates may not operate as trigger factors; alternatively, they could indicate premonitory symptoms. For instance, rather than being the causal pathway, the perception of sensitivity to strong smells or bright lights may actually be a part of the premonitory phase prior to the attack (Pavlovic et al. 2014). For management of migraine trigger factors, it is very essential to identify and modify the trigger factor (Fig. 11.2).

11.4.1.2 Dietary Changes

Tension-type headaches (TTHs) are indeed more prevalent than migraines, affecting around 3% of the global population compared to 1% for migraine. Research has indeed illustrated that a significant portion of people with migraines experience sensitivity to cheese, chocolate, and citrus fruits. The percentages you mentioned (19.2% for cheese, 18.2% for chocolate, and 11.1% for citrus fruits) are consistent with findings in the medical literature. The sensitivity to wine, particularly red wine, is also well-documented in migraine sufferers, with women appearing to be more susceptible than men. This could be due to hormonal factors or other differences in metabolism. Other potential triggers for some individuals include sausage, salami, MSG, milk, aspartame, soft drinks, ice cream, and nuts. However, the evidence for these is generally less robust compared to the ones you mentioned initially (Hauge et al. 2010). Individuals experiencing headache after exposure to a chemical have a high probability in developing spontaneous unprovoked migraine headaches also. A multitude of diets and dietary interventions have been suggested to improve symptoms. The most common researched dietary interventions are discussed below.

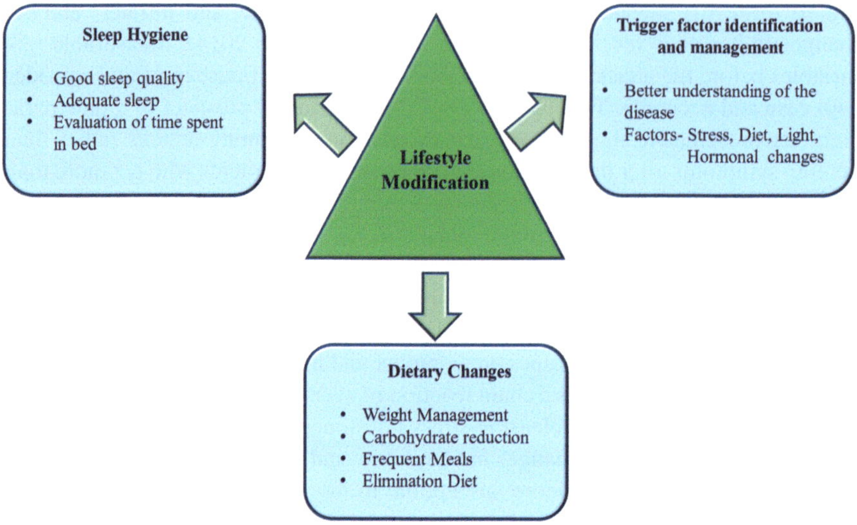

Fig. 11.2 Lifestyle modifications included trigger identification and management

Elimination Diets

Various dietary and nutritional treatments have been suggested to improve symptoms. The most studied dietary interventions are discussed below.

The classic elimination diet aims to reduce the recurrence and persistence of headaches by identifying foods that cause headaches and gradually reintroducing them into the diet while reducing all foods that are unlikely to be triggers. That's what I mean. Elimination diets are based on the idea that certain foods can trigger migraines in migraine-prone people (Houle and Turner 2013). IgE and IgG, and nonspecific immune cell responses to a series of preselected food samples have been measured in clinical laboratory blood tests to test the predictive power of immunoassays for individual food stimuli (Mullin et al. 2010).

Weight Management

There's a growing body of evidence suggesting a link between obesity and migraines, with obesity acting as a risk factor for both the development and severity of migraine. Here's a deeper dive into the connection. Studies have shown that obese individuals have two to three times more chances of experiencing migraine compared to those with a healthy weight. The risk of developing chronic migraine (15 or more headache days per month) is also significantly higher in people with obesity, as compared to those at a healthy weight. The exact mechanisms linking obesity and migraine are still being explored, but several factors might play a role in inflammation, which are linked to obesity chronic low-grade inflammation, that could contribute to migraine pain. Hormonal changes: fat cells produce hormones that can influence pain pathways and migraine susceptibility. Moreover, changes in brain structure and function-associated studies suggest that obesity may lead to alterations in brain regions involved in pain processing.

Frequent Meals

One well-known migraine cause is fasting. The longer the fast, the higher the chance of having a headache attack. As a result, it is advised to follow a regular eating schedule that involves eating roughly every 3 h and possibly five to six small meals throughout the day to maintain a healthy intake of calories and nutrients. When selecting meals, attention is paid to those that will satisfy hunger and maintain a stable blood sugar level (Song et al. 2016). This suggestion hasn't been fully put to the test through experimentation to see if it can prevent anything, nonetheless, it's likely that underlying triggers such as stressful situations interfere with regularly planned eating times (Peroutka 2014).

Carbohydrate Restriction

More rigorous research is needed to establish whether these diets offer actual preventive benefits beyond potential effects related to weight loss. The evidence for ketogenic diets (KD) in migraine prevention is limited to small observational studies and case reports. These lack the randomized controlled design necessary for definitive conclusions. Separating the effects of weight loss from the KD itself is challenging. Losing weight can improve migraines independently, making it difficult to determine whether the benefits observed in KD studies are due to the diet or the weight loss it often induces. KD requires significant patient commitment and lifestyle changes. Patients need proper screening and counselling to ensure they can adhere to the strict dietary restrictions and potential side effects.

11.4.1.3 Sleep Hygiene

Sleep hygiene refers to the setup and activities that facilitate regular and efficient sleep. This covers routine wake-up and bedtime schedules. The total amount of time spent in bed is equivalent to the amount of time needed for stable, sufficient sleep. Minimizing caffeine, alcohol, and other substances may help to ensure sleep quality. Moreover, the improved dietary habits, physical activity, and surrounding factors that imparts quality sound sleep must be encouraged (Calhoun and Ford 2007). It is possible to have good sleep quality by living according to sleep hygiene instructions. Someone younger may experience a headache or migraine due to inadequate sleep and their discomfort or suffering is mistakenly classified as a headache rather than insufficient sleep. In addition, adolescents and teens are particularly likely to experience the syndrome of delayed sleep phases and other sleep problems brought on by social factors (Bruni et al. 1999).

11.4.2 Behavioral and Psychological Interventions

11.4.2.1 Cognitive Behavioral Therapy

Cognitive-behavioral therapy focuses on the cognitive, affective, and behavioral precipitants and components of headache (Holroyd et al. 2009). Cognitive behavioral therapy (CBT) has emerged as a powerful tool for managing migraine symptoms beyond reliance on pain medication. While traditionally linked to treating psychological disorders, CBT offers significant benefits for migraine sufferers in several ways, addressing triggers and behaviors. CBT helps patients identify specific triggers that can initiate or worsen migraines, such as stress, sleep disturbances, certain foods, or environmental factors. Also, it assists in recognizing unhelpful behaviors that may contribute to migraines, like medication overuse or avoidance of physical activity. Cognitive restructuring CBT focuses on reframing negative thought patterns associated with migraines. These patterns can amplify stress and anxiety, which in turn can exacerbate headache symptoms. By challenging and replacing these thoughts with more adaptive and realistic beliefs, patients can learn to manage stress more effectively and improve their coping mechanisms. Developing behavioral strategies and CBT equips the patients with practical tools and techniques to manage stress and migraine episodes. Examples include relaxation techniques like deep breathing and meditation, mindfulness exercises, and cognitive distraction strategies. By building a personalized toolkit of coping skills, patients gain a sense of control over their migraine and overall well-being (Seng and Holroyd 2010). The goal of CBT is to treat the psychological, stress-related, and lifestyle variables that lead to the onset of migraine. CBT improves a patient's quality of life and psychological well-being by teaching them stress management, relaxation techniques, and pain coping abilities. These skills help patients better control their migraine. Long-term advantages arise from incorporating CBT into migraine treatment since it gives patients more control over how they manage and lessen the effects of migraine attacks (Bae et al. 2021).

Another benefit of CBT is that it can effectively treat other illnesses such as anxiety, panic disorder, depression, and sleep disorders. All these are diseases that often occur together with migraines. Assertiveness training, improving coping and problem-solving skills, stress management, and other factors associated with headache onset, exacerbation, and maintenance can be helpful in cognitive behavioral therapy (CBT), for migraine or severe headache patients. The goal of behavioral techniques is to help patients identify and modify behaviors that may cause, worsen, or maintain undesirable or harmful conditions. This usually involves keeping a diary (paper or electronic) and using other behavioral techniques to identify and modify the target behavior. Keeping detailed records of daily headaches can help patients learn how their thoughts, behaviors, emotions, and lifestyle choices relate to their headache characteristics. Relaxation training and biofeedback are often included in CBT for migraines (Fig. 11.3) (Russo et al. 2012).

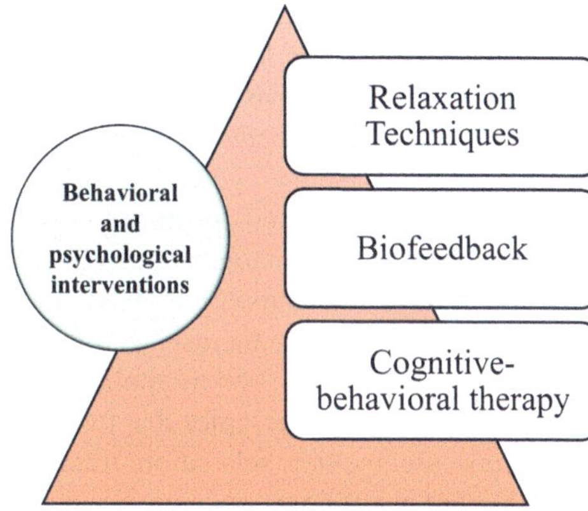

Fig. 11.3 A list of behavioral and psychological interventions

11.4.2.2 Biofeedback

Biofeedback provides a way for individuals to gain control over physiological processes known to contribute to migraines, such as muscle tension, blood flow, and brain activity. By using monitoring equipment to measure these processes and providing real-time feedback, individuals can learn to self-regulate them, potentially reducing the frequency, duration, and intensity of their migraine. Some benefits of biofeedback are that biofeedback offers a non-invasive and non-pharmacological approach to migraine management, which can be appealing to individuals seeking alternatives to medication. It can lead to increased self-awareness and empower individuals to actively participate in their own pain management. Artificial intelligence (AI) has the potential to enhance biofeedback by providing more personalized and actionable feedback, further improving patient care and treatment outcomes (Fig. 11.3).

A third approach, pulse volume biofeedback (BVP), is specifically designed for migraines. The patient is trained to restrict blood flow in the temporal artery. (Think of it as the non-drug counterpart of the abortion pill.) Another migraine-specific approach focuses on specific EEG parameters. Early research is promising, but the number of studies is too small to evaluate (Andrasik and Rime 2007).

11.4.2.3 Relaxation Techniques

Reducing sympathetic arousal and retraining the stress response are the primary objectives of relaxation therapies (Fig. 11.3). In order to elicit a relaxation response, patients are usually taught a range of methods for relaxation, such as stimulating the parasympathetic nervous system and lowering the sympathetic nervous system (Benson et al. 1975). Jacobson initially presented progressive muscle relaxation training (PMRT) in 1938 (Conrad and Roth 2007). Training in muscular awareness is used in this method. The patient is instructed to deliberately tighten and release certain muscle groups while observing opposing feelings. Through a homeostatic process, the muscles instinctively relax once the forceful contractions cease, which improves discrimination learning even more. Patients are advised to employ this technique on a regular basis and to do so whenever they feel their levels of stress or tension rising. Several methods of inducing the relaxation response that are frequently provided include guided imagery, cue-controlled relaxation, breathing through the diaphragm, self- or therapist-performed hypnosis, meditation, and any other approach that the patient finds advantageous. Several medical practitioners can teach sedative methods if they have the appropriate knowledge and education. Patients can also educate themselves on relaxation methods by employing paper, audio, or web materials (such as applications, iTunes, and YouTube). The majority of behavioral migraine therapies that psychologists and other mental health professionals provide include relaxation training. It takes consistent practice in the patient's regular setting to attain maximal efficacy and use the procedure as needed. Patients are usually encouraged to spend 20–30 min each day practicing a series of gradual relaxation exercises. Patients can eventually learn to incorporate basic relaxation techniques into their everyday activities by integrating daily rest intervals with them. Whether the majority of the training takes place in a work environment or outside of one, the benefits frequently correspond with practice, especially during the first month of acquiring the skill.

11.5 Advanced Interventional Procedures for Migraine Treatment

It affects a large proportion of the population, with a 12% prevalence in the United States, resulting in millions of primary care visits each year. Chronic migraine (CM) are defined as having at least 15 headache days per month, with at least 8 of those days satisfying migraine criteria with or without aura. Status migrainosus (SM) develops when a migraine attack lasts for more than 72 h and causes severe disability. Individuals with EM have less than 15 headache days per month, whereas those with CM have more intense, longer-lasting migraine and greater functional impairment.

Low-frequency episodic migraine (LFEM) and high-frequency episodic migraine (HFEM) are two more subtypes of EM that are further differentiated by a range of headache days per month; HFEM is commonly thought to lie between 8 and 14 or 10 and 14 headache days per month. Limitations remain despite notable progress in the pharmacological and neuromodulator treatments for migraine. Treatment options may be limited by coexisting disorders such as mental comorbidities, pregnancy, renal or hepatic impairment, cardiovascular diseases, or possible drug interactions. Remarkably, less research and interventions focus on the most refractory aspects of CM, with less attention going to the EM population. This bias in research focus highlights the need for more thorough knowledge and customized strategies for various migraine subtypes, recognizing the limitations of available treatments in certain patient groups.

11.5.1 Occipital Nerve Blocks

11.5.1.1 Greater Occipital Nerve

Among the peripheral nerves frequently targeted in injectable therapy for episodic migraine (EM) is the greater occipital nerve (GON). The GON, which receives its origin from the C2 nerve, provides feeling to the posterior scalp extending from the vertex to the occipital protuberance. It is frequently located next to the occipital artery, which is occasionally palpable or about 2 cm lateral to the occipital protuberance, however individual anatomy may differ. In order to perform a GON block, a solution is injected along the GON's pathway (Akerman et al. 2011). As for the ideal method for GON block preparation, there isn't much agreement. Many medical practitioners choose to use either bupivacaine 0.25–0.5% (which has a delayed onset but longer-lasting effects) or lidocaine 1–2% (which has a quicker onset but shorter duration), occasionally combining the two. Some practitioners also use a corticosteroid with the local anesthetic. However, when comparing the effectiveness of utilizing a corticosteroid with a local anesthetic alone, a randomized, double-blind controlled research found no additional benefit (Fig. 11.4) (Hepp et al. 2015).

11.5.1.2 Lesser Occipital Nerve

The C2 and C3 nerves give rise to the lesser occipital nerve (LON), which supplies feeling to the posterior scalp and extends laterally to the upper neck. The normal location of the LON injection site is 1–2 cm medially of the mastoid process. This injection usually uses a combination of lidocaine and bupivacaine, much like the procedure for the greater occipital nerve (GON). While there aren't many systematic studies that explicitly address LON blockades in episodic migraine (EM),

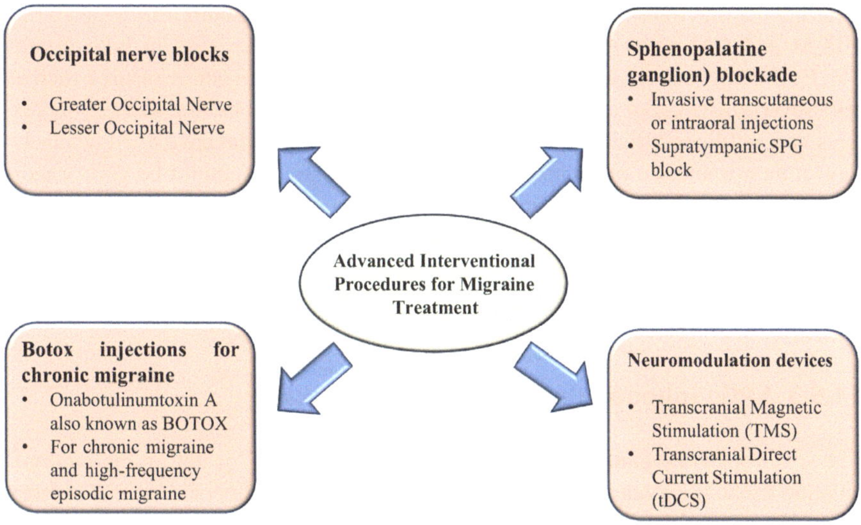

Fig. 11.4 Advanced interventional procedures for migraine treatment

practitioners frequently combine LON blockades with GON and other peripheral nerve blocks (PNBs) in their treatment plans. This integrated strategy attempts to treat several neural pathways linked to migraine without specific studies concentrating on LON blockades' effectiveness in EM (Plato and Whitt 2020).

11.5.2 Botox Injections Treatment for Chronic Migraine (CM)

Known for its effectiveness in treating disorders involving excessive muscular contraction, botulinum toxin was approved by the US Food and Drug Administration (FDA) in 2010 for the treatment of chronic migraine using the preparation onabotulinumtoxin A, also known as Botox. This approval was based on strong data from two noteworthy trials that demonstrated the medication's ability to relieve persistent migraine discomfort. Preliminary investigations into the use of onabotulinumtoxin A in patients suffering from episodic migraine (EM) did not yield strong evidence supporting its use in this specific subgroup. On the other hand, because chronic migraine (CM) and high-frequency episodic migraine (HFEM) have similar degrees of disability, HFEM has received more attention recently. In both the CM and HFEM populations, a prospective trial evaluated the immediate and long-term advantages of onabotulinumtoxin A (Silberstein et al. 2000). Those with 8–14 headache days per month were classified as having HFEM, which was described as having CM-like treatment and injection frequency using the PREEMPT protocol. Observations made 6, 12, 18, and 24 months after the baseline produced some interesting results. By the time the 6-month period rolled around, HFEM patients had reduced their monthly headache days from 13.4 to 9.4, which translated into a 34.2% drop in migraine incidence (Binder et al. 2000). Additionally, the number of analgesic pills taken per month dropped from 21.4 to 14.4. Patients with HFEM saw a further decline to 8.7 headache days per month at the 12-month follow-up. When HFEM and CM were compared, headache frequency decreased in both groups (44.7% for CM and 41.2% for HFEM), as were other significant measures such as pain severity, the use of analgesic and preventative medications, and MIDAS scores. This new unblinded trial suggests that those with HFEM, who have 8–14 headache days a month and are not responsive to existing treatments, could be good candidates for onabotulinumtoxin A as a preventive medication. However, more investigation and controlled trials are necessary to confirm these results and provide more precise instructions for using it to this patient population (Fig. 11.4) (Diener et al. 2010).

11.5.3 SPG (Sphenopalatine Ganglion) Blockade

The sphenopalatine ganglion (SPG) is a group of neurons located in the pterygopalatine fossa. It includes parasympathetic and sympathetic fibers as well as sensory fibers innervating the V2 branch. SPG blocking has long been a mainstay in the treatment of headache disorders, particularly in situations of cluster headaches

(Khan et al. 2014). Research has, however, also investigated its use in migraine, mostly for acute care. There are several ways to block SPG, from less invasive procedures to invasive transcutaneous or intraoral injections. A local anesthetic is applied to the mucosa of the lateral posterior nasal cavity during the less invasive procedures. Some patients may self-administer this technique at home by slowly inserting a cotton-tipped applicator soaked in lidocaine or a syringe filled with the drug into their nostrils. As an alternative, healthcare professionals can use flexible catheters to do a minimally invasive SPG blockage in-office (Maizels et al. 1996).

Most studies have investigated how well SPG blocking works to cure acute migraine. Studies using intranasal lidocaine that were placebo-controlled showed that patients treated with lidocaine had headache relief more quickly than those given with a placebo; nonetheless, many of them had headaches again after the first episode of relief. When compared to individuals who received a placebo, a sizable portion of patients in a follow-up research who received intranasal lidocaine at home experienced headache alleviation within 15 min (Binfalah et al. 2018). Furthermore, a substantial majority of patients experienced headache freedom following a single lidocaine-assisted trans nasal SPG block treatment, according to an uncontrolled retrospective investigation. Furthermore, a transcutaneous method for treating status migraineurs was investigated in a recent study. In this method, patients had supratympanic SPG block using a dexamethasone and ropivacaine combination. Within 30 min, the process significantly reduced the intensity of the migraine, especially in those situations where previous therapies for acute migraine had not worked. SPG blocking, all things considered, shows promise in reducing migraine symptoms, especially when they are acute. To confirm its efficacy, investigate its long-term effects, and standardize its application techniques, more research is needed (Fig. 11.4) (Khan et al. 2014).

11.5.4 Neuromodulation Devices: TMS (Transcranial Magnetic Stimulation), tDCS (Transcranial Direct Current Stimulation)

11.5.4.1 Transcranial Magnetic Stimulation (TMS)

TMS targets and modifies neuronal activity by applying magnetic pulses to specific areas of the brain. It is mostly used in repetitive TMS (rTMS), a technique in which a coil is applied to the scalp and creates magnetic fields that cause electrical currents to be induced in the brain. TMS targets aberrant brain activity associated with migraine and attempts to modify cortical excitability (Hallett 2007). TMS treatment for acute and preventive migraine has shown potential in studies. When a migraine occurs for the first time, the administration of TMS may be able to reduce the intensity of the pain and relieve symptoms in acute situations. Regular TMS sessions may also eventually reduce the frequency and intensity of migraine as a preventative measure. However, individual responses may vary and the best procedure for TMS therapy of migraine is still unknown (Naro et al. 2016).

11.5.4.2 Transcranial Direct Current Stimulation (tDCS)

This modifies the activity of neurons in specific areas of the brain by passing a low-intensity electrical current through electrodes placed on the head. The goal of this method is to change neural networks, and cortical excitability, that are involved in the pathophysiology of migraine. tDCS for migraine has been studied and results have been encouraging in both acute and preventive contexts (Elder and Taylor 2014). When used as an acute therapy, transcranial magnetic stimulation (tDCS) can relieve pain and associated symptoms during a migraine attack. Frequent tDCS sessions have shown promise in preventing migraines by reducing their frequency and intensity. However, individual responses vary, similar to TMS, and further research is needed to develop standardized procedures and maximize efficacy (Liu et al. 2017). Thus, TMS and tDCS are two examples of neuromodulator therapies that may offer a non-pharmacological alternative for the acute therapy and prevention of various headache conditions. In addition to their minimal adverse effect profile, these technologies may prove advantageous for individuals who show resistance to conventional drugs or who are not suitable for oral pharmacological care (Fig. 11.4) (Baloh 1997).

11.6 Integrative Approaches

11.6.1 Acupuncture for Migraine Relief

Acupuncture is a healing technique that is rooted in Traditional Chinese Medicine (TCM) and has a rich history dating back to 100 BC (Linde et al. 2005). The practice is influenced by Confucianism and Taoism, and over centuries, its techniques have evolved with varying governmental support. However, under Communist rule in China in 1949, there was a notable resurgence of acupuncture, leading to the formation of TCM, which is a collection of different acupuncture practices. Acupuncture was gradually introduced to Western nations, but it gained traction in the USA during the Nixon administration (Li et al. 2012). The turning point for acupuncture's acceptance in the US medical community came in 1998, when the National Institutes of Health recognized its efficacy. Specifically, it was acknowledged as effective in relieving postoperative dental pain and treating nausea and vomiting induced by postoperative and chemotherapy procedures (Da Silva 2015). This endorsement paved the way for increased credibility and inclusion of acupuncture into comprehensive treatment plans at academic hospitals across the United States. Since then, acupuncture has gained popularity nationwide, with many academic hospitals integrating it as part of a holistic approach to patient care (Bäcker et al. 2008). Additionally, insurance companies have become more inclined to cover the costs of acupuncture treatments. As a result, an estimated 1.5% of the US population has sought acupuncture at least once, often as a complementary therapy alongside established standard care regimens. This widespread acceptance and incorporation into mainstream healthcare highlight the evolving landscape of acupuncture as a respected and recognized form of therapy in the western medical context (Li et al. 2012).

Migraine pose a formidable challenge in terms of treatment, often accompanied by unwanted side effects from conventional options, especially for those with chronic migraine (CM) who may experience exacerbated headache pain. While drug therapy remains the primary approach, the limitations and side effects have prompted exploration of alternative treatments, with acupuncture emerging as a promising adjunct to migraine management (Zhao et al. 2021). Numerous studies indicate that acupuncture demonstrates measurable effects on both the duration and frequency of migraine attacks, positioning it as a safe and beneficial option for certain migraine patients.

Despite the positive findings, there is a consensus in the medical community that further research is essential to solidify the role of acupuncture in migraine treatment. Larger scale randomized controlled trials (RCTs) are warranted to provide more robust evidence and enhance the clinical support for incorporating acupuncture into standard migraine care. Although existing studies have investigated the effects of acupuncture on migraines, ongoing research is needed to ensure the positive outcomes observed are consistent and applicable across diverse patient populations. Addressing this need for comprehensive investigation, a recently developed Acupuncture Expectancy Scale holds promise in refining future research efforts. Incorporating this scale in studies may contribute to a more nuanced understanding of patient expectations and outcomes, related to acupuncture treatment for migraine (Song et al. 2022). Additionally, future research could benefit from regular follow-up treatments and interviews to gather insights into the sustained effects of acupuncture over time. Long-term follow-up studies may offer valuable information about the enduring benefits and potential for acupuncture to provide lasting relief for migraine sufferers. In summary, while acupuncture shows promise in alleviating migraines, ongoing research and the development of standardized assessment tools are crucial for further establishing its clinical efficacy and optimizing its integration into migraine treatment protocols (Chen et al. 2022).

11.6.2 Herbal Supplements and Nutraceuticals

Individuals who experience migraine are often frustrated by the inefficacy, high costs, and toxic action associated with conventional therapeutic drugs (D'Onofrio et al. 2017). As a result, they are increasingly exploring alternative avenues for relief, such as nutraceuticals. Coined by DeFelice in 1979, the term "nutraceutical" blends "nutrition" and "pharmaceutical" and refers to foods or components of diet which offer pharmacological or well-being advantages, including the prevention and treatment of diseases. Nutraceuticals fall within the category of biological-based therapies in the realm of complementary and alternative medicine (CAM) (Kalra 2003). There should have to be a surge in interest in nutraceuticals for migraine relief, as evidenced by the findings from the 2007 NHIS (National Health Interview Study) (Rajapakse and Pringsheim 2016). The study revealed that among mature persons dealing with migraine or extreme headaches, 49.5% reported using CAM, including nutraceuticals, compared to 33.9% among those without migraine or headaches. This shift towards CAM, particularly nutraceuticals, highlights a significant movement away from traditional pharmaceutical approaches, as individuals seek alternatives that are perceived as more natural and potentially associated with fewer adverse effects. As more patients incorporate nutraceuticals into their migraine management strategies, it is important to conduct comprehensive research to evaluate the efficacy, safety, and specific benefits of these approaches. This will provide a more informed foundation for their integration into mainstream migraine care (D'Onofrio et al. 2017).

A common neurological condition that primarily affects women who are fertile, migraine peak in frequency in the fourth decade and afflict about 25% of people in a year. The frequency and intensity of migraine attacks are significantly influenced by hormonal fluctuations associated with several reproductive events, including menarche, the menstruation, gestation, menopause, and use of oral contraceptives and hormone replacement therapy (Rajapakse and Pringsheim 2016).

11.6.2.1 Riboflavin

As a member of the vitamin B2 family, riboflavin gets its name from the combination of the ring-moiety of "flavin" and "ribitol," the reduced sugar form that gives the vitamin its structure. When this moiety oxidizes, it provides the urine of those who consume this vitamin its distinctive yellow color. It is thought that riboflavin alters the irregularities in cerebral information processing and deficiencies in mitochondrial energy metabolism that may contribute to the pathophysiology of migraine. As a preventative medication, it might aid in migraine management (Maizels et al. 2004).

11.6.2.2 Coenzyme Q10

Coenzyme Q10 (coQ10) serves as a cofactor in the ETC (Electron Transport Chain). CoQ10 safeguards against mitochondrial collapse or degeneration by maintaining proper electron transport down the respiratory chain. It also prevents the opening of mitochondrial membrane transition pores, thereby hindering the translocation of large molecules that could lead to mitochondrial functional decline. Additionally, CoQ10 counteracts endothelial dysfunction by stimulating the release of nitric oxide and exhibits anti-inflammatory effects (Taylor 2009).

11.6.2.3 Magnesium

Magnesium is proposed to influence the threshold for migraine occurrence through various mechanisms. These include neuroinflammatory blockades, calcium channel blocking effects, modulation of *N*-methyl-D-aspartate (NMDA)-glutamate receptors, and effects on glutamate and nitric oxide synthesis, release, and activity (Maizels et al. 2004). It also affects serotonin receptor affinities and activities, as well as endogenous hormone regulation, thereby regulating various vascular and neuronal mechanisms.

11.6.2.4 Feverfew

Feverfew (*Tanacetum parthenium*) is a herbal supplement that comprises dried chrysanthemum leaves. It exhibits a 400% variation in dosage strength of the active ingredient parthenolide. Additionally, it contains melatonin and chrysanthenyl acetate. It is unclear, however, whether parthenolide is the major active ingredient in available feverfew preparations (Jin et al. 2008). Clinical studies on feverfew's efficacy and safety in migraine prevention have yielded insufficient evidence to suggest its superiority over a placebo. This uncertainty highlights the need for further research to establish the role and effectiveness of feverfew in managing migraine (Heptinstall et al. 1992).

11.6.3 Yoga and Mindfulness-Based Interventions

Yoga is a form of complementary and alternative medicine that has gained popularity worldwide. It is rooted in ancient Indian practices and is based on mind-body medicine principles (Anheyer et al. 2020). Research has shown that yoga can positively impact various measures related to migraine, including headache frequency, intensity, attack duration, medication scores, and the overall nature of pain experienced by individuals. The main focus of one study was to estimate the effectiveness of yoga as the adjuvant therapy for individuals with episodic migraine (Kumar et al. 2020). Yoga's benefits in managing migraine are attributed to its potential to improve the dysfunctional hypothalamic-pituitary-adrenal system, which regulates the body's autonomic functions. By positively influencing this system, yoga may enhance the overall quality of life for people with migraine (Anheyer et al. 2020). Additionally, there is a reported reduction in headache frequency among those who incorporate yoga into their therapeutic regimen. Yoga's holistic approach, which combines physical postures, breath control, and mindfulness, likely plays a role in addressing both the physiological and psychological aspects of migraine, making it a promising adjuvant therapy (Sharma et al. 2018). In summary, the positive effects of yoga on migraine measures suggest that it could be a valuable and holistic complementary approach for individuals with episodic migraine. The potential improvement in the hypothalamic-pituitary-adrenal system and overall quality of life further support the inclusion of yoga as part of a comprehensive migraine management strategy (Sharma et al. 2020).

Yoga is a comprehensive practice that includes breathing techniques, meditation, mindfulness, and physical development through postures and routines (Mooventhan and Nivethitha 2017). It originated in ancient India and has been found to be beneficial in managing tension-type headaches (TTH) and migraine. Controlled trials have shown positive outcomes, indicating that yoga can be a valuable non-drug intervention for headache management (Kumar et al. 2020). However, it is important to exercise caution in advocating yoga as the first-line treatment for headaches until further studies are conducted. While existing research has provided encouraging results, more rigorous studies are needed to establish a comprehensive understanding of yoga's efficacy in treating different types of headaches (Yardi 2001).

Further research will help determine the specific conditions and populations for which yoga may be most beneficial, ultimately leading to informed integration into headache treatment plans (Shashikiran et al. 2022).

11.7 Patient-Specific Considerations

11.7.1 Pediatric Migraine Treatment

According to epidemiological research, the prevalence of headaches in youngsters varies greatly, ranging from 5.9% to 82%. Even in childhood and adolescence, migraine is the most prevalent and severely incapacitating type of primary headache in children (Patniyot and Gelfand 2016). There is a discernible age-related variation in the average prevalence of pediatric migraine, which begins around 3% in younger children and rises to about 20% in teenagers. Chronic migraine, which are characterized by headaches occurring more than 15 days per month, is a major societal concern that affects 0.6–1.8% of children and adolescents (Sartory et al. 1998).

11.7.1.1 Dopamine Receptor Antagonists

Dopamine receptor antagonists are commonly used in emergency department (ED) settings for the immediate management of migraine (Teleanu et al. 2016). This class includes metoclopramide, butyrophenones, and phenothiazines such as prochlorperazine, chlorpromazine, and promethazine (Peroutka 1997). Although these medications are useful in treating migraine symptoms like nausea and vomiting, they can also cause side effects like akathisia and dystonic reactions. Additionally, because these medications have the potential to prolong QTc intervals, it is important to evaluate children's ECGs before administering them (Lewis et al. 2004).

11.7.1.2 Beta-Blockers and Calcium-Channel Blockers

Beta-blockers, particularly propranolol, have been historically used for migraine prophylaxis in children. Propranolol a non-selective beta adrenoceptor antagonist that blocks b1 and b2 receptors. Some studies have shown its effectiveness, but follow-up research has yielded more controversial results. Despite being used for over 50 years in treating migraine in children, its efficacy remains a subject of ongoing discussion (Rao et al. 2000).

Calcium channel blockers have been extensively studied for migraine prophylaxis in adults. However, there is a lack of substantial evidence for their use in children and adolescents, except for flunarizine. Flunarizine, a non-selective calcium channel blocker with specific effects on the cerebrovascular system, has demonstrated effectiveness in preventing migraine. However, the broader evidence for calcium-channel blockers in this population remains limited. Overall, managing pediatric migraines involves careful consideration of acute and prophylactic treatment options, taking into account potential side effects and the evolving nature of research in this field (Sorge et al. 1988).

11.7.2 Migraine Treatment During Pregnancy and Lactation

One interesting finding is that 80–90% of pregnant women who suffer from migraine report gradually improving symptoms and some even report total remission (Cassina et al. 2010). This encouraging tendency, however, is less common among people who get aura-related migraine or migraine unrelated to the menstrual cycle. Interestingly, 4–8% of migraineurs may have an aggravation of their symptoms when pregnant, and occasionally migraine may appear for the first time during pregnancy, usually in the first trimester (Hutchinson et al. 2013). More than half of women discover that their migraine return to pre-pregnancy levels within a month of giving birth, despite improvements throughout pregnancy (Jügens et al. 2009). Stress and lack of sleep are two further variables that could precipitate or exacerbate headaches during the puerperium, the postpartum phase (Lewis et al. 2004). Although there is conflicting evidence on whether breastfeeding provides protective effects against migraine recurrence after delivery, the interplay of hormonal changes, reproductive events, and external factors during this complex period underscores the dynamic nature of migraine experiences in women.

Nonsteroidal anti-inflammatory drugs (NSAIDs) are commonly used to alleviate migraine symptoms by regulating prostaglandins. Prostaglandins play important roles in various reproductive processes such as ovulation, implantation, cervical ripening, and uterine contractility (Lucas 2009). Therefore, the use of NSAIDs during pregnancy requires careful consideration due to their impact on these processes. Paracetamol (acetaminophen) is considered a safe treatment option for migraine during all stages of pregnancy. However, aspirin (acetylsalicylic acid) is often consumed without a prescription and with less detailed product labeling. Despite some general cautions related to upper gastric intolerability, asthma, and Reye's syndrome, there is no widespread public warning against the use of aspirin during pregnancy. It's worth noting that Reye's syndrome has only been reported in association with viral illness and not specifically linked to migraine. Therefore, clear and comprehensive information dissemination is crucial regarding the use of medications, especially during pregnancy. The divergence in the perceived safety of different medications underlines the importance of informed decision-making and effective communication between healthcare providers and pregnant individuals to make sure of the well-being of both the mother and the developing fetus.

Despite its accessibility, large exposures to codeine over extended periods, even when used for many years, have not been associated with any reported teratogenic hazard to either the fetus or the mother. This observation underscores the importance of weighing the potential risks and benefits of opioid use during pregnancy, emphasizing the need for cautious and informed decision-making by healthcare providers and pregnant individuals. While codeine may be accessible and has not shown teratogenic effects, its use should still be managed judiciously, especially during pregnancy, to make sure of the safety of both the mother and the growing fetus.

11.7.3 Geriatric Considerations

The common neurological condition of migraine has not been sufficiently studied in the elderly, with little attention paid to its features, attack patterns in relation to past migraine history, factors influencing severity, and the effects of coexisting pathologies, especially in menopausal women and women receiving hormone replacement therapy. The thorough examination of pain location, quality, and related symptoms like phonophobia and photophobia in this population is frequently overlooked in the research that is currently in publication. In spite of this research vacuum, studies conducted recently on older persons show that the prevalence of these migraine symptoms is still similar to what has been found in younger groups, according to earlier epidemiological studies (Wijeratne et al. 2019).

Migraine can be difficult to manage in elderly patients due to changes in their body with age and other health issues they may have. Although migraine are usually associated with younger patients, they can also occur in older adults. Elderly patients may experience different symptoms, such as non-pulsatile headaches, and accompanying neurological symptoms. It can be challenging to diagnose migraine in the elderly because there are other medical conditions that have similar symptoms, such as vascular disorders or neurodegenerative diseases. When treating migraine in the elderly, it is necessary to consider their overall health, existing medications, and potential drug interactions. In addition, the management of migraine in the elderly requires a team approach, involving neurologists, geriatricians, and other healthcare professionals. Overall, proper care for elderly patients with migraine requires a thorough understanding of their medical history, a careful symptom evaluation, and tailored therapeutic strategies that balance effectiveness with the potential risks associated with age-related factors (Riggins and Ehrlich 2022).

11.8 Challenges and Future Directions

While significant strides have been made in the management of migraine pain, several challenges persist, necessitating a forward-looking perspective to drive future advancements in the field. One prominent challenge lies in the limited efficacy and tolerability of existing interventions for certain patient populations. The variability in individual response to medications underscores the need for more personalized and targeted therapeutic approaches.

Moreover, the potential for medication overuse headache (MOH), particularly with acute treatments, poses a substantial hurdle. Striking a delicate balance between providing relief and avoiding the risk of MOH remains a clinical conundrum. Addressing these challenges calls for a deeper understanding of the underlying pathophysiology of migraines, which remains incompletely elucidated.

The future directions in the management of migraine pain revolve around embracing emerging therapies and innovative research avenues. The advent of calcitonin gene-related peptide (CGRP) inhibitors has marked a transformative phase, signaling the potential for more targeted and specific pharmacological interventions. Continued exploration of CGRP pathway modulation and the development of novel classes of drugs offer hope for enhanced efficacy and reduced side effects. Additionally, the paradigm shift towards personalized medicine holds immense promise. Genetic and environmental factors influencing migraine susceptibility and response to treatments are becoming increasingly apparent. Integrating this knowledge into clinical practice can pave the way for tailored interventions, optimizing outcomes for individuals with migraine.

Collaboration between researchers, clinicians, and pharmaceutical industries is paramount for overcoming current challenges and propelling the field forward. By addressing limitations, embracing emerging therapies, and harnessing the power of personalized medicine, the future of migraine pain management holds the potential to provide more effective, tolerable, and individualized care, ultimately improving the lives of millions affected by migraine.

11.9 Conclusion

The landscape of migraine treatment is evolving, propelled by advancements in our understanding of the disorder and innovative therapeutic approaches. Acknowledging the limitations of current interventions and embracing emerging therapies and personalized medicine positions the field on a trajectory toward more effective, patient-centric, and tailored migraine management strategies. As research continues to unravel the intricate facets of migraine pathophysiology, the promise of transforming challenges into opportunities for improved patient outcomes remains at the forefront of ongoing efforts in the field. In the exploration of migraine treatment, this chapter has traversed the spectrum of current interventions, shedding light on their efficacy, limitations, and the evolving landscape of therapeutic options. Key findings underscore the complexity of migraines as a neurological disorder, emphasizing the need for tailored and comprehensive approaches to address the multifaceted nature of the condition. The chapter has elucidated the challenges inherent in current interventions, including incomplete efficacy, the risk of Medication Overuse Headache (MOH), and limitations in certain populations. However, it also illuminates promising avenues, such as emerging therapies like CGRP inhibitors and the potential for personalized medicine to revolutionize migraine treatment. The future of migraine treatment holds immense promise, propelled by ongoing research, technological advancements, and a deeper understanding of the intricate mechanisms underlying the disorder. Continued exploration of the genetic basis of migraine is anticipated to unveil new discoveries and potential biomarkers. Identifying genetic markers associated with susceptibility and treatment response will contribute to more targeted and precise interventions. Additionally, technological innovations in neuromodulation advancements in neuromodulation techniques, including transcranial magnetic stimulation (TMS) and transcranial direct current stimulation (tDCS), offer exciting prospects for non-pharmacological migraine management. As technology evolves, these interventions may become more accessible and refined. The integration of personalized medicine into clinical practice prompts reflection on ethical considerations. Striking a balance between harnessing the benefits of genetic and molecular insights and safeguarding patient privacy and equitable access will be paramount. Patient-reported outcomes and quality of life future research should prioritize the evaluation of patient-reported outcomes and the impact of interventions on the overall quality of life. Understanding the holistic effects of treatments beyond headache reduction will contribute to a more comprehensive assessment of therapeutic efficacy. Global accessibility and equity in the advancement of migraine treatment should not be limited to specific demographics or geographic regions. Ensuring global accessibility to innovative therapies and personalized interventions requires a concerted effort to address disparities in healthcare resources.

References

Abu-Arafeh I et al (2010) Prevalence of headache and migraine in children and adolescents: a systematic review of population-based studies. Dev Med Child Neurol 52(12):1088–1097

Akerman S, Holland PR, Goadsby PJ (2011) Diencephalic and brainstem mechanisms in migraine. Nat Rev Neurosci 12(10):570–584

Amery WK, Vandenbergh V (1987) What can precipitating factors teach us about the pathogenesis of migraine? Headache 27(3):146–150

Amin FM et al (2013) Magnetic resonance angiography of intracranial and extracranial arteries in patients with spontaneous migraine without aura: a cross-sectional study. Lancet Neurol 12(5):454–461

Aminoshariae A, Kulild JC, Donaldson M (2016) Short-term use of nonsteroidal anti-inflammatory drugs and adverse effects: an updated systematic review. J Am Dent Assoc 147(2):98–110

Andrasik F, Rime C (2007) Can behavioural therapy influence neuromodulation? Neurol Sci 28(Suppl 2):S124–S129

Anheyer D et al (2020) Yoga for treating headaches: a systematic review and meta-analysis. J Gen Intern Med 35:846–854

Argoff C (2011) Mechanisms of pain transmission and pharmacologic management. Curr Med Res Opin 27(10):2019–2031

Bäcker M et al (2008) Acupuncture in migraine: investigation of autonomic effects. Clin J Pain 24(2):106–115

Bae JY et al (2021) Cognitive behavioral therapy for migraine headache: a systematic review and meta-analysis. Medicina (Kaunas) 58(1):44

Baloh RW (1997) Neurotology of migraine. Headache 37(10):615–621

Becker WJ (2015) Acute migraine treatment in adults. Headache 55(6):778–793

Benson H, Greenwood MM, Klemchuk H (1975) The relaxation response: psychophysiologic aspects and clinical applications. Int J Psychiatry Med 6(1–2):87–98

Binder WJ et al (2000) Botulinum toxin type A (BOTOX) for treatment of migraine headaches: an open-label study. Otolaryngol Head Neck Surg 123(6):669–676

Binfalah M et al (2018) Sphenopalatine ganglion block for the treatment of acute migraine headache. Pain Res Treat 2018:2516953

Blumenfeld AM et al (2011) Disability, HRQoL and resource use among chronic and episodic migraineurs: results from the International Burden of Migraine Study (IBMS). Cephalalgia 31(3):301–315

Bruni O, Galli F, Guidetti V (1999) Sleep hygiene and migraine in children and adolescents. Cephalalgia 19(Suppl 25):57–59

Burch R (2019) Migraine and tension-type headache: diagnosis and treatment. Med Clin North Am 103(2):215–233

Burch R (2021) Preventive migraine treatment. Continuum (Minneap Minn) 27(3):613–632

Burstein R, Noseda R, Borsook D (2015) Migraine: multiple processes, complex pathophysiology. J Neurosci 35(17):6619–6629

Buse DC et al (2013) Psychiatric comorbidities of episodic and chronic migraine. J Neurol 260(8):1960–1969

Buse DC et al (2016) Impact of migraine on the family: perspectives of people with migraine and their spouse/domestic partner in the CaMEO study. Mayo Clin Proc. https://doi.org/10.1016/j.mayocp.2016.02.013

Calhoun AH, Ford S (2007) Behavioral sleep modification may revert transformed migraine to episodic migraine. Headache 47(8):1178–1183

Cassina M et al (2010) Migraine therapy during pregnancy and lactation. Expert Opin Drug Saf 9(6):937–948

Chen Y et al (2022) Therapeutic applications and potential mechanisms of acupuncture in migraine: a literature review and perspectives. Front Neurosci 16:1022455

Chiang CC, Schwedt TJ (2020) Calcitonin gene-related peptide (CGRP)-targeted therapies as preventive and acute treatments for migraine—the monoclonal antibodies and gepants. Prog Brain Res 255:143–170

Close LN et al (2019) Cortical spreading depression as a site of origin for migraine: role of CGRP. Cephalalgia 39(3):428–434

Conrad A, Roth WT (2007) Muscle relaxation therapy for anxiety disorders: it works but how? J Anxiety Disord 21(3):243–264

Cristofoletti R, Dressman JB (2016) Bridging the gap between in vitro dissolution and the time course of ibuprofen-mediating pain relief. J Pharm Sci 105(12):3658–3667

Croop R et al (2021) Oral rimegepant for preventive treatment of migraine: a phase 2/3, randomised, double-blind, placebo-controlled trial. Lancet 397(10268):51–60

D'Onofrio F et al (2017) Usefulness of nutraceuticals in migraine prophylaxis. Neurol Sci 38(1):117–120

Da Silva AN (2015) Acupuncture for migraine prevention. Headache 55(3):470–473

Dakhale GN et al (2019) Low-dose sodium valproate versus low-dose propranolol in prophylaxis of common migraine headache: a randomized, prospective, parallel, open-label study. Indian J Pharmacol 51(4):255–262

Danesh A, Gottschalk PCH (2019) Beta-blockers for migraine prevention: a review article. Curr Treat Options Neurol 21(4):20

de Boer I et al (2023) Place of next generation acute migraine specific treatments among triptans, non-responders and contraindications to triptans and possible combination therapies. Cephalalgia 43(2):3331024221143773

Derry CJ, Derry S, Moore RA (2012) Sumatriptan (oral route of administration) for acute migraine attacks in adults. Cochrane Database Syst Rev 2012(2):Cd008615

Diener HC et al (2010) OnabotulinumtoxinA for treatment of chronic migraine: results from the double-blind, randomized, placebo-controlled phase of the PREEMPT 2 trial. Cephalalgia 30(7):804–814

Diener HC et al (2015) New therapeutic approaches for the prevention and treatment of migraine. Lancet Neurol 14(10):1010–1022

Diener HC et al (2016) Medication-overuse headache: risk factors, pathophysiology and management. Nat Rev Neurol 12(10):575–583

Dobrek L, Głowacka K (2023) Depression and its phytopharmacotherapy—a narrative review. Int J Mol Sci 24(5):4772

Dodick DW et al (2019) Ubrogepant for the treatment of migraine. N Engl J Med 381(23):2230–2241

Dos Santos JBR, da Silva MRR (2022) Small molecule CGRP receptor antagonists for the preventive treatment of migraine: a review. Eur J Pharmacol 922:174902

Elder GJ, Taylor JP (2014) Transcranial magnetic stimulation and transcranial direct current stimulation: treatments for cognitive and neuropsychiatric symptoms in the neurodegenerative dementias? Alzheimers Res Ther 6(9):74

El-Metwally A et al (2020) The epidemiology of migraine headache in Arab countries: a systematic review. ScientificWorldJournal 2020:4790254

Fumal A, Schoenen J (2008) Current migraine management—patient acceptability and future approaches. Neuropsychiatr Dis Treat 4(6):1043–1057

GBD 2016 Headache Collaborators (2018) Global, regional, and national burden of migraine and tension-type headache, 1990–2016: a systematic analysis for the Global Burden of Disease Study 2016. Lancet Neurol 17(11):954–976

Giffin NJ et al (2003) Premonitory symptoms in migraine: an electronic diary study. Neurology 60(6):935–940

Goadsby PJ et al (2017) Pathophysiology of migraine: a disorder of sensory processing. Physiol Rev 97(2):553–622

Graham GG et al (2013) The modern pharmacology of paracetamol: therapeutic actions, mechanism of action, metabolism, toxicity and recent pharmacological findings. Inflammopharmacology 21(3):201–232

Hallett M (2007) Transcranial magnetic stimulation: a primer. Neuron 55(2):187–199

Hansen JM, Charles A (2019) Differences in treatment response between migraine with aura and migraine without aura: lessons from clinical practice and RCTs. J Headache Pain 20(1):96

Hareendran A et al (2018) Development of a new tool for evaluating the benefit of preventive treatments for migraine on functional outcomes—the Migraine Functional Impact Questionnaire (MFIQ). Headache 58(10):1612–1628

Hauge AW, Kirchmann M, Olesen J (2010) Trigger factors in migraine with aura. Cephalalgia 30(3):346–353

Hepp Z et al (2015) Adherence to oral migraine-preventive medications among patients with chronic migraine. Cephalalgia 35(6):478–488

Heptinstall S et al (1992) Parthenolide content and bioactivity of feverfew (Tanacetum parthenium (L.) Schultz-Bip.). Estimation of commercial and authenticated feverfew products. J Pharm Pharmacol 44(5):391–395

Holm JE, Lokken C, Myers TC (1997) Migraine and stress: a daily examination of temporal relationships in women migraineurs. Headache 37(9):553–558

Holroyd KA, Labus JS, Carlson B (2009) Moderation and mediation in the psychological and drug treatment of chronic tension-type headache: the role of disorder severity and psychiatric comorbidity. Pain 143(3):213–222

Houle TT, Turner DP (2013) Natural experimentation is a challenging method for identifying headache triggers. Headache 53(4):636–643

Houle TT et al (2017) Forecasting individual headache attacks using perceived stress: development of a multivariable prediction model for persons with episodic migraine. Headache 57(7):1041–1050

Hutchinson S et al (2013) Use of common migraine treatments in breast-feeding women: a summary of recommendations. Headache 53(4):614–627

Hutchinson S et al (2021) Ubrogepant for the acute treatment of migraine: pooled efficacy, safety, and tolerability from the ACHIEVE I and ACHIEVE II phase 3 randomized trials. Neurol Ther 10(1):235–249

Jin P, Madieh S, Augsburger LL (2008) Selected physical and chemical properties of Feverfew (Tanacetum parthenium) extracts important for formulated product quality and performance. AAPS PharmSciTech 9(1):22–30

Johannessen Landmark C (2008) Antiepileptic drugs in non-epilepsy disorders: relations between mechanisms of action and clinical efficacy. CNS Drugs 22(1):27–47

Joshi S et al (2021) A narrative review of the importance of pharmacokinetics and drug-drug interactions of preventive therapies in migraine management. Headache 61(6):838–853

Jügens TP, Schaefer C, May A (2009) Treatment of cluster headache in pregnancy and lactation. Cephalalgia 29(4):391–400

Kalra EK (2003) Nutraceutical—definition and introduction. AAPS PharmSci 5(3):25

Karsan N, Goadsby PJ (2020) Imaging the premonitory phase of migraine. Front Neurol 11:140

Khan S, Schoenen J, Ashina M (2014) Sphenopalatine ganglion neuromodulation in migraine: what is the rationale? Cephalalgia 34(5):382–391

Kumar A et al (2020) Effect of yoga as add-on therapy in migraine (CONTAIN): a randomized clinical trial. Neurology 94(21):e2203–e2212
Láinez MJ, García-Casado A, Gascón F (2013) Optimal management of severe nausea and vomiting in migraine: improving patient outcomes. Patient Relat Outcome Meas 4:61–73
Larik MO et al (2023) Nasal spray (Zavegepant) for migraine: a mini-review. Ann Med Surg (Lond) 85(6):2787–2790
Laurell K et al (2016) Premonitory symptoms in migraine: a cross-sectional study in 2714 persons. Cephalalgia 36(10):951–959
Leonardi M et al (2005) The global burden of migraine: measuring disability in headache disorders with WHO's Classification of Functioning, Disability and Health (ICF). J Headache Pain 6(6):429–440
Lewis DW et al (2004) Prophylactic treatment of pediatric migraine. Headache 44(3):230–237
Li Y et al (2012) Acupuncture for migraine prophylaxis: a randomized controlled trial. CMAJ 184(4):401–410
Linde K et al (2005) Acupuncture for patients with migraine: a randomized controlled trial. JAMA 293(17):2118–2125
Lipton RB et al (2016) A comparison of the Chronic Migraine Epidemiology and Outcomes (CaMEO) study and American Migraine Prevalence and Prevention (AMPP) study: demographics and headache-related disability. Headache 56(8):1280–1289
Lipton RB et al (2018) Estimating the clinical effectiveness and value-based price range of erenumab for the prevention of migraine in patients with prior treatment failures: a US societal perspective. J Med Econ 21(7):666–675
Liu S et al (2017) Recent advances in non-invasive brain stimulation for major depressive disorder. Front Hum Neurosci 11:526
Löscher W, Schmidt D (2006) New Horizons in the development of antiepileptic drugs: innovative strategies. Epilepsy Res 69(3):183–272
Lucas S (2009) Medication use in the treatment of migraine during pregnancy and lactation. Curr Pain Headache Rep 13(5):392–398
Maizels M et al (1996) Intranasal lidocaine for treatment of migraine: a randomized, double-blind, controlled trial. JAMA 276(4):319–321
Maizels M, Blumenfeld A, Burchette R (2004) A combination of riboflavin, magnesium, and feverfew for migraine prophylaxis: a randomized trial. Headache 44(9):885–890
Martin VT, Behbehani MM (2001) Toward a rational understanding of migraine trigger factors. Med Clin North Am 85(4):911–941
Martirosov AL et al (2023) Zavegepant intranasal spray for migraine. Ann Pharmacother 2023:10600280231209439
Mayans L, Walling A (2018) Acute migraine headache: treatment strategies. Am Fam Physician 97(4):243–251
Mechtler L et al (2021) Real-world evidence for the safety and efficacy of CGRP monoclonal antibody therapy added to onabotulinumtoxinA treatment for migraine prevention in adult patients with chronic migraine. Front Neurol 12:788159
Messali A et al (2016) Direct and indirect costs of chronic and episodic migraine in the United States: a web-based survey. Headache 56(2):306–322
Minen MT et al (2016) Migraine and its psychiatric comorbidities. J Neurol Neurosurg Psychiatry 87(7):741–749
Mooventhan A, Nivethitha L (2017) Evidence based effects of yoga in neurological disorders. J Clin Neurosci 43:61–67
Muddam MR et al (2023) Efficacy and safety of anti-calcitonin gene-related peptide (CGRP) monoclonal antibodies in preventing migraine: a systematic review. Cureus 15(9):e45560
Mullin GE et al (2010) Testing for food reactions: the good, the bad, and the ugly. Nutr Clin Pract 25(2):192–198
Murray AM et al (2022) Real-world patient experience of CGRP-targeting therapy for migraine: a narrative review. Curr Pain Headache Rep 26(10):783–794

Naro A et al (2016) Non-invasive brain stimulation, a tool to revert maladaptive plasticity in neuropathic pain. Front Hum Neurosci 10:376
Olesen J, Ashina M (2011) Emerging migraine treatments and drug targets. Trends Pharmacol Sci 32(6):352–359
Ong JJY, De Felice M (2018) Migraine treatment: current acute medications and their potential mechanisms of action. Neurotherapeutics 15(2):274–290
Patniyot IR, Gelfand AA (2016) Acute treatment therapies for pediatric migraine: a qualitative systematic review. Headache 56(1):49–70
Pavlovic JM et al (2014) Trigger factors and premonitory features of migraine attacks: summary of studies. Headache 54(10):1670–1679
Peroutka SJ (1997) Dopamine and migraine. Neurology 49(3):650–656
Peroutka SJ (2014) What turns on a migraine? A systematic review of migraine precipitating factors. Curr Pain Headache Rep 18(10):454
Peters GL (2019) Migraine overview and summary of current and emerging treatment options. Am J Manag Care 25(2 Suppl):S23–S34
Plato BM, Whitt M (2020) Interventional procedures in episodic migraine. Curr Pain Headache Rep 24(12):75
Pleş H et al (2023) Migraine: advances in the pathogenesis and treatment. Neurol Int 15(3):1052–1105
Rajapakse T, Pringsheim T (2016) Nutraceuticals in migraine: a summary of existing guidelines for use. Headache 56(4):808–816
Rao BS et al (2000) A double blind controlled study of propranolol and cyproheptadine in migraine prophylaxis. Neurol India 48(3):223–226
Riggins N, Ehrlich A (2022) Episodic migraine and older adults. Curr Pain Headache Rep 26(4):331–335
Roberto G et al (2015) Adverse cardiovascular events associated with triptans and ergotamines for treatment of migraine: systematic review of observational studies. Cephalalgia 35(2):118–131
Rose FC (1995) The history of migraine from Mesopotamian to Medieval times. Cephalalgia 15(Suppl 15):1–3
Russo AF, Hay DL (2023) CGRP physiology, pharmacology, and therapeutic targets: migraine and beyond. Physiol Rev 103(2):1565–1644
Russo A et al (2012) Executive resting-state network connectivity in migraine without aura. Cephalalgia 32(14):1041–1048
Sartory G et al (1998) A comparison of psychological and pharmacological treatment of pediatric migraine. Behav Res Ther 36(12):1155–1170
Saxena PR, Ferrari MD (1989) 5-HT(1)-like receptor agonists and the pathophysiology of migraine. Trends Pharmacol Sci 10(5):200–204
Schoenen J et al (2020) Monoclonal antibodies blocking CGRP transmission: an update on their added value in migraine prevention. Rev Neurol (Paris) 176(10):788–803
Seng EK, Holroyd KA (2010) Dynamics of changes in self-efficacy and locus of control expectancies in the behavioral and drug treatment of severe migraine. Ann Behav Med 40(3):235–247
Sharma VM et al (2018) Combination of Ayurveda and Yoga therapy reduces pain intensity and improves quality of life in patients with migraine headache. Complement Ther Clin Pract 32:85–91
Sharma SK et al (2020) Yoga as an add-on therapy in the management of migraine. Indian J Community Med 45(2):244–245
Shashikiran H et al (2022) Effect of Yoga Nidra on the brain activity in individuals with migraine. Yoga Mimamsa 54(1):18–23
Silberstein S et al (2000) Botulinum toxin type A as a migraine preventive treatment. For the BOTOX Migraine Clinical Research Group. Headache 40(6):445–450
Song TJ et al (2016) Characteristics of elderly-onset (≥65 years) headache diagnosed using the International Classification of Headache Disorders, Third Edition Beta Version. J Clin Neurol 12(4):419–425

Song Y et al (2022) Comparative efficacy of acupuncture-related therapy for migraine: a systematic review and network meta-analysis. Front Neurol 13:1010410
Sorge F et al (1988) Flunarizine in prophylaxis of childhood migraine. A double-blind, placebo-controlled, crossover study. Cephalalgia 8(1):1–6
Steiner TJ (2004) Lifting the burden: the global campaign against headache. Lancet Neurol 3(4):204–205
Steiner TJ et al (2020) Migraine remains second among the world's causes of disability, and first among young women: findings from GBD2019. J Headache Pain 21(1):137
Stewart WF et al (2010) Employment and work impact of chronic migraine and episodic migraine. J Occup Environ Med 52(1):8–14
Stovner LJ et al (2014) The methodology of population surveys of headache prevalence, burden and cost: principles and recommendations from the Global Campaign against Headache. J Headache Pain 15(1):5
Straube A et al (2011) Combined analgesics in (headache) pain therapy: shotgun approach or precise multi-target therapeutics? BMC Neurol 11:43
Suthisisang CC et al (2010) Meta-analysis of the efficacy and safety of naproxen sodium in the acute treatment of migraine. Headache 50(5):808–818
Szeto CC et al (2020) Non-steroidal anti-inflammatory drug (NSAID) therapy in patients with hypertension, cardiovascular, renal or gastrointestinal comorbidities: joint APAGE/APLAR/APSDE/APSH/APSN/PoA recommendations. Gut 69(4):617–629
Tajti J et al (2015) Drug safety in acute migraine treatment. Expert Opin Drug Saf 14(6):891–909
Taylor FR (2009) Lifestyle changes, dietary restrictions, and nutraceuticals in migraine prevention. Techn Regional Anesth Pain Manag 13(1):28–37
Teleanu RI et al (2016) Treatment of pediatric migraine: a review. Maedica (Bucur) 11(2):136–143
Ungrungseesopon N, Wongtanasarasin W (2022) Pain reduction and adverse effects of intravenous metoclopramide for acute migraine attack: a systematic review and meta-analysis of randomized-controlled trials. World J Methodol 12(4):319–330
Urits I et al (2020) Pharmacological options for the treatment of chronic migraine pain. Best Pract Res Clin Anaesthesiol 34(3):383–407
van Dongen RM, Haan J (2019) Symptoms related to the visual system in migraine. F1000Res 8:F1000 Faculty Rev-1219
Varrassi G et al (2020) Ibuprofen safety at the golden anniversary: are all NSAIDs the same? A narrative review. Adv Ther 37(1):61–82
Waeber C (2003) Emerging drugs in migraine treatment. Expert Opin Emerg Drugs 8(2):437–456
Wammes-van der Heijden EA et al (2006) Risk of ischemic complications related to the intensity of triptan and ergotamine use. Neurology 67(7):1128–1134
Wang J, McDonagh DL, Meng L (2021) Calcium channel blockers in acute care: the links and missing links between hemodynamic effects and outcome evidence. Am J Cardiovasc Drugs 21(1):35–49
Wijeratne T et al (2019) Prevalence of migraine in the elderly: a narrated review. Neuroepidemiology 52(1–2):104–110
Yardi N (2001) Yoga for control of epilepsy. Seizure 10(1):7–12
Zhang X et al (2011) Activation of central trigeminovascular neurons by cortical spreading depression. Ann Neurol 69(5):855–865
Zhao T et al (2021) A bibliometric analysis of research trends of acupuncture therapy in the treatment of migraine from 2000 to 2020. J Pain Res 14:1399–1414
Zobdeh F et al (2021) Pharmacological treatment of migraine: drug classes, mechanisms of action, clinical trials and new treatments. Br J Pharmacol 178(23):4588–4607

Drugs Under Clinical Trials for the Treatment of Migraine 12

Thangaraj Devadoss and Ravindra Babu Pingili

12.1 Introduction

Migraine is one of the headache disorders, also called one side headache. It affects 12% of the population (Arulmozhi et al. 2005). Based on the frequency and other symptoms associated with headaches, migraine disorders can be classified into migraine with aura, migraine without aura, episodic migraine, and chronic migraine.

The symptoms of the aura are related to disturbance of the neurological, autonomic, and gastrointestinal systems. The blind spots, flashes of light, and tingling in the hands or face are the characteristic symptoms of migraine with aura (Arulmozhi et al. 2005). The recurrent pulsating headache appearing for 4–72 h with unilateral location and associated with nausea, photophobia, and phonophobia are the characteristics of migraine without aura (Arulmozhi et al. 2005). Furthermore, physical activities worsen the migraine conditions. Migraine without aura category account for 70% of migraine. A migraine is episodic when headaches appear for less than 15 days per month (Lipton and Silberstein 2015). In a month, 15 days, or more than 15 days of headache appearing with the symptoms of migraine for a minimum of 8 days in a month, and occurring for more than 3 months is considered chronic migraine (Lipton and Silberstein 2015).

The exact mechanism involved in the etiology of migraine is not known (Garcia-Martinez et al. 2020). However, both the central and peripheral nervous systems are implicated. The development of migraine occurs due to exposure to or withdrawal from several factors. It includes stress, hormonal changes, skipped meals, weather

T. Devadoss (✉)
Department of Pharmaceutical Chemistry, Faculty of Pharmacy, Karpagam Academy of Higher Education, Coimbatore, Tamil Nadu, India

R. B. Pingili
School of Pharmacy and Technology Management, SVKM's NMIMS (Deemed to be University), Shirpur, Maharashtra, India

S. Bhatt et al. (eds.), *Management of Migraine Pain*,
https://doi.org/10.1007/978-981-97-4529-6_12

changes, excessive or inadequate sleep, odor, exposure to light, consumption of alcohol, smoking, late sleeping, heat, exercise, etc. (Haghdoost and Togha 2022). The prevalence of migraine rise is three times higher in relatives with migraine disorders than in relatives without migraine disorders (Devoto et al. 1986; Merikangas et al. 1988). This indicates genetics are also one of the risk factors for migraine disorder.

The types of treatment or management of migraine can be classified into two types, namely abortive and prophylactic treatment (Aguilar-Shea et al. 2022). The objective of the abortive treatment is to stop the progression of the headache. This type of treatment is also known as acute treatment. The list of most widely-used drugs is presented in Table 12.1 (Eigenbrodt et al. 2021). The chemical structure of drugs used for the abortive treatment is represented in Fig. 12.1a, b.

The second type of treatment is prophylactic, also known as preventive treatment. The prime objective of the treatment is to reduce the attack frequency, improve responsiveness to acute attacks, and reduce disability. The drug used under this category is represented in Table 12.2, and the chemical structure of small molecules is presented in Fig. 12.2 (Eigenbrodt et al. 2021).

The adverse effects and ineffectiveness of the available therapy urges scientists to discover novel therapies or new drugs or to repurpose the existing drug molecules as antimigraine drugs (Arulmozhi et al. 2005). This chapter discusses the drugs under clinical trials for the prevention and treatment of migraine disorders in a phase-wise manner. It also focuses on the possible mechanism involved in the antimigraine effect of the molecules under the clinical trials.

12.2 Drugs Under Clinical Trials

The molecules under clinical trials for the treatment and management of migraine disorder are discussed in the following subsections. The ID of the clinical trial and study start date (YYYY-MM-DD) are presented in parenthesis.

Table 12.1 List of most widely-used abortive treatment drugs for migraine

S. No.	Drug category	Example
1.	NSAID's	Diclofenac, Naproxen, Ibuprofen, Aspirin, Paracetamol
2.	Triptans	Sumatriptan, Zolmitriptan, Eletriptan, Rizatriptan, Almotriptan
3.	Antiemetics	Metoclopramide, Chlorpromazine, Prochlorperazine
4.	Gepants	Rimegepant, Ubrogepant
5.	Selective 5-HT_{1F} receptor agonist	Lasmiditan
6.	Ergot alkaloids	Ergotamine, Dihydroergotamine
7.	Steroids	Prednisolone, Dexamethasone

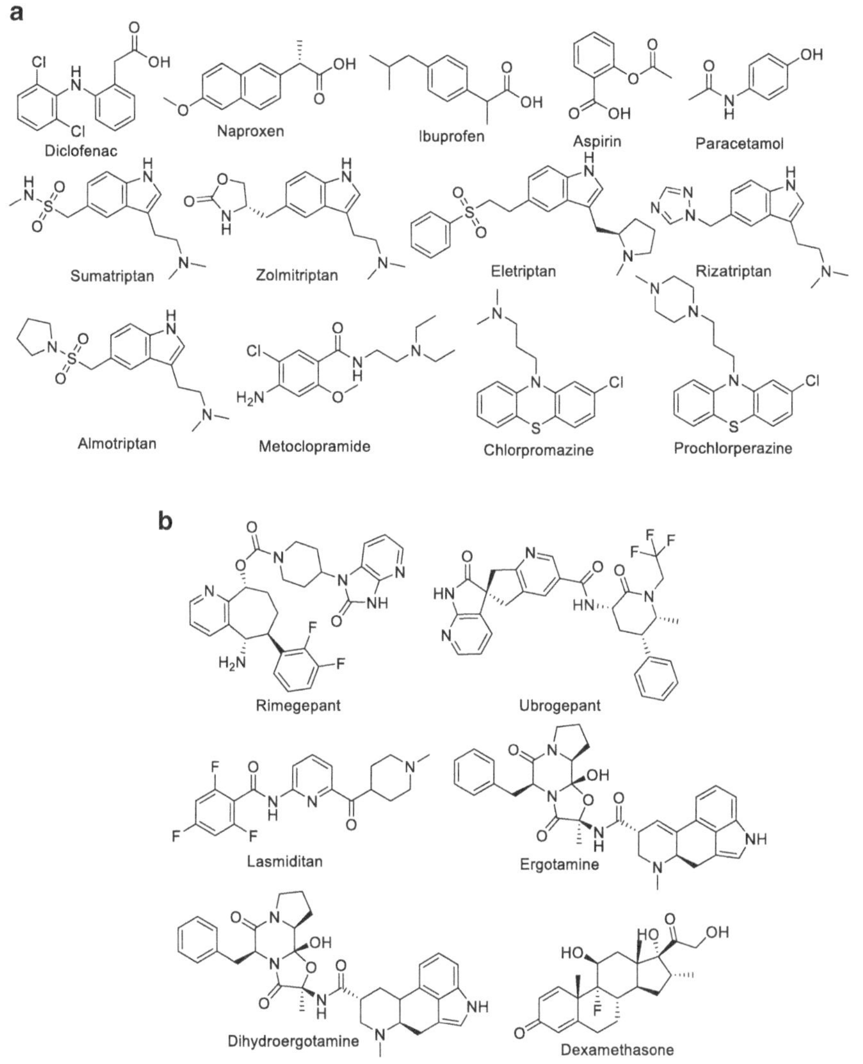

Fig. 12.1 (**a**) Chemical structures of NSAID's, triptans, and antiemetics. (**b**) Chemical structures of CGRP receptor antagonists, selective serotonin type-1 receptor agonist, ergot alkaloids and steroid

12.2.1 Early Phase 1 Clinical Trial

The peripheral nerve block method is a treatment option for various headache disorders, including migraines (Ashkenazi et al. 2010). This method involves injecting local anesthetic agents around peripheral nerve branches to provide prompt relief that can last up to a month (Blumenfeld et al. 2013). Texas Tech University Health Sciences Center recently registered a clinical study (ID: NCT06016465) in

Table 12.2 List of most widely-used preventive treatment drugs for migraine

S. No.	Drug category	Example
1.	β-Blockers	Metoprolol, Propranolol, Atenolol, Bisoprolol
2.	Antidepressants	Amitriptyline, Venlafaxine
3.	Anticonvulsants	Sodium valproate, Topiramate
4.	Calcium channel blockers	Verapamil, Flunarizine
5.	Monoclonal antibodies, targeting CGRP	Erenumab, Fremanezumab, Galcanezumab

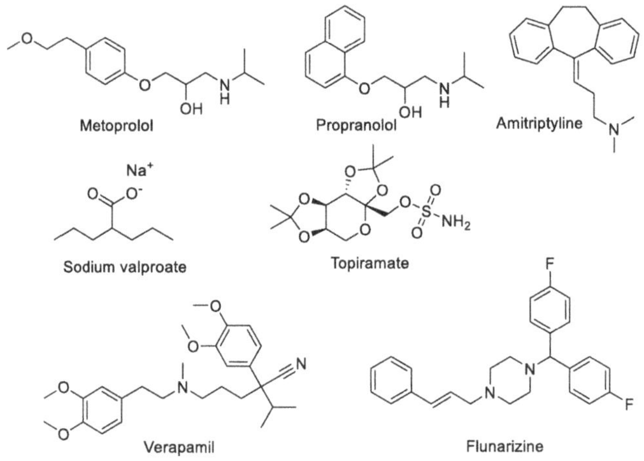

Fig. 12.2 Chemical structures of some small molecules of the prophylactic category

September 2023. The study aims to reduce the recurrence of headaches within 72 h by using a combination of nerve blockers and steroids (dexamethasone). This study is based on previous research indicating that adding steroids to conventional migraine treatment can reduce the recurrence of headaches (Kashyap and Chabri 2023). Prednisolone and dexamethasone are the most commonly used steroids for this purpose, which can be administered via the intravenous or oral route (Kashyap and Chabri 2023). While peripheral nerve blocking is the standard treatment for migraine, dexamethasone is used as an adjuvant treatment.

12.2.2 Phase 1 Clinical Trials

St. Justine's Hospital registered a clinical trial (NCT02903680; 2014-05) to observe the effectiveness of intravenous dexamethasone in preventing migraine recurrence in children. This study also aimed to study the risk factors for migraine relapse after discharge from the hospital emergency department (ED).

Yale University started the clinical study (NCT04218539; 2021-08-10) to evaluate the efficacy of single and repeated dosing of psilocybin, the maximum being two

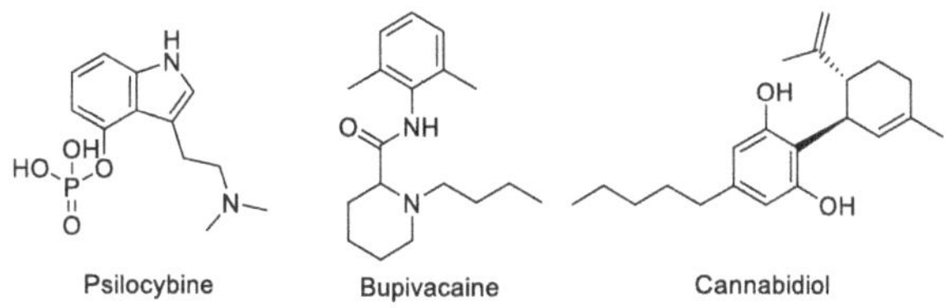

Fig. 12.3 Chemical structure of psilocybin, bupivacaine and cannabidiol

doses. The objective of this study is to identify the mechanism of action of psilocybin and to measure the neuroinflammatory markers of the migraine disorder. Psilocybin is an indole nucleus containing a hallucinogen having substituent at second and fourth position of indole, namely dimethylaminoethyl and the phosphoric acid group, respectively (Daniel and Haberman 2018). The chemical structure of psilocybin is shown in Fig. 12.3. In 1957, Albert Hofmann isolated the psilocybin from the mushroom, *Psilocybe mexicana* (Daniel and Haberman 2018). Now its presence is identified in wide ranges of mushroom species. Some recent studies indicate the clinical applications of psilocybin in anxiety, obsessive-compulsive disorder (Harlow et al. 2024), tobacco cessation (Johnson et al. 2014) and alcohol dependence (Bogenschutz et al. 2015). The pharmacological properties exhibited by psilocybin are due to agonists at 5-HT_{2A}, 5-HT_{1A}, and 5-HT_{2C} receptors (Daniel and Haberman 2018).

Recently, Montefiore Medical Center initiated clinical trials (NCT05707754, 2023-03-06) to answer the following questions: whether a high dose (3 mL) of bupivacaine will give more relief over a low dose (1 mL) of bupivacaine and comparing bilateral administration over unilateral administration. Bupivacaine is a local anesthetic agent and was discovered in 1957 (Vardanyan and Hruby 2006; Bowdle et al. 2007). It is a piperidine-2-carboxamide derivative, the chemical structure of bupivacaine is presented in Fig. 12.3.

Shanghai Junshi Bioscience Co., Ltd., registered a phase I clinical trial (NCT06158737; 2023-05-09) to study the safety, treatment-emergent adverse events, tolerability, pharmacokinetics, pharmacodynamics properties, and immunogenicity of a single dose (3 mg or 30 mg or 150 mg or 300 mg or 600 mg or 900 mg) of JS010 in healthy subjects. The JS010 is a type of monoclonal antibody that acts as an antagonist of CRGP. This study is expected to be completed by September 2024.

The term "cannabinoids" represents 80 chemicals obtained from the *Cannabis sativa* plant (Pereira and Wiegand 2014). Tetrahydrocannabinol (THC) is a well-known compound present in *Cannabis sativa*. It has psychiatric effects, but cannabidiol (CBD) exhibits anti-inflammatory activity without psychoactive effects (Atalay et al. 2020). The University of Manitoba recently registered a clinical study (NCT05337033, 2023-11-27) to evaluate the tolerability of cannabidiol-enriched cannabis herbal extracts (MPC-001) in adolescent populations with chronic headaches. Like other phase 1 trials, the prime objective of this study is to examine the cannabis-related adverse events produced by MPC-001. The secondary objective is to measure the beneficial effects of cannabis herbal extracts in migraine patients (14–17 years old) in comparison with baseline. The chemical structure of CBD is depicted in Fig. 12.3. This study is expected to be completed by December, 2025.

12.2.3 Phase 2 Clinical Trials

Olavas Hospital registered a clinical study (NCT04574713; 2020-04-26) to assess the efficacy of candesartan as a preventive drug for episodic migraine. Candesartan is an angiotensin II receptor blocker. It was approved for the treatment of hypertension and heart failure (Bulsara and Makaryus 2024). The angiotensin II blocking property of candesartan inhibits the release of catecholamine and thus provides beneficial effects in migraines (Sprenger et al. 2018). It is a benzimidazole derivative, its chemical structure is depicted in Fig. 12.4.

The vestibular migraine (VM) is one of the types of migraine characterized by episodic vertigo/dizziness, sometimes with or without a headache. The VM patients preferred to take treatment for the dizziness over the headache, as the headache is less common. Galcanezumab is a monoclonal antibody that exhibits antimigraine effects by blocking the calcitonin gene-related peptide pathway, by targeting the CGRP (Scott 2020a). Galcanezumab was approved in the USA for the prevention of migraine, particularly for episodic or chronic migraine (Lamb 2018) but was not tested for the vestibular migraine. Therefore, the University of California registered the clinical trial in the year 2023.

AEON Biopharma Inc., is investigating (NCT04845178, 2021-03-01) the efficacy and safety of ABP-450 as a migraine preventive drug in adults who suffer from 6 or more migraine days per month. ABP-450 is a botulinum toxin type-A complex having the molecular weight of 900 kDa, obtained from *Clostridium botulinum*, a Gram-positive anaerobic bacterium (Ray et al. 2021). The botulinum toxin expresses the beneficial effects in migraine by blocking the release of CGRP, glutamate and substance P.

Bangabandhu Sheikh Muih Medical University, Dhaka, Bangladesh, registered a clinical study to assess the efficacy of nicotinic acid as a prophylactic agent for moderate to severe migraine (NCT05846373, 2022-11-25). Nicotinic acid is also known as niacin, or vitamin B3. Nicotinamide adenine dinucleotide (NAD) and nicotinamide adenine dinucleotide phosphate (NADP) are coenzymes, derived from nicotinic acid (Gasperi et al. 2019). The chemical structure of niacin is depicted in Fig. 12.4. The level of serotonin is low in migraine patients (Ren et al. 2018). Serotonin is a neurotransmitter biosynthesized from the tryptophan in a sequence of reactions. In the normal state, 1% of tryptophan is transformed into serotonin in the brain (Sandyk 1992). In another biosynthetic pathway (kynurenine pathway), tryptophan is converted to NAD (Richard et al. 2009).

Fig. 12.4 Chemical structure of candesartan and nicotinic acid

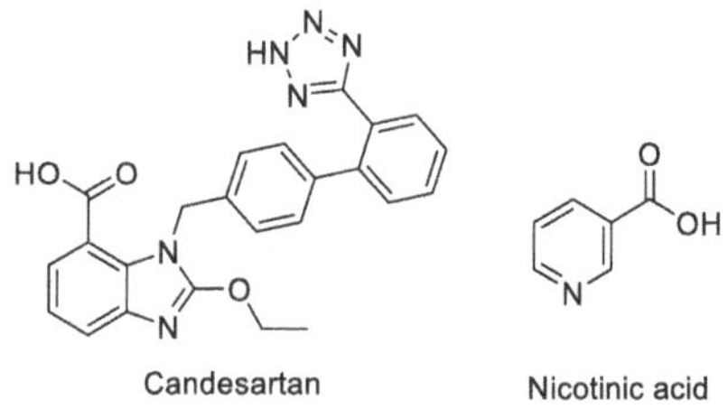

The supplementation of niacin will provide an adequate quantity of NAD, which suppresses the kynurenine pathway and facilitates the biosynthesis of serotonin from tryptophan. The sufficiently available serotonin produces vasoconstrictions by acting on the serotonin type 1 receptor. It also inhibits the synthesis and release of nitric oxide, glutamate, and CGRP. This leads to the prevention of neuroinflammation. On the other hand, nicotinic acid itself displays anti-inflammatory-like effects by decreasing the levels of proinflammatory cytokines like IL-6, IL-11b, TNF-α, and high sensitivity c-creative protein (NCT05846373). Further, an increased level of nicotinic acid improves brain energy deficiency and has potent antioxidant properties.

Tonix Pharmaceuticals, Inc. started the phase 2 clinical trial on 2022-12-06 (NCT05679908) to assess the efficacy and safety of two different dosages (30 IU daily and 60 IU daily), of intranasal oxytocin (TNX-1900) in chronic migraine patients. Many studies indicated that a high level of oxytocin showed beneficial effects on migraine patients (Richard et al. 2009; Tzabazis et al. 2017). Oxytocin blocks the release of CGRP from trigeminal afferent neurons (Tzabazis et al. 2017). CGRP is one of the critical components in the pathogenesis of chronic migraine headaches (Richard et al. 2009).

Bangabandhu Sheikh Mujib Medical University, Dhaka, Bangladesh, initiated the clinical study (NCT06158230; 2023-03-19) to compare the efficacy and observe common adverse effects of a combination of amitriptyline-propranolol and pizotifen as prophylactic treatments for migraines. The prime aim of the study is to answer whether pizotifen is more effective than the combination of amitriptyline-propranolol in migraine prophylaxis.

Pizotifen has structural and pharmacological similarities with cyproheptadine, an antihistamine drug (Lee and Abrahams 2012; Speight and Avery 1972). It also possesses antagonistic properties against serotonergic and cholinergic systems. These antagonistic effects are related to the drug's mode of action in reducing the occurrence of migraine attacks. In pizotifen, a tricyclic system is linked with the piperdine ring. The structure of pizotifen and cyproheptadine are shown in Fig. 12.5.

Zavegepant (BHV-3500) is clinically approved for the treatment of acute migraine, administered in the form of nasal spray. Zavegepant is a small molecule that produces beneficial effects by antagonizing the CGRP receptor (Dhillon 2023). It reduces the symptoms of both migraine with or without aura. However, it is not indicated for the prevention of migraine. Recently, Pfizer initiated a clinical study (NCT04804033; 2021-03-26) to determine the efficacy of the oral route of zavegepant in preventing migraine. The chemical structure of zavegepant is depicted in Fig. 12.6.

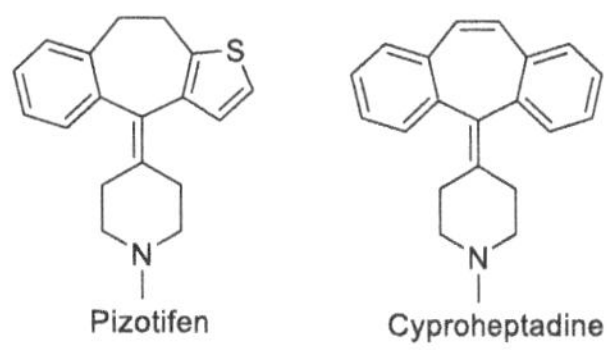

Fig. 12.5 Chemical structure of pizotifen and cyproheptadine

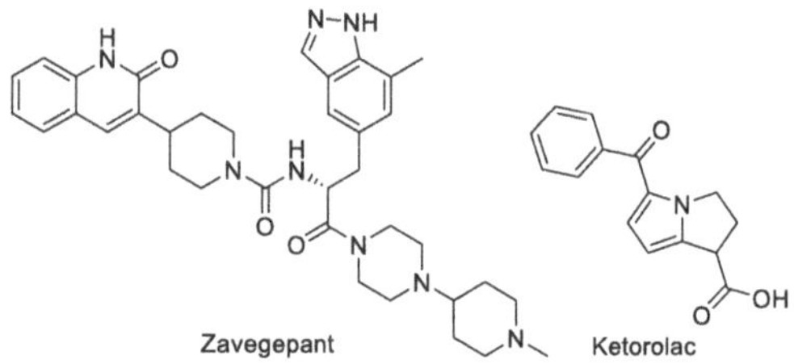

Fig. 12.6 Chemical structure of zavegepant and ketorolac

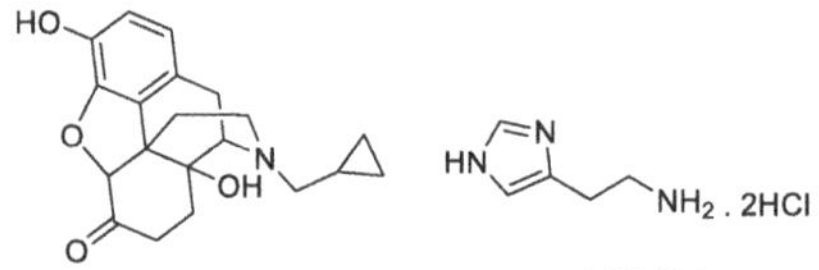

Fig. 12.7 Chemical structure of naltrexone and AGX-201

Very recently, Pfizer registered another clinical study (ID: NCT05989048; 2023-11-29) to determine the safety and efficacy of intranasal zavegepant in the Asian adult population as an acute treatment for migraine (NCT05989048).

The Washington University of Medicine registered a clinical study (NCT06083571) to check the efficacy of intranasal ketorolac in combination with oral prochlorperazine and diphenhydramine and to compare it with intravenous ketorolac in combination with oral prochlorperazine and diphenhydramine. Ketorolac is a non-steroidal anti-inflammatory agent that exhibits pharmacological action by inhibiting the enzyme cyclooxygenase (Mahmoodi and Kim 2024). It is commercially available in various formulations like tablets, injectables, nasal sprays, and ophthalmic solutions. Ketorolac is a derivative of pyrrolizine-1-carboxylic acid (Chandrasekharan 2007), its chemical structure depicted in Fig. 12.6.

A clinical study was registered (NCT05685225) to determine the efficacy of naltrexone-acetaminophen in acute migraine as well as migraine associated with anxiety. Naltrexone acts as an opioid receptor antagonist as well as a toll-like receptor-4 (TLR-4) antagonist. It is approved for treating opioid and alcohol addiction due to its opioid receptor antagonistic property (Singh and Saadabadi 2024). The TLR-4 antagonistic property of naltrexone decreases the levels of pro-inflammatory cytokines, which are responsible for pain generation. It occurs mainly at the trigeminal ganglion and dorsal root ganglia. The acetaminophen is a well-known antipyretic agent and non-opioid analgesic, also known as paracetamol. The chemical structure of paracetamol and naltrexone is depicted in Figs. 12.1a and 12.7, respectively.

Naltrexone exhibited effectiveness in animal and clinical studies. However, confirmation trials for treating pain in humans are still lacking. Further, half of the migraine population is comorbid with anxiety. Therefore, this study is planned to assess the effectiveness of combination therapy for migraine comorbid with anxiety.

AgoneX biopharmaceutical registered the clinical study of AGX-201 for migraine prophylaxis (NCT02021474; 2022-10). The objective is to evaluate the efficacy and safety of subcutaneous AGX-201 to prevent migraine. The AGX-201 is chemically 2-(4-imidazolyl)ethylamine dihydrochloride, refer to Fig. 12.7 for the chemical structure. AGX-201 is a dihydrochloride salt of histamine. In the previous study (Millán-Guerrero et al. 2003), metabolite of histamine selectively interacted with histamine type 3 receptor, thereby decreasing neurological inflammation.

12.2.4 Phase 3 Clinical Trials

In this section, drugs under clinical trials are discussed in a year-wise manner, from 2018 to 2024.

12.2.4.1 2018

Eli Lilly and Company performed the clinical trial (NCT03432286; 2018-03-14) to evaluate the safety and effectiveness of galcanezumab as a preventive drug in children and adolescents. The core objective of this study was to establish the superiority of galcanezumab over placebo treatment in chronic migraine.

12.2.4.2 2019

In May 2018, the USFDA approved erenumab, a monoclonal antibody, as a preventive treatment of migraine in adults (Markham 2018). In the subsequent month (June 2018), this drug got approval from the European Medicines Agency. It produces beneficial effects by blocking the activity of CGRP. It is available in two doses (70 and 140 mg), recommended once a month, and administered through a subcutaneous route (Andreou et al. 2020). In 2019, Amgen started a clinical study (NCT03836040; 2019-07-19) to assess the safety and effectiveness of erenumab in preventing migraine attacks in children and adolescents with episodic migraine. A few months later, they initiated another study (NCT03832998) with the same objectives in children and adolescents with chronic migraine.

Females are more prone to migraine than males, which indicates that sex hormones play a major role in the difference in prevalence (Delaruelle et al. 2018). Further, clinical studies indicated that hormonal changes in pregnancy resulted in a reduction in the frequency of migraine attacks. However, the involvement of sex hormones in the pathophysiology of migraine is not clearly understood. Some recent studies indicated the beneficial effect of oral contraceptives on migraine (Lyall et al. 2023). Based on this, Leiden University Medical Center initiated the clinical study (NCT04007874; 2019-09-10) to compare the efficacy of continuous daily use of the oral contraceptive [ethinylestradiol/levonorgestrel (30/150 μg/day)] pill over vitamin E (400 IU/day) in the treatment of women with menstrually-related and perimenopausal migraine. The chemical structure of ethinylestradiol and levonorgestrel is presented in Fig. 12.8.

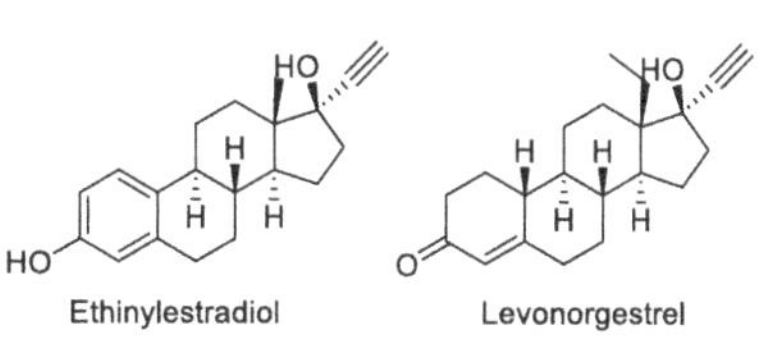

Fig. 12.8 Chemical structure of ethinylestradiol, and levonorgestrel

12.2.4.3 2020

In October 2019, the USFDA gave approval to use lasmiditan for the acute treatment of migraine with and without aura in the adult population (Lamb 2019). The selective serotonin 1F receptor agonistic property of lasmiditan blocks the release of CGRP and the neurotransmitter glutamate, thus providing beneficial effects in migraine (Clemow et al. 2020). The chemical structure of lasmiditan is depicted in Fig. 12.1a. Eli Lilly undertaking a clinical study (NCT04396236; 2020-06-15) to determine the safety and efficacy of lasmiditan in children aged between 6 and 17 years having migraine disorder.

Teva Branded Pharmaceutical Products R&D, Inc. is investigating (NCT04458857; 2020-07-15) the efficacy of fremanezumab in preventing migraine in episodic migraineurs in the age group of 6–17 years. While the safety and tolerability are assessed as a secondary objective of the study. A few months later, they also registered another clinical study (NCT04530110; 2020-09-16) to determine the long-term safety and tolerability of fremanezumab in children and adolescents for the prevention of both episodic and chronic migraine. Fremanezumab is a monoclonal antibody, that produces beneficial effects in migraine by binding with CGRP and thereby preventing the ligand from binding with the CGRP receptor (Silberstein et al. 2017).

Eli Lilly and Company initiated the clinical study (ID: NCT04616326; 2020-11-25) to determine the safety and effectiveness of galcanezumab as a preventive treatment for chronic migraine patients of ages 12–17 years. The core objective of this study is to establish the superiority of galcanezumab over placebo treatment in chronic migraine patients (NCT04616326).

12.2.4.4 2021

Pfizer started a clinical trial (ID: NCT04649242; 2021-01-15) to evaluate the safety and efficacy of BHV-3000 (rimegepant) as an acute treatment for moderate to severe migraine patients in children and adolescents (6–18 years). In February 2020, rimegepant was approved by the USFDA to treat migraine patients with or without aura in adult populations (Scott 2020b). Rimegepant is a small molecule antagonist of the CGRP receptor. Rimegepant is a derivative of piperdine-1-carboxylate, its chemical structure is represented in Fig. 12.1b. A few months later, they initiated another clinical study (NCT04743141; 2021-04-28) to evaluate the long-term safety and tolerability of orally dispersible tablets (75 mg) of BHV-3000 (rimegepant) in migraine patients of the age group 6–18 years.

Atogepant is an orally active calcitonin gene-related peptide receptor antagonist that was approved by the USFDA in September 2021 (Deeks 2022). Alternatively, atogepant is identified with the following codes MK-8081, and AGN-241689. Atogepant is a pyrrolo pyridine derivative, the chemical structure of atogepant is presented in Fig. 12.9. Allergan registered a clinical study (ID: NCT04686136; 2021-02-19) to determine the long-term safety and tolerability profile of atogepant as a preventive treatment in chronic or episodic migraine adult patients.

H. Lundbeck A/S conducted a clinical study (NCT04921384; 2021-05-29) to determine the efficacy of eptinezumab in reducing the number of days in adults

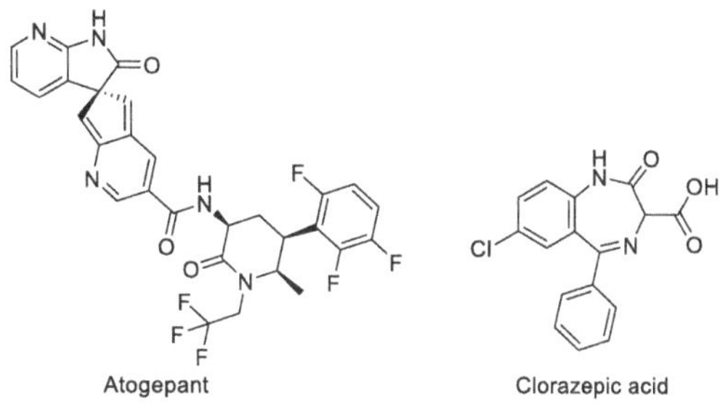

Fig. 12.9 Chemical structure of atogepant and clorazepic acid

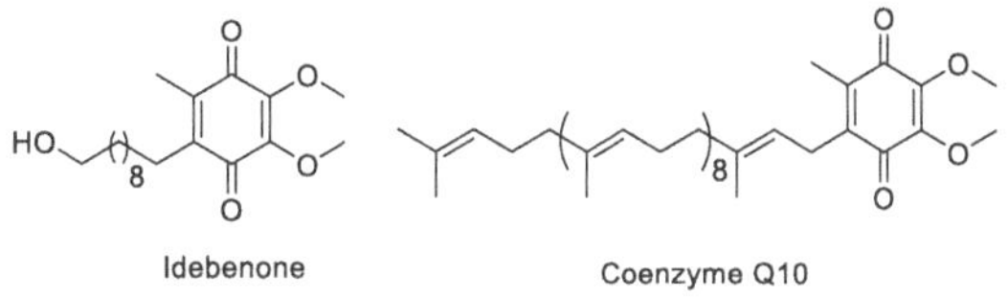

Fig. 12.10 Chemical structure of idebenone and coenzyme Q10

(18–75 years) with chronic migraine. Keeping the same objectives, Lundbeck also initiated another clinical study (NCT04965675; study start date: 2021-06-30) in adolescent patients (12–17 years) with chronic migraine.

Clorazepic acid is a 1,4-benzodiazepine derivative. The salt form of the drug is freely soluble in water. The chemical structure of clorazepic acid is shown in Fig. 12.9. Like other benzodiazepine derivatives, it possesses muscle relaxant, anti-epileptic, and anxiolytic properties (Nikfarjam et al. 2022). Assistance Publique-Hôpitaux de Paris is studying (NCT04726592; 2021-07-08) the efficacy of clorazepate in migraine patients, who are already consuming usual medication to treat migraine attacks in the emergency room.

The School of Medicine from Zhejiang University initiated the clinical trial (NCT04151472; 2021-12-08) to evaluate the efficacy of idebenone in preventing migraine with or without aura in adult patients. Idebenone is a synthetic analog of coenzyme Q10. It has a short-chain saturated hydrocarbon residue with a hydroxyl group instead of a long-chain alkenyl system (Yu-Wai-Man and Chinnery 2016). The chemical structure of both idebenone and coenzyme Q10 is given in Fig. 12.10. The antioxidant property of idebenone is due to the presence of a quinone system. Idebenone penetrates the biological membrane rapidly, compared to coenzyme Q10, due to the short side chain (Filla and Rinaldi 2010). Clinically, it is used to treat a wide range of neurological disorders, including Friedreich ataxia (FRDA), mitochondrial encephalomyopathies, senile dementia, and Huntington's disease (Filla and Rinaldi 2010). Many antioxidants display beneficial effects on migraine, reduce the duration of episodes, or decrease the frequency of migraine headaches. Further, many drugs acting as a prophylactic in migraine possess antioxidant properties (Goschorska et al. 2020).

12.2.4.5 2022

AbbVie, a pharmaceutical company, registered a clinical study (NCT05125302; 2022-01-13) to assess the safety, efficacy, tolerability, and pharmacokinetic properties of oral ubrogepant in children and adolescent population (6–17 years) for acute treatment of migraine. In December 2019, the USFDA approved ubrogepant for the acute treatment of migraine in adults in oral dosage form (Scott 2020c). 50 mg or 100 mg are the recommended doses of ubrogepant for the acute treatment in a single dose, however a second dose is recommended if needed (Dodick et al. 2019). Ubrogepant exhibits antimigraine activity by antagonizing the CGRP receptor. The chemical structure of ubrogepant is shown in Fig. 12.11, it is a pyridine-3-carboxamide derivative.

Pfizer is studying (NCT05156398; 2022-02-28) the efficacy and safety of rimegepant in preventative treatment migraine in children and adolescents (6–18 years) with episodic migraine. A few months later, Pfizer initiated another clinical study to investigate the (ID: NCT05371652, Study start date: 2022-05-19) long-term safety and tolerability of orally dispersible tablets (75 mg) of BHV-3000 (rimegepant) in Chinese subjects. In August 2022, keeping the same objectives, Pfizer registered a clinical study (NCT05399485; 2022-08-09) in Japanese populations.

Kafrelsheikh University is investigating (ID: NCT05632133; 2022-06-01) the effect of lacosamide on serum concentration of calcitonin gene-related peptide in migraine patients with episodic migraine. Lacosamide is an antiepileptic drug; it is a derivative of propanoic acid, chemically known as (2*R*)-2-acetamido-*N*-benzyl-3--methoxypropanamide (Beydoun et al. 2009). The chemical structure of lacosamide is depicted in Fig. 12.11. This drug produces its beneficial effects by inhibiting the production and release of calcitonin gene-related peptide (Greco et al. 2016).

Axsome Therapeutics, Inc. registered a clinical study (NCT05550207; 2022-08-31) to evaluate the safety and efficacy of AXS-07 in treating migraine patients, those who do not respond to calcitonin gene-related peptide inhibitors. AXS-07 is a molecular solubility-enhanced inclusion complex of meloxicam and rizatriptan (Jones et al. 2022). Meloxicam is a non-steroidal anti-inflammatory agent, while rizatriptan belongs to the category of 5-$HT_{1B/1D}$ receptor agonist (Ogino et al. 2002; Dooley and Faulds 1999). Meloxicam preferentially inhibits the COX2 enzyme, which leads to the inhibition of prostaglandin synthesis, which results in reduced neuroinflammation (Ogino et al. 2002). On the other hand, rizatriptan inhibits the release of CGRP and thereby reverses the CGRP-mediated vasodilation

Fig. 12.11 Chemical structure of ubrogepant and lacosamide

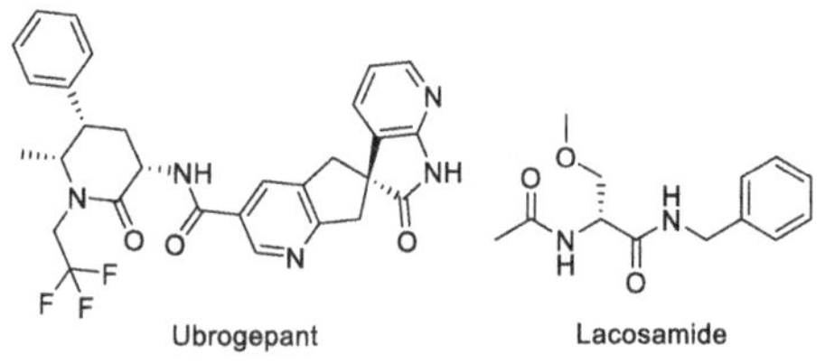

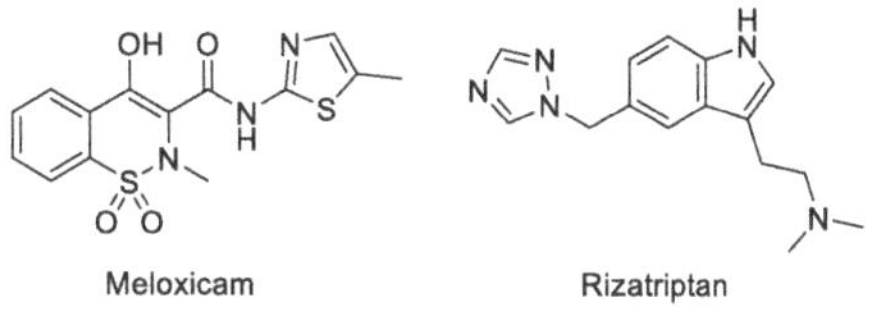

Fig. 12.12 Chemical structure of meloxicam and rizatriptan

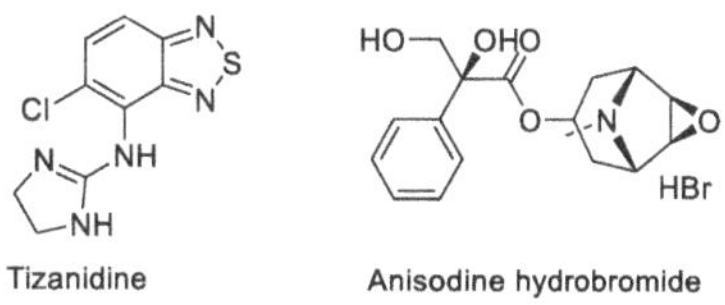

Fig. 12.13 Chemical structure of tizanidine and anisodine hydrobromide

(Hargreaves 2000). The chemical structure of meloxicam and rizatriptan are given in Fig. 12.12.

12.2.4.6 2023

AbbVie started the clinical study (ID: NCT05711394; 2023-05-01) to evaluate the adverse effects and changes in disease activity of oral atogepant tablets in pediatric patients of ages 6–17 years. Atogepant was approved in the USA for the treatment of episodic migraine in adults (Deeks 2022). The main objective of this study is to assess the safety and efficacy of a low-dose and high-dose of atogepant in paediatric participants in the above stated age categories. In August 2023, AbbVie registered another clinical study on atogepant, to assess the safety and effectiveness of three different doses of atogepant in adult Japanese participants (NCT05861427; 2023-07-12). A few months later, they initiated a comparative study (NCT05748483; 2023-10-07) between oral dosage forms of atogepant and topiramate in tablet and capsule formulation respectively, for treating migraine patients. This comparative study was aimed to assess the safety and tolerability of atogepant and topiramate in treating migraine patients between the age of 18 and 80 years.

H. Lundbeck A/S conducted a clinical study (ID: NCT05897320; 2023-06-08) to determine the efficacy of eptinezumab in reducing the number of days in paediatric patients (6–17 years) with episodic migraine. Eptinezumab is a monoclonal antibody clinically approved by the USFDA in February 2020 for preventive treatment of migraines in adults (Dhillon 2020). It produces beneficial effects by binding with human calcitonin gene-related peptide and thus prevents its binding with CGRP receptors (Dhillon 2020).

Zhejiang University (Second Affiliated Hospital, School of Medicine) is conducting a clinical study (NCT05484349; 2023-08-31) to determine the safety and effectiveness of oral tizanidine hydrochloride in preventing episodic migraine in the adult population. Tizanidine is a benzothiadiazole derivative, its chemical structure is presented in Fig. 12.13. The drug tizanidine is commonly used to treat spasms and is a muscle relaxant, displaying the agonistic effect at α2-adrenergic receptor sites (Jewell 2007). The α2-adrenergic agonistic activity of tizanidine blocks the release of norepinephrine at both the spinal cord and brain which results in beneficial effects on migraine. The analgesic effect of tizanidine is independent of the endogenous opioid system (Saper et al. 2002).

The School of Medicine from Zhejiang University started the clinical study (ID: NCT05416476; 2023-09-01) to assess the safety, efficacy, and tolerability of anisodine hydrobromide in preventing episodic migraine in adults. Anisodine hydrobromide is a tropane alkaloid, also known as α-hydroxyscopolamin hydrobromide and daturamine hydrobromide, obtained from the plant *Anisodus tanguticus* (Liu et al. 2020). The chemical structure of anisodine hydrobromide is depicted in Fig. 12.13. The antispasmodic property of anisodine hydrobromide is due to muscarinic antagonism (PubChem 2004). The analgesic, antipyretic, and antiplatelet effects of anisodine hydrobromide are due to cyclooxygenase enzyme inhibitor property, which leads to inhibition of prostaglandins synthesis (PubChem 2004).

Dysport® is a formulation that consists of Botulinum toxin type A as an active pharmaceutical ingredient. Two clinical studies are simultaneously initiated by Ipsen, a global biopharmaceutical company, to evaluate the safety and efficacy of this formulation in two different clinical conditions, the episodic migraine, and the chronic migraine. The episodic (NCT06047457) and chronic migraine study (NCT06047444) were started on 2023-09-29 and 2023-10-12, respectively.

12.2.5 Phase 4 Clinical Trials

The Phase 4 clinical study is conducted to ensure the safety of drugs in the long run. This stage of study is initiated once the drug has been approved and is available in the market. The primary objectives of these studies are to evaluate the drug's side effects, safety profile, and long-term risks and benefits. It also aims to determine the effectiveness of the drug in populations that were excluded in phase 3 trials. Additionally, identifying rare side effects is also one of the key objectives of the phase 4 study. The migraine drugs under phase 4 clinical studies are presented in Table 12.3.

12.3 Conclusions

Most of the antimigraine drugs under clinical trials exhibit their beneficial effects by any one of the following properties, antagonizing the calcitonin gene-related peptide, CGRP (monoclonal antibodies like JS010, galcanezumab, erenumab, fremanezumab, eptinezumab), antagonizing the CGRP receptors (zavegepant, rimegepant, atogepant, ubrogepant), inhibiting the cyclooxygenase enzyme (anisodine, ketorolac), selective agonistic activity on histamine type 3 receptor (AGX-201), blocking the release of CGRP (oxytocin, lasmiditan, lacosamide, onabotulinumtoxinA), antioxidant properties (idebenone), sedative and hypnotic properties (clorzepic acid), facilitating the biosynthesis of serotonin (niacin), local anesthetic effects (bupivacaine), agonistic effects on the serotonin receptors (psilocybin, rizatriptan), toll-like receptor-4 (TLR-4) antagonist properties (naltrexone) and inhibiting the release of

Table 12.3 List of drugs under phase 4 clinical trials

S. No.	Clinical Trial ID/study start date	Name of the drug(s)/Sponsor	Category of the drug(s)	Purpose of the study
1.	NCT04271202/2020-07-29	Galcanezumab/Beth Israel Deaconess Medical Center	Monoclonal antibody, targeting Calcitonin gene-related peptide	To understand the mechanism of action of galcanezumab in migraine prevention
2.	NCT04674020/2020-10-22	Erenumab/Danish Headache Center	Monoclonal antibody, targeting CGRP	To identify the cerebral changes in migraine patients before and after treatment with erenumab
3.	NCT04748601/2021-02-19	Qudexy XR (Topiramate extended release)/Upsher-Smith Laboratories	Anticonvulsant	Prevention of migraine in children with the age of 6–11 years
4.	NCT04693533/2021-07-29	Fremanezumab/Beth Israel Deaconess Medical Center	Monoclonal antibody, targeting CGRP	To identify the relationship between ability of fremazenumab in migraine prevention and improved sleep quality in migraine patients
5.	NCT05217927/2022-03-04	Rimegepant/Pfizer	CGRP receptor antagonist	To determine the safety and effectiveness of rimegepant in preventing episodic migraine
6.	NCT05207865/2022-03-15	Rimegepant/Pfizer	CGRP receptor antagonist	To identify the long-term safety and tolerability of rimegepant as a preventive drug for the episodic migraine patients
7.	NCT05211154/2022-05-05	Diclofenac Potassium and Rimegepant/Danish Headache Center	Diclofenac Potassium: NSAID; Rimegepant: CGRP receptor antagonist	To compare the efficacy of diclofenac potassium with rimegepant for treating acute, mild, and moderate migraine attack
8.	NCT05411978/2022-05-09	Oral Idebenone/Beijing Tiantan Hospital	Antioxidant	To determine the safety and efficacy of orally administered idebenone for preventive migraine in adults
9.	NCT05214001/2022-06-30	Almotriptan and Ubrogepant/Messoud Ashina, MD	Almotriptan: Triptans; Ubrogepant: CGRP receptor antagonist	To evaluate and compare the efficacy of almotriptan over ubrogepant for treating acute, mild, and moderate migraine attack in adults

(continued)

Table 12.3 (continued)

S. No.	Clinical Trial ID/study start date	Name of the drug(s)/Sponsor	Category of the drug(s)	Purpose of the study
10.	NCT05452239/2022-07-01	Eptinezumab/H. Lundbeck A/S	Monoclonal antibody, targeting CGRP	To check the effectiveness of eptinezumab in patients with medication overuse headache
11.	NCT05503082/2022-09-06	Ubrogepant/Chicago Headache Center & Research Institute	CGRP receptor antagonist	To evaluate the safety and effectiveness of ubrogepant in migraine patients, who presently treated with monoclonal antibody, targeting CGRP
12.	NCT04989413/2022-10-18	Cannabidiol + Cannabigerol + Tetrahydrocannabinol/ Hospital Israelita Albert Einstein	Cannabinoids	To determine the efficacy of combination of cannabinoids as adjuvant for chronic migraine
13.	NCT05509400/2022-10-18	Rimegepant/Pfizer	CGRP receptor antagonist	To determine the safety and tolerability of rimegepant in adult patients who are not compatible or lack of efficacy with triptan
14.	NCT05518123/2022-11-07	Rimegepant/Pfizer	CGRP receptor antagonist	To study the effectiveness and tolerability of rimegepant as a preventive migraine drug in adults who failed to respond for oral preventive drugs
15.	NCT06089356/2023-01-01	Zolmitriptan/Tanta University	5-$HT_{1B/1D/1F}$ receptor agonists	To study the prophylactic efficacy of zolmitriptan in childhood patients
16.	NCT05724771/2023-01-26	Botox + Fremanezumab/Chicago Headache Center & Research Institute	OnabotulinumtoxinA: Block the release of CGRP + Fremanezumab: Monoclonal antibody, targeting CGRP	To study the beneficial effects of fremanezumab in migraine patients who currently under the monotherapy treatment of OnabotulinumtoxinA
17.	NCT06173661/2024-01	Fremanezumab/Brigham and Women's Hospital	Monoclonal antibody, targeting CGRP	To determine the safety and effectiveness of fremanezumab in menstrual migraine patient with the age of 18–45 years

catecholamine (candesartan). More than one mechanism is applicable, particularly when two or more drugs (naltrexone-acetaminophen; botox + fremanezumab; cannabidiol + cannabigerol + tetrahydrocannabinol) are used in combination to get the synergistic effect or better treatment compared to a single drug or existing treatment. Though many biological targets are discovered, identifying the exact mechanism involved in the pathogenesis of migraines will help researchers to develop safe and potent drugs for migraine disorders.

References

Aguilar-Shea AL, Membrilla Md JA, Diaz-de-Teran J (2022) Migraine review for general practice. Aten Primaria 54(2):102208. https://doi.org/10.1016/j.aprim.2021.102208

Andreou AP, Fuccaro M, Lambru G (2020) The role of erenumab in the treatment of migraine. Ther Adv Neurol Disord 13:1756286420927119. https://doi.org/10.1177/1756286420927119

Arulmozhi DK, Veeranjaneyulu A, Bodhankar SL (2005) Migraine: current concepts and emerging therapies. Vasc Pharmacol 43(3):176–187. https://doi.org/10.1016/j.vph.2005.07.001

Ashkenazi A, Blumenfeld A, Napchan U, Narouze S, Grosberg B, Nett R, DePalma T, Rosenthal B, Tepper S, Lipton RB, Interventional Procedures Special Interest Section of the American (2010) Peripheral nerve blocks and trigger point injections in headache management—a systematic review and suggestions for future research. Headache 50(6):943–952

Atalay S, Jarocka-Karpowicz I, Skrzydlewska E (2020) Antioxidative and anti-inflammatory properties of cannabidiol. Antioxidants 9(1):21. https://doi.org/10.3390/antiox9010021

Beydoun A, D'Souza J, Hebert D, Doty P (2009) Lacosamide: pharmacology, mechanisms of action and pooled efficacy and safety data in partial-onset seizures. Expert Rev Neurother 9(1):33–42. https://doi.org/10.1586/14737175.9.1.33

Blumenfeld A, Ashkenazi A, Napchan U, Bender SD, Klein BC, Berliner R, Ailani J, Schim J, Friedman DI, Charleston L IV, Young WB, Robertson CE, Dodick DW, Silberstein SD, Robbins MS (2013) Expert consensus recommendations for the performance of peripheral nerve blocks for headaches—a narrative review. Headache 53(3):437–446

Bogenschutz MP, Forcehimes AA, Pommy JA, Wilcox CE, Barbosa PC, Strassman RJ (2015) Psilocybin-assisted treatment for alcohol dependence: a proof-of-concept study. J Psychopharmacol 29(3):289–299. https://doi.org/10.1177/0269881114565144

Bowdle TA, Knutsen LJS, Williams M (2007) 6.15—Local and adjunct anesthesia. In: Taylor JB, Triggle DJ (eds) Comprehensive medicinal chemistry II. Elsevier, pp 351–367. ISBN 9780080450445. https://doi.org/10.1016/B0-08-045044-X/00175-9

Bulsara KG, Makaryus AN (2024) Candesartan. In: StatPearls [Internet]. StatPearls Publishing, Treasure Island, FL. PMID: 30137786

Chandrasekharan NV (2007) Ketoralac. In: Enna SJ, Bylund DB (eds) xPharm: the comprehensive pharmacology reference. Elsevier, pp 1–6. ISBN 9780080552323. https://doi.org/10.1016/B978-008055232-3.61985-3

Clemow DB, Johnson KW, Hochstetler HM, Ossipov MH, Hake AM, Blumenfeld AM (2020) Lasmiditan mechanism of action—review of a selective 5-HT1F agonist. J Headache Pain 21(1):71. https://doi.org/10.1186/s10194-020-01132-3

Daniel J, Haberman M (2018) Clinical potential of psilocybin as a treatment for mental health conditions. Ment Health Clin 7(1):24–28. https://doi.org/10.9740/mhc.2017.01.024

Deeks ED (2022) Atogepant: first approval. Drugs 82(1):65–70. https://doi.org/10.1007/s40265-021-01644-5

Delaruelle Z, Ivanova TA, Khan S, Negro A, Ornello R, Raffaelli B, Terrin A, Mitsikostas DD, Reuter U, European Headache Federation School of Advanced Studies (EHF-SAS) (2018)

Male and female sex hormones in primary headaches. J Headache Pain 19(1):117. https://doi.org/10.1186/s10194-018-0922-7
Devoto M, Lozito A, Staffa G, D'Alessandro R, Sacquegna T, Romeo G (1986) Segregation analysis of migraine in 128 families. Cephalalgia 6(2):101–105. https://doi.org/10.1046/j.1468-2982.1986.0602101.x
Dhillon S (2020) Eptinezumab: first approval. Drugs 80(7):733–739. https://doi.org/10.1007/s40265-020-01300-4
Dhillon S (2023) Zavegepant: first approval. Drugs 83(9):825–831. https://doi.org/10.1007/s40265-023-01885-6. Erratum in: Drugs. 2023;83(11):1063
Dodick DW, Lipton RB, Ailani J, Lu K, Finnegan M, Trugman JM, Szegedi A (2019) Ubrogepant for the treatment of migraine. N Engl J Med 381(23):2230–2241. https://doi.org/10.1056/NEJMoa1813049
Dooley M, Faulds D (1999) Rizatriptan. Drugs 58:699–723. https://doi.org/10.2165/00003495-199958040-00013
Eigenbrodt AK, Ashina H, Khan S, Diener HC, Mitsikostas DD, Sinclair AJ, Pozo-Rosich P, Martelletti P, Ducros A, Lantéri-Minet M, Braschinsky M, Del Rio MS, Daniel O, Özge A, Mammadbayli A, Arons M, Skorobogatykh K, Romanenko V, Terwindt GM, Paemeleire K, Sacco S, Reuter U, Lampl C, Schytz HW, Katsarava Z, Steiner TJ, Ashina M (2021) Diagnosis and management of migraine in ten steps. Nat Rev Neurol 17(8):501–514. https://doi.org/10.1038/s41582-021-00509-5
Filla A, Rinaldi C (2010) Idebenone and Friedreich ataxia. In: Kompoliti K, Metman LV (eds) Encyclopedia of movement disorders. Academic, pp 63–66. ISBN 9780123741059. https://doi.org/10.1016/B978-0-12-374105-9.00184-2
Garcia-Martinez LF, Raport CJ, Ojala EW, Dutzar B, Anderson K, Stewart E, Kovacevich B, Baker B, Billgren J, Scalley-Kim M, Karasek C, Allison D, Latham JA (2020) Pharmacologic characterization of ALD403, a potent neutralizing humanized monoclonal antibody against the calcitonin gene-related peptide. J Pharmacol Exp Ther 374(1):93–103. https://doi.org/10.1124/jpet.119.264671
Gasperi V, Sibilano M, Savini I, Catani MV (2019) Niacin in the central nervous system: an update of biological aspects and clinical applications. Int J Mol Sci 20(4):974. https://doi.org/10.3390/ijms20040974. PMID: 30813414; PMCID: PMC6412771
Goschorska M, Gutowska I, Baranowska-Bosiacka I, Barczak K, Chlubek D (2020) The use of antioxidants in the treatment of migraine. Antioxidants (Basel) 9(2):116. https://doi.org/10.3390/antiox9020116
Greco MC, Capuano A, Navarra P, Tringali G (2016) Lacosamide inhibits calcitonin gene-related peptide production and release at trigeminal level in the rat. Eur J Pain 20(6):959–966. https://doi.org/10.1002/ejp.820
Haghdoost F, Togha M (2022) Migraine management: non-pharmacological points for patients and health care professionals. Open Med (Wars) 17(1):1869–1882. https://doi.org/10.1515/med-2022-0598
Hargreaves RJ (2000) Pharmacology and potential mechanisms of action of rizatriptan. Cephalalgia 20(Suppl 1):2–9. https://doi.org/10.1046/j.1468-2982.2000.020s1002.x
Harlow AF, Hendricks PS, Leventhal AM, Barrington-Trimis JL (2024) Psychedelic microdosing among young adults from Southern California. J Psychoactive Drugs 10:1–12. https://doi.org/10.1080/02791072.2024.2313684
Jewell R (2007) Tizanidine. In: Enna SJ, Bylund DB (eds) xPharm: the comprehensive pharmacology reference. Elsevier, pp 1–6. ISBN 9780080552323. https://doi.org/10.1016/B978-008055232-3.62773-4
Johnson MW, Garcia-Romeu A, Cosimano MP, Griffiths RR (2014) Pilot study of the 5-HT2AR agonist psilocybin in the treatment of tobacco addiction. J Psychopharmacol 28(11):983–992. https://doi.org/10.1177/0269881114548296
Jones A, Tepper S, Lipton R, Tabuteau H (2022) Efficacy and safety of AXS-07 (MoSEIC Meloxicam-Rizatriptan) for the acute treatment of migraine: results from the INTERCEPT

phase 3, randomized, double-blind, placebo-controlled trial (P14-2.004). Neurology 98(18_Suppl). https://doi.org/10.1212/WNL.98.18_supplement.1112

Kashyap PV, Chabri S (2023) Steroids in headache: a comprehensive review of recent research. Ann Neurosci 30(4):256–261. https://doi.org/10.1177/09727531231173286

Lamb YN (2018) Galcanezumab: first global approval. Drugs 78(16):1769–1775. https://doi.org/10.1007/s40265-018-1002-7

Lamb YN (2019) Lasmiditan: first approval. Drugs 79(18):1989–1996. https://doi.org/10.1007/s40265-019-01225-7. PMID: 31749059

Lee MC, Abrahams M (2012) Chapter 18—Pain and analgesics. In: Bennett PN, Brown MJ, Sharma P (eds) Clinical pharmacology, 11th edn. Churchill Livingstone, pp 278–294. ISBN 9780702040849. https://doi.org/10.1016/B978-0-7020-4084-9.00057-4

Lipton RB, Silberstein SD (2015) Episodic and chronic migraine headache: breaking down barriers to optimal treatment and prevention. Headache 55(Suppl 2):103–122; quiz 123–126

Liu Y, Wang L, Wan F, Yang N (2020) Effects of anisodine hydrobromide on the cardiovascular and respiratory functions in conscious dogs. Drug Des Devel Ther 14:4263–4276. https://doi.org/10.2147/DDDT.S268113

Lyall M, de Oliveira BR, Mody SK (2023) Considerations for contraceptive use among patients with migraines. Curr Obstet Gynecol Rep 12:57–63. https://doi.org/10.1007/s13669-023-00349-8

Mahmoodi AN, Kim PY (2024) Ketorolac. In: StatPearls [Internet]. StatPearls Publishing, Treasure Island, FL. PMID: 31424756

Markham A (2018) Erenumab: first global approval. Drugs 78(11):1157–1161. https://doi.org/10.1007/s40265-018-0944-0

Merikangas KR, Risch NJ, Merikangas JR, Weissman MM, Kidd KK (1988) Migraine and depression: association and familial transmission. J Psychiatr Res 22(2):119–129. https://doi.org/10.1016/0022-3956(88)90076-3

Millán-Guerrero RO, Pineda-Lucatero AG, Hernández-Benjamín T, Tene CE, Pacheco MF (2003) Nalpha-methylhistamine safety and efficacy in migraine prophylaxis: phase I and phase II studies. Headache 43(4):389–394

Nikfarjam Z, Doustkhah E, Zamani F, Brown RW (2022) Chapter 5—Pharmaceutical applications of 1,4-benzodiazepines. In: Zamani F, Doustkhah E (eds) Heterocyclic drug discovery, benzodiazepine-based drug discovery. Elsevier, pp 125–182. ISBN 9780128245163. https://doi.org/10.1016/B978-0-12-824516-3.00009-4

Ogino K, Saito K, Osugi T, Satoh H (2002) Meloxicam (Mobic): a review of its pharmacological and clinical profile. Nihon Yakurigaku Zasshi 120(6):391–397. Japanese. https://doi.org/10.1254/fpj.120.391

Pereira J, Wiegand T (2014) Marijuana. In: Wexler P (ed) Encyclopedia of toxicology, 3rd edn. Academic, pp 157–159. ISBN 9780123864550. https://doi.org/10.1016/B978-0-12-386454-3.00745-4

PubChem [Internet] (2004) PubChem compound summary for CID 11972243, Anisodine hydrobromide. National Library of Medicine (US), National Center for Biotechnology Information, Bethesda, MD. https://pubchem.ncbi.nlm.nih.gov/compound/Anisodine-hydrobromide

Ray JC, Hutton EJ, Matharu M (2021) OnabotulinumtoxinA in migraine: a review of the literature and factors associated with efficacy. J Clin Med 10(13):2898. https://doi.org/10.3390/jcm10132898

Ren C, Liu J, Zhou J, Liang H, Wang Y, Sun Y, Ma B, Yin Y (2018) Low levels of serum serotonin and amino acids identified in migraine patients. Biochem Biophys Res Commun 496(2):267–273. https://doi.org/10.1016/j.bbrc.2017.11.203

Richard DM, Dawes MA, Mathias CW, Acheson A, Hill-Kapturczak N, Dougherty DM (2009) L-Tryptophan: basic metabolic functions, behavioral research and therapeutic indications. Int J Tryptophan Res 2:45–60. https://doi.org/10.4137/ijtr.s2129

Sandyk R (1992) L-Tryptophan in neuropsychiatry disorders: a review. Int J Neurosci 67(1–4):127–144. https://doi.org/10.3109/00207459208994781

Saper JR, Lake AE III, Cantrell DT, Winner PK, White JR (2002) Chronic daily headache prophylaxis with tizanidine: a double-blind, placebo-controlled, multicenter outcome study. Headache 42(6):470–482. https://doi.org/10.1046/j.1526-4610.2002.02122.x

Scott LJ (2020a) Galcanezumab: a review in the prevention of migraine and treatment of episodic cluster headache. Drugs 80(9):893–904. https://doi.org/10.1007/s40265-020-01329-5

Scott LJ (2020b) Rimegepant: first approval. Drugs 80(7):741–746. https://doi.org/10.1007/s40265-020-01301-3

Scott LJ (2020c) Ubrogepant: first approval. Drugs 80(3):323–328. https://doi.org/10.1007/s40265-020-01264-5

Silberstein SD, Dodick DW, Bigal ME, Yeung PP, Goadsby PJ, Blankenbiller T, Grozinski-Wolff M, Yang R, Ma Y, Aycardi E (2017) Fremanezumab for the preventive treatment of chronic migraine. N Engl J Med 377(22):2113–2122. https://doi.org/10.1056/NEJMoa1709038

Singh D, Saadabadi A (2024) Naltrexone. In: StatPearls [Internet], Treasure Island, FL, StatPearls Publishing. PMID: 30521232

Speight TM, Avery GS (1972) Pizotifen (BC-105): a review of its pharmacological properties and its therapeutic efficacy in vascular headaches. Drugs 3:159–203. https://doi.org/10.2165/00003495-197203030-00002

Sprenger T, Viana M, Tassorelli C (2018) Current prophylactic medications for migraine and their potential mechanisms of action. Neurotherapeutics 15(2):313–323. https://doi.org/10.1007/s13311-018-0621-8

Tzabazis A, Kori S, Mechanic J, Miller J, Pascual C, Manering N, Carson D, Klukinov M, Spierings E, Jacobs D, Cuellar J, Frey WH II, Hanson L, Angst M, Yeomans DC (2017) Oxytocin and migraine headache. Headache 57(Suppl 2):64–75. https://doi.org/10.1111/head.13082

Vardanyan RS, Hruby VJ (2006) 2—Local anesthetics. In: Vardanyan RS, Hruby VJ (eds) Synthesis of essential drugs. Elsevier, pp 9–18. ISBN 9780444521668. https://doi.org/10.1016/B978-044452166-8/50002-9

Yu-Wai-Man P, Chinnery PF (2016) Chapter 7—Leber hereditary optic neuropathy. In: Saneto RP, Parikh S, Cohen BH (eds) Mitochondrial case studies. Academic, pp 55–64. ISBN 9780128008775. https://doi.org/10.1016/B978-0-12-800877-5.00007-3

Clinical Trials

NCT02021474. https://www.clinicaltrials.gov/study/NCT02021474
NCT02903680. https://www.clinicaltrials.gov/study/NCT02903680
NCT03432286. https://www.clinicaltrials.gov/study/NCT03432286
NCT03832998. https://www.clinicaltrials.gov/study/NCT03832998
NCT03836040. https://www.clinicaltrials.gov/study/NCT03836040
NCT04007874. https://www.clinicaltrials.gov/study/NCT04007874
NCT04151472. https://www.clinicaltrials.gov/study/NCT04151472
NCT04218539. https://www.clinicaltrials.gov/study/NCT04218539
NCT04271202. https://www.clinicaltrials.gov/study/NCT04271202
NCT04396236. https://www.clinicaltrials.gov/study/NCT04396236
NCT04417361. https://www.clinicaltrials.gov/study/NCT04417361
NCT04458857. https://www.clinicaltrials.gov/study/NCT04458857
NCT04530110. https://www.clinicaltrials.gov/study/NCT04530110
NCT04574713. https://www.clinicaltrials.gov/study/NCT04574713
NCT04616326. https://www.clinicaltrials.gov/study/NCT04616326
NCT04649242. https://www.clinicaltrials.gov/study/NCT04649242
NCT04674020. https://www.clinicaltrials.gov/study/NCT04674020
NCT04686136. https://www.clinicaltrials.gov/study/NCT04686136
NCT04693533. https://www.clinicaltrials.gov/study/NCT04693533
NCT04726592. https://www.clinicaltrials.gov/study/NCT04726592

NCT04743141. https://www.clinicaltrials.gov/study/NCT04743141
NCT04748601. https://www.clinicaltrials.gov/study/NCT04748601
NCT04804033. https://www.clinicaltrials.gov/study/NCT04804033
NCT04845178. https://www.clinicaltrials.gov/study/NCT04845178
NCT04921384. https://www.clinicaltrials.gov/study/NCT04921384
NCT04965675. https://www.clinicaltrials.gov/study/NCT04965675
NCT04989413. https://www.clinicaltrials.gov/study/NCT04989413
NCT05125302. https://www.clinicaltrials.gov/study/NCT05125302
NCT05156398. https://www.clinicaltrials.gov/study/NCT05156398
NCT05207865. https://www.clinicaltrials.gov/study/NCT05207865
NCT05211154. https://www.clinicaltrials.gov/study/NCT05211154
NCT05214001. https://www.clinicaltrials.gov/study/NCT05214001
NCT05217927. https://www.clinicaltrials.gov/study/NCT05217927
NCT05337033. https://www.clinicaltrials.gov/study/NCT05337033
NCT05371652. https://www.clinicaltrials.gov/study/NCT05371652
NCT05399485. https://www.clinicaltrials.gov/study/NCT05399485
NCT05411978. https://www.clinicaltrials.gov/study/NCT05411978
NCT05416476. https://www.clinicaltrials.gov/study/NCT05416476
NCT05452239. https://www.clinicaltrials.gov/study/NCT05452239
NCT05484349. https://www.clinicaltrials.gov/study/NCT05484349
NCT05503082. https://www.clinicaltrials.gov/study/NCT05503082
NCT05509400. https://www.clinicaltrials.gov/study/NCT05509400
NCT05518123. https://www.clinicaltrials.gov/study/NCT05518123
NCT05550207. https://www.clinicaltrials.gov/study/NCT05550207
NCT05632133. https://www.clinicaltrials.gov/study/NCT05632133
NCT05679908. https://www.clinicaltrials.gov/study/NCT05679908
NCT05685225. https://www.clinicaltrials.gov/study/NCT05685225
NCT05707754. https://www.clinicaltrials.gov/study/NCT05707754
NCT05711394. https://www.clinicaltrials.gov/study/NCT05711394
NCT05724771. https://www.clinicaltrials.gov/study/NCT05724771
NCT05748483. https://www.clinicaltrials.gov/study/NCT05748483
NCT05846373. https://www.clinicaltrials.gov/study/NCT05846373
NCT05861427. https://www.clinicaltrials.gov/study/NCT05861427
NCT05897320. https://www.clinicaltrials.gov/study/NCT05897320
NCT05989048. https://www.clinicaltrials.gov/study/NCT05989048
NCT06016465. https://www.clinicaltrials.gov/study/NCT06016465
NCT06047444. https://www.clinicaltrials.gov/study/NCT06047444
NCT06047457. https://www.clinicaltrials.gov/study/NCT06047457
NCT06083571. https://www.clinicaltrials.gov/study/NCT06083571
NCT06089356. https://www.clinicaltrials.gov/study/NCT06089356
NCT06158230. https://www.clinicaltrials.gov/study/NCT06158230
NCT06158737. https://www.clinicaltrials.gov/study/NCT06158737
NCT06173661. https://www.clinicaltrials.gov/study/NCT06173661

Future Perspectives for the Management of Migraine Pain

13

Riya Patil, Aashka Thakur, Tania Purswani, Bappaditya Chatterjee, and Saritha Shetty

13.1 Introduction

An estimated 50% of adults worldwide are reported to suffer from migraine condition. In the past year, half to three quarters of persons worldwide between the ages of 18 and 65 reported having a headache, and at least 30% of those people reported having a migraine (Headache disorders 2023). A primary headache illness called migraine is characterised by recurrent attacks. The disorder is caused by the activation of a deep brain function that produces inflammatory chemicals that causes pain around the blood vessels and nerves in the head. Migraine is classified mainly by those who have "suffered from at least eight terrible headache days per month for at least three consecutive months without improvement and having failed at least three classes of migraine preventatives" and are known to have resistant migraines. Secondly, by those that "have failed all available preventatives and suffer from at least eight debilitating headache days per month for at least six consecutive months". Recurrent episodes of disabling headaches, frequently coupled with sensory and motor abnormalities, are the hallmark of migraine disease. The result is affected by a number of factors, including genetics, age, sex, and environment. It is generally acknowledged that migraine is food-sensitive and that certain foods, such as chocolate, dairy products, alcoholic beverages, nuts, ice cream, tomatoes, onions, tyramine, histamine, phenyl ethylamine, nitrites, aspartame, sucralose, and gluten can cause migraine attacks (Gazerani 2020). Globally, young adults and women are disproportionately affected by migraine. Comorbidities associated with migraine can range from sleep difficulties and stress to suicidal thoughts and actions. Numerous biological and social risk

R. Patil · A. Thakur · T. Purswani · S. Shetty (✉)
Department of Pharmaceutics, Shobhaben Pratapbhai Patel School of Pharmacy & Technology Management, SVKMS NMIMS University, Mumbai, India

B. Chatterjee
School of Pharmacy, GITAM University, Hyderabad, India

S. Bhatt et al. (eds.), *Management of Migraine Pain*,
https://doi.org/10.1007/978-981-97-4529-6_13

factors such as hormonal abnormalities, genetic and epigenetic modifications, neurological, immunological, and cardiovascular illnesses, have been proposed as potential causes of migraines due to the intricate and largely unknown processes that underlie their development. The trigeminovascular system (TGVS) is thought to be the cause of discomfort at this time. To date, it is believed that the trigeminovascular system (TGVS) is the source of this discomfort. The TGVS is a crucial channel for the transmission of pain and serves as a significant control centre for cerebral blood flow. Vasoactive chemicals such as substance P, neurokinin A, and calcitonin gene-related peptide (CGRP) are released in response to the stimulation of trigeminal sensory nerve endings. This leads to vasodilation and dural plasma extravasation, which ultimately results in neurogenic inflammation (Tardiolo et al. 2019). Modern migraine treatment includes medication designed to reduce the frequency, severity, and duration of migraine attacks. Triptans or dihydroergotamine (DHE) are administered for moderate-to-severe migraine attacks; acetaminophen and aspirin are used for mild attacks. Additionally, there are brand-new FDA-approved devices (nerivio®, cefaly®, reliviion®, saphire™) that can both reduce the frequency of episodes and end them once they start (Tiwari and Agrawal 2022). These are known as neuromodulation devices and they prevent the progression of migraines by applying electric or magnetic pulses. They can be obtained with a prescription and are quite simple to use. Numerous dietary factors for migraine have been found, which has led to the development of helpful diets including ketogenic diets, exclusion diets, and comprehensive diets to avoid migraine. The current therapies for episodic migraine are frequently ineffective in treating chronic migraineurs and may even increase the frequency and severity of migraine attacks. This chapter aims to provide an overview of translational challenges for the current migraine therapeutic strategies and novel therapeutic targets in migraine, with future prospects for the treatment of migraines, like currently investigating additional novel neuromodulation tools and present various pharmacoeconomic models to improve the quality of life (QOL).

13.2 Epidemiology

Globally, migraine is a prevalent neurological condition with a high incidence and morbidity rate, particularly in young adults and women as shown in Fig. 13.1 (Gazerani 2020). Over the past three decades, migraine prevalence has significantly increased worldwide (Hindiyeh et al. 2020). Between the ages of 3 and 7, males had a higher prevalence of migraines. Between the ages of 7 and 11, both genders had an equal prevalence. Only in girls did the prevalence of migraines increase beyond the age of 11. It varies between 3.3% and 21.9% for women, and 0.7% and 16.1% for males among adults (Sahithi et al. 2020). The information at hand supports the co-occurrence of other neurological conditions with migraine, emphasising the part that the same molecular pathways play in the pathophysiology of these conditions (Schembri et al. 2022). In order to evaluate other aspects and data on pain characteristics, co-morbidities, impact on quality of life, impact on family, and migraine risk factors, epidemiological research is crucial in assessing the prevalence and incidence of migraine, especially chronic migraine (Schembri et al. 2022).

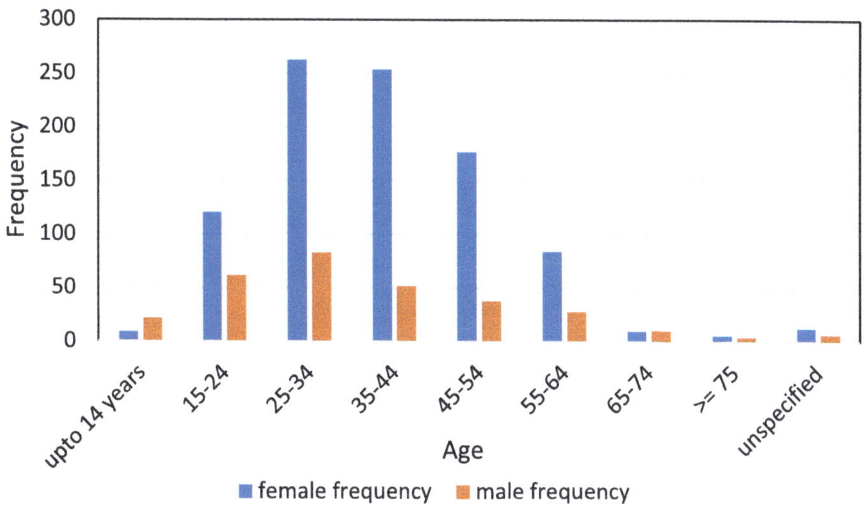

Fig. 13.1 Age and sex-wise distribution of patients suffering from migraines (Amiri et al. 2022)

13.3 Neuromodulation in Migraine

To activate, alter, or modify the actions within the central nervous system (CNS) and peripheral nervous system (PNS), neuromodulation involves the use of electrical, magnetic, or chemical stimulation. Neuromodulation is typically a non-harmful and flexible procedure. It is becoming increasingly clear that this technology can offer more effective and safe treatment for migraines when compared to conventional medication therapies. The neuromodulation techniques and their site of action are given in Fig. 13.2. Neuromodulation devices can be invasive, non-invasive, portable, or require surgical placement (Tiwari and Agrawal 2022).

Some examples of FDA approved devices are (Hindiyeh et al. 2020):

1. The Gamma Core Sapphire™, which is a portable, non-invasive vagus nerve stimulator (nVNS) designed for migraineurs 12 years of age and up. It can be used for both acute and preventive pain management.
2. The FDA has approved Nerivio®, a remote electronic neuromodulation (REN) device, for the acute treatment of episodic and chronic migraines in patients 12 years of age and older.
3. The electrical trigeminal nerve stimulation (eTNS) device, Cefaly® Dual, is used to treat adult patients' migraines both acutely and proactively.
4. For patients 12 years of age and older, the sTMS mini™ single pulse. Transcranial magnetic stimulation is used for both acute and preventative therapy of migraines.
5. Relivion®, an Combined Occipital and Trigeminal Nerve Stimulation device for acute treatment of migraine (Yuan and Chuang 2021).

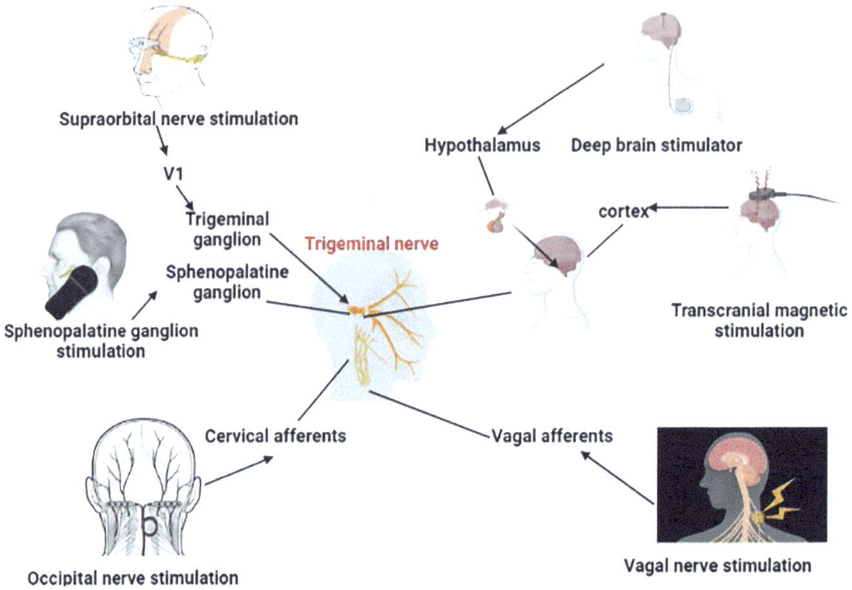

Fig. 13.2 Neuromodulation techniques and their site of action (Tardiolo et al. 2019)

13.3.1 Mechanism of Action

Neuromodulation employs electrical or magnetic signals to alter or activate the central or peripheral nervous pathways. This treatment attempts to modulate the pain systems in an effort to reduce the pain ratings. These electrical or magnetic pulsations of pain pathways can cause changes in central neurotransmitters. These modulations could be applied to treat acute migraine attacks by interfering with the mechanisms that set off the attack. As a preventive approach, these changes aim to decrease the central sensitisation that causes chronic headaches (Tiwari and Agrawal 2022).

13.3.1.1 An Individual-Pulse Transcranial Magnetic Stimulator

The single-pulse transcranial magnetic stimulator is a console device that modifies electrical signals in the brain by means of producing a magnetic impulse. Recently, it has also been approved to prevent migraines. It results in a varying magnetic field, which causes an electric current to flow, and changes the excitability of thalamocortical networks and cortical neurons. Meningeal nociceptors are hypothesised to be activated in migraine patients due to a condition of brain hyper excitability that lowers the threshold for cortical spreading depression, either directly or indirectly via descending assistance channels. In migraineurs, this stimulation has been linked to the onset of aura. Repeated transcranial magnetic stimulation reduces cortical hyper excitability by controlling glutamate and dopamine levels in the caudate and hippocampal regions. Therefore, long-term changes in neuronal excitability, a decrease in central sensitization, and a reduction in the frequency of headaches are all possible with sustained transcranial magnetic stimulation (Andreou et al. 2016).

13.3.1.2 Vagus Nerve Stimulator Without Invasion

The cervical branch of the tenth cranial nerve receives transcutaneous stimulation from the gamma core device, a portable electrical stimulator. The limbic system, thalamus, locus coeruleus, nucleus tractus solitarius, dorsal pons, and other areas that imaging studies have revealed as components of the composition of headache pain have all been demonstrated to be inhibited by continuous VNS. For people with chronic headaches, it may reduce pain and potentially even reverse central sensitisation by lowering glutamate levels in the trigeminal nucleus caudalis. The primary aim of this treatment is the cervical branch of the tenth cranial nerve (Silberstein et al. 2017).

13.3.1.3 Supraorbital Nerve Stimulator

Supraorbital nerve stimulation produces an immediate sedative effect in healthy persons. The current understanding of the pathophysiology of migraines suggests that central sensitisation plays a role in the disease's progression from intermittent to persistent, and that the headache phase is caused by stimulation of the trigeminovascular system and neuronal hyper excitability of the pain pathways (Urits et al. 2020). Activation of the supraorbital nerve may block the trigeminal sensory pain channels by regulating both peripheral and central activity in the trigeminal vascular system (Vikelis et al. 2017).

13.4 Diagnostic Accuracy of Artificial Intelligence in Migraine (Silberstein et al. 2017)

Artificial intelligence (AI) has been increasingly used to improve the diagnostic accuracy of migraine pain. Promising outcomes by Nicolas Vandenbussche et al. have been observed in studies utilising artificial intelligence (AI) and machine learning (ML) algorithms to diagnose migraines and differentiate them from other headache diseases. Using the International Classification of Headache Disorders, Third Edition (ICHD-3) criteria, a multi-centre study evaluated the concordance in migraine diagnosis between a semi-structured interview (SSI) conducted by a headache specialist and an online, self-administered computer-based diagnostic engine (CDE) (Tardiolo et al. 2019). The above conducted study found that the SSI and CDE had excellent concordance in diagnosing migraine/probable migraine. The CDE showed high specificity and a positive likelihood ratio for ruling in migraine/probable migraine and high sensitivity and a low-negative likelihood ratio for ruling out migraine/probable migraine. Hence, the above study concludes that machine learning and natural language processing (NLP) could accurately identify between cluster headache and migraine, discovering patterns in the words patients used to describe their headache symptoms with an accuracy rate of nearly 86%. These studies suggest that AI and ML have the potential to improve the diagnostic accuracy of migraine pain and help healthcare professionals in making more informed decisions.

13.5 Dietary Triggers and Dietary Modifications in Migraine

Certain dietary elements might cause migraine attacks, as they are sensitive to food. Medication can prevent and treat migraines, but its effectiveness is limited, and if taken excessively, it can have harmful side effects such as headaches and increased cardiovascular risks. Dietary changes may be a viable treatment for migraine because the identification of migraine triggers is influenced by a number of internal (e.g., genetic and epigenetic) and external (e.g., culture, ethnicity, geography, and eating habits) factors (Gazerani 2020). Although it is well recognised that food triggers affect migraine, less research has been done on the underlying mechanisms. According to research involving aspartame or MSG, certain foods or ingredients may need to be consumed in high quantities in order to cause a headache. Headaches can be enhanced with intake of few foods, while overindulging in some ingredients like caffeine—can also result in headaches. Certain patient subgroups, such as those with celiac sprue or those who experience specific food-related immune reactions, such as those who test positive for IgG antibodies, may only experience headaches in response to specific foods or substances (Gazerani 2020). Common diet-related triggers such as alcohol, fasting, food allergens and caffeine, chocolate, cheese, and citrus have a substantial correlation between migraine and changes in plasma oestrogen concentrations during menstrual periods (Johnson et al. 2013). In addition to food, a variety of additional triggers, such as stress, certain types of weather, and dehydration can initiate the migraine process (Cortés et al. 2011).

For migraineurs, there are two primary approaches to dietary modification: eliminating trigger foods and implementing general dietary adjustments to potentially lessen the frequency or severity of migraine attacks. Regretfully, no one can promise that either strategy will be effective for them. Regular migraine attacks may be avoided if you prioritise anti-inflammatory foods in your diet and are aware of potential dietary triggers. In particular, it has been shown in multiple trials that eating meals high in omega-3 fatty acids helps migraineurs experience fewer headache days. These consist of plant-based foods like walnuts, chia seeds, and flaxseed as well as fatty cold-water fish like salmon, cod, and lake trout. Furthermore, some research suggests that some migraineurs can benefit from a high-fat, low-carb ketogenic diet. An IgG antibody test, which examines a particular immune reaction in your body to minute amounts of numerous food items, may be a helpful technique to detect migraine food triggers (Migraine diet modifications: what really works? 2023). SEEDS (Sleep, Exercise, Eat, Diary, and Stress) are best practices that primary care physicians have lately recommended to aid migraine patients by helping them to make small lifestyle changes.

13.6 Gut Microbiota and Migraine

Recent research has shown that gut microbiota is important in the pathogenesis of migraine. The gut microbiota may have a role in neurological conditions like migraines by mediating gut-brain communication (Crawford et al. 2022). Research published in the Journal of Headache and Pain suggests that migraine, migraine with aura (MA), and migraine without aura (MO) may be caused by differences in the gut microbiomes. The study provided additional evidence that abnormalities in the gut-brain axis are the cause of migraines (Yong et al. 2023). The makeup of the gut microbiota was different in migraineurs than in normal individuals. It is clearly shown that a certain genus linked to headache frequency and severity demonstrate the possible influence of gut microbiota on migraine features. A metagenome-wide association study was published by Frontiers in Cellular and Infection Microbiology to look at the connection between gut microbiota and migraine. It demonstrated a notable decline in metabolic processes and species diversity in the gut microbiota of migraineurs, suggesting a possible gut-brain axis connection between gut microbiota and migraine (Migraine diet modifications: what really works? 2023).

Comprehending these alterations could result in customised, microbiome-driven, non-opioid migraine therapies, such as dietary modifications or faecal microbiota transplantation (He et al. 2023). Figure 13.3 represents the relationship between the gut-brain axis and gut microbiota in migraine.

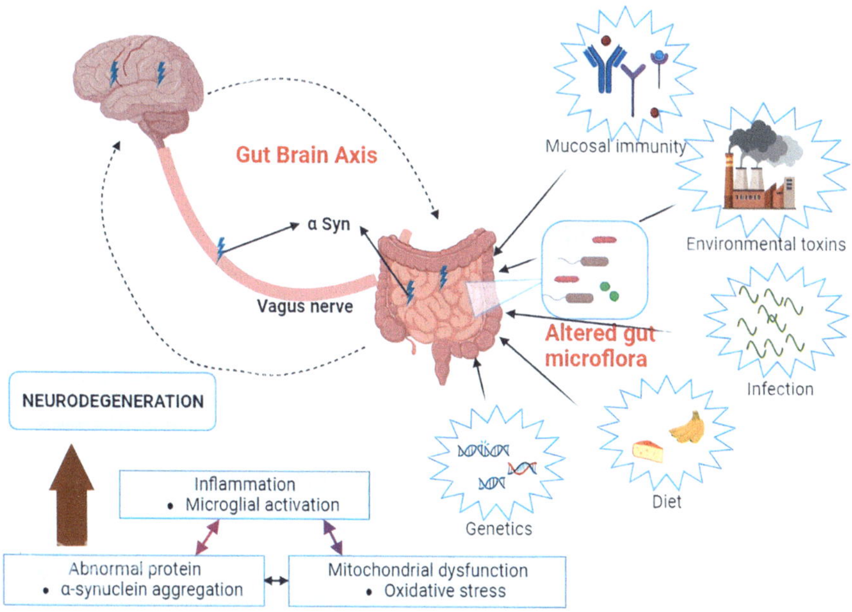

Fig. 13.3 Interpretation of gutmicrobiota axis in migraine

13.7 Translational Challenges for the Current Migraine Therapeutic Strategies

Numerous neurological and neuropsychiatric conditions have responded well to therapies such as deep brain stimulation (DBS), intracranial cortical stimulation (ICS), transcranial direct current stimulation (tDCS), and transcranial magnetic stimulation (TMS). A few fundamental translational challenges are often pertinent to improving neuromodulation efficacy.

One problem is the lack of understanding regarding the neurophysiological principles underlying DBS, which restricts the potential for improving stimulation and targeting, for more reliable and efficient treatment. This is partly because of a lack of animals available to allow therapists to more accurately shape the electric fields surrounding a DBS lead. Hence, to overcome these, certain parameters are to be taken into consideration to increase the consistency of therapy across patients like careful patient selection, imaging to identify cortical regions suitable for neuromodulation, placement of neurosurgical arrays, consistency of stimulation parameters, and patient-specific computational models to confirm that the stimulation settings are appropriate for activation of the desired cortical region. But a point to remember is that not all mental illnesses respond well to cortical activation, especially those with subcortical pathophysiology. Therefore, it is essential that cortical stimulation be thoroughly explored in animal models going forward in order to inform the translation of DBS method. More characterisation of the physiological alterations linked to therapeutic repetitive transcranial magnetic stimulation (rTMS) is necessary to enable the creation of stimulation parameters and TMS machine designs that are more effective. Despite being used as a neuromodulation therapy on a large number of people, TMS' physiological mechanisms of action still need to be further investigated in order to increase therapeutic efficacy and explore new avenues. There are also many obstacles in the way of functional neuroimaging of transcranial direct current stimulation (tDCS). There haven't been many attempts to simultaneously record electroencephalogram (EEG) when tDCS stimulation is being applied. EEG was only obtained prior to and following a tDCS stimulation period in the majority of tDCS-EEG experiments. The main reason for this is that simultaneous EEG recordings are not a good fit for conventional sponge electrodes. These electrodes cause broad artefacts in the EEG signal and interfere with EEG recordings at electrode positions that coincide with tDCS electrode locations. They are also heavy and challenging to place beneath an EEG cap. High-density tDCS electrodes, on the other hand, are far more appropriate for use during simultaneous EEG. EEG data was successfully recorded at 24 electrode locations during high density tDCS in pilot research that was recently published by Mathew D. Johnson et al. (2013).

13.8 Novel Therapeutic Targets in Migraine

Recent studies have identified numerous novel therapeutic targets for migraine, such as pituitary adenylate cyclase-activating polypeptide (PACAP), δ-opioid receptor (DOR), transient receptor potential ion channels (TRP), adenosine, and acid-sensing ion channels (ASIC) (He et al. 2023).

A study conducted by Areeba Nisar et al. looked into the relationship between gut microbiota and migraine development and was published in The Journal of Headache and Pain. It revealed a number of bacterial taxonomies, linked causally to migraines, MA, and MO, offering further proof of the disruption of the gut-brain axis in migraines (Nisar et al. 2023). The study by Thien Phu Do et al. demonstrated the possible influence of gut microbiota on migraine features by identifying particular genera linked to headache frequency and intensity (Nisar et al. 2023). A meta genome-wide association study of gut microbiome in patients with multiple sclerosis revealed novel disease pathology was carried out by Toshihiro Kishikawa et al. in Frontiers in Cellular and Infection Microbiology to look at the connection between gut microbiota and migraine. The gut microbiota of migraineurs showed a marked reduction in species diversity and metabolic activities, suggesting a possible gut-brain axis connection between gut microbiota and migraine (Do et al. 2019). Further research is needed to elucidate the underlying mechanisms and validate the potential of targeting the gut microbiota for managing migraines, given its potential substantial role in migraine pathophysiology. Understanding these changes could enable the tailoring of non-opioid migraine treatments based on the microbiome, such as fecal microbiota transplantation or dietary adjustments (Do et al. 2019). Apart from the gut microbiota, gepants, which are antagonists of the calcitonin gene-related peptide (CGRP) receptor, anti-CGRP monoclonal antibodies (mAbs), and 5-HT1F receptor agonists (alniditans and lasmiditans) are other innovative treatment approaches for migraines (Do et al. 2019).

13.9 Emerging Drugs for Migraine Treatment

These days, migraine treatment is developing at a very quick pace. These findings have led to the development of novel, targeted, acute, and preventative medications, including ditans (5-HT1F receptor agonists), gepants (calcitonin gene-related peptide (CGRP) receptor antagonists), and anti-CGRP monoclonal antibodies (mAbs) (Do et al. 2019). Migraine treatment can be divided into two categories: preventative and abortive. While abortive therapy consists of medications that are typically given promptly to halt or minimize symptoms, preventive therapies seek to reduce the frequency, intensity, or both of migraine attacks (Ogunlaja et al. 2021). List of recent drugs used in migraine treatment are given in Table 13.1.

Table 13.1 Emerging drugs for migraine treatment (Drugs.com 2023)

Sr. no.	Drug name	Drug class	Dosage form	Administration
1.	Zavegepant	Calcitonin gene-related peptide (CGRP) receptor antagonist (gepant)	Nasal spray	One dosage (10 mg into one nostril), if necessary; It is recommended to take no more than one dose in a 24-h period (the safety of taking more than eight doses in a 30-day period has not been established)
2.	Rizatriptan	Serotonin $(5\text{-HT})_{1B/1D}$ receptor agonist (triptan)	Oral film	Adults are not allowed to take more than three 10-mg oral films in a 24-h period, but they are allowed to take a second dose of 10 mg if needed, at least 2 h after the first dose
3.	Dihydroergotamine mesylate	Ergotamine derivative	Nasal spray	Repeat a dose (0.725 mg into each nostril) as needed, but do not use more than two doses in a 24-h period or three doses in a 7-day period. Do this at least 1 h after the first dose
4.	Rimegepant	Calcitonin gene-related peptide (CGRP) receptor antagonist (gepant)	Orally disintegrating tablets	When necessary, take one dose (75 mg) no more than once per 24 h (it has not been determined if taking more than 18 doses in a 30-day period is safe)
5.	Atogepant	Calcitonin gene-related peptide (CGRP) receptor antagonist (gepant)	Tablets	Once daily
6.	Eptinezumab-jjmr	Calcitonin-gene related peptide antagonist	Intravenous infusion	Once every 3 months
7.	Lasmiditan	Serotonin $(5\text{-HT})_{1F}$ receptor agonist (ditan)	Tablets	As needed; a maximum of one dose should be taken in a 24-h period
8.	Sumatriptan	Serotonin $(5\text{-HT})_{1B/1D}$ receptor agonist (triptan)	Nasal spray	As needed; doses should be spaced out by at least an hour, with a maximum of 30 mg to be taken in a 24-h period

13.10 Emerging Pharmacological Treatments for Migraine in Different Populations

Based on new findings in migraine research, these focused medicines have surfaced and represent a paradigm shift in migraine treatment, giving patients and physicians renewed hope. A large percentage of people worldwide suffer from migraines, a common and crippling neurological condition. Acute and preventive medicines are among the available treatment options for migraine, but they are frequently insufficient, therefore new therapeutic targets are required to enhance migraine management (Agostoni et al. 2019).

Several innovative pharmacological treatments for migraines have been identified in recent times. These comprise gepants, acting as antagonists of the calcitonin gene-related peptide (CGRP) receptor, monoclonal antibodies (mAbs) targeting anti-CGRP, and agonists for the 5-HT1F receptor (Nisar et al. 2023). Based on new findings in migraine research, these medications have emerged as a revolutionary shift in the treatment of migraines, providing both patients and physicians with a sense of renewed optimism. The study addresses the management of chronic migraines in light of the latest advancements in evidence-based pharmaceutical treatments. Apart from delving into the possibilities offered by monoclonal antibodies targeting the CGRP pathway in migraine pathogenesis, the research underscores the necessity for enhanced preventive medications to alleviate the substantial burden associated with chronic migraines (Do et al. 2019). Scientific reports have investigated variations in gut microbiota between individuals without migraines (controls) and those with episodic migraines (EM) and chronic migraines (CM). The study identified a specific genus that showed associations with the frequency and intensity of headaches, indicating a potential influence of gut microbiota on migraine characteristics (Iannone et al. 2022).

The gut microbiota of migraineurs showed a marked reduction in species diversity and metabolic activities, suggesting a possible gut-brain axis connection between gut microbiota and migraine (Tajti et al. 2023). Pituitary adenylate cyclase-activating polypeptide (PACAP), adenosine, δ-opioid receptor (DOR), potassium channels, transient receptor potential ion channels (TRP), and acid-sensing ion channels (ASIC) are further novel pharmacological treatments for migraine. Through current study by Qiang He et al., several targets have been discovered as potential processes in migraine signalling and nociception. Newly developed pharmaceutical medicines for migraine may enhance migraine care and provide patients and medical professionals fresh hope (Agostoni et al. 2019) as shown in Table 13.1.

13.11 Evaluation of the Pharmacoeconomic Model

When assessing whether migraine medications are cost-effective, pharmacoeconomic models are essential (Fig. 13.4). Economic analyses of pharmaceutical therapy for persons with chronic migraines have been the subject of numerous studies. For example, model-based cost-utility studies were found through a systematic

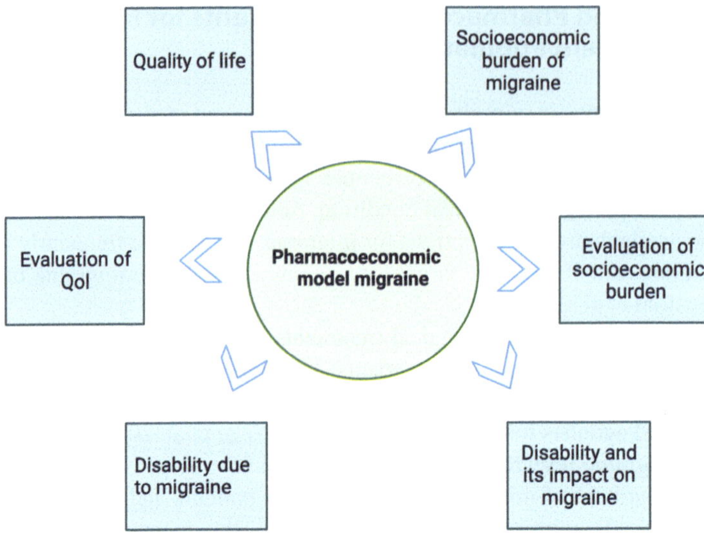

Fig. 13.4 Representation of the pharmacoeconomic model

review, indicating that Erenumab could be a viable treatment for chronic migraine, particularly for those patients who are not responding to other treatments like Botox (Khanal et al. 2022). Furthermore, few literature has also shown, how to calculate the quality of life and the number of migraine days per month. These figures may then be connected to other variables in an economic model to determine how much money will need to be spent on new migraine medications in order to achieve a particular degree of treatment impact (Hu et al. 1999). The aforementioned study supplied valuable information for analysing cost-effectiveness and utilising longitudinal regression models to estimate how the average monthly migraine days are distributed (Hu et al. 1999). Moreover, a Canadian study on sumatriptan justified the financial viability of employing the medication for migraine treatment, emphasising significant cost savings linked to decreased time lost as a result of migraine symptoms (Agostoni et al. 2019).

13.11.1 Evaluation of Disability and the Impact of Migraine

Migraine can impact a person's quality of life (QoL). Many studies have assessed migraines' effects on the economy and disability (Hu et al. 1999). Using population-based estimates of migraine prevalence, attack frequency, disability, and healthcare resource utilization, one study calculated the direct and indirect costs of migraine. According to the study, the majority of a migraine's effects are linked to disability and a decline in functional status, which results in indirect expenses for employers. Overall, migraines are quite expensive for society. The costs are higher than those associated with asthma and similar to those associated with diabetes (Hu et al. 1999).

Model-based cost-utility studies employing fictitious cohorts of chronic migraineurs were found through a thorough examination of economic evaluations of pharmaceutical therapies for individuals with the condition (Khanal et al. 2022). According to the studies by Saval Khanal et al., those who require a change in preventative therapy experience a greater burden from migraines. Average reductions in presenteeism and absenteeism, along with anticipated declines in visits to neurologist and healthcare providers, were linked to an increase of one headache-free day (Khanal et al. 2022). In conclusion, much research has shed light on the financial and social cost associated with migraines, and has validated the assessment of illness impact and incapacity. These results highlight the importance of developing effective treatments for migraines and the need for further research to ascertain the financial viability of these interventions.

13.12 Conclusion

Migraine affects a large portion of the population worldwide. It is a multifaceted disorder that affects the brain and the nerves throughout the body. Migraine is associated with various symptoms such as a throbbing headache, sensitivity to sound, light, smell etc. To minimise these symptoms, advancements in curative techniques become essential. Hence the environmental, genetic, and nutritional factors that were chief precipitaters of migraine need to be detected and studied. Development of (1) instruments used for the diagnosis of migraine that were driven by artificial intelligence, and (2) devices that modulated the nervous system were among these advanced curative techniques for treatment. The segment of the population suffering from migraine has been increasing disturbingly over the globe as stated by the Global Burden of Diseases study that was published in 2019. Thus, there was an urgent need to consider the socioeconomic aspects as well while developing medications for its treatment. The medical treatment for migraine has to be combined with several lifestyle modifications to help alleviate the pain of patients suffering from migraine.

Various innovative routes i.e. the advancement of several neuromodulation methods, which include non-invasive vagus nerve stimulation and transcranial magnetic stimulation, have proven beneficial in the treatment of migraine to a great extent. Artificial intelligence plays a significant role in the diagnosis of migraine in patients. It makes the process of diagnosis more specific and assists medical practitioners in the crucial process of making decisions.

There are various other aspects such as changes in the nutritional intake of the patient which help mitigate the symptoms of migraine and have added great value to the treatment strategy. The dietary consumption of the individual significantly impacts the functions of the gut bacteria as well. Certain distinct diets such as a low-carbohydrate diet and the administration of probiotics lower the occurrence of migraine attacks substantially. This has gained the attention of various scientists globally as an important adjunct in the therapy. The Gut-Brain Axis (GBA), which is a bidirectional molecular signalling that occurs between the central nervous system and the gastrointestinal tract, serves as a novel path for the development of several medications concerning migraine treatment. The uncovering of information related to specific receptors such as the CGRP receptor (Calcitonin gene-related peptide) has provided us with the knowledge that such receptors can act as effective targets for several different medications for the duration of the therapy. The newer medications such as the anti-CGRP antibodies derived from a single cell (monoclonal) and the first and second-generation calcitonin gene-related peptide (CGRP) receptor antagonists have attested to be a revolutionary development in the management of migraine.

Pharmacoeconomic models contemplate various factors of the medication, such as the quality of life, disability, and the impact of the treatment on the social and economic facets of a society. Thus, these models are vital for the evaluation of the efficiency of the treatment. There is a need for prompt and feasible solutions to reduce the tremendous financial burdens linked to migraine. The goals of the novel medications, pharmacoeconomic models, improved instruments for diagnosis, and newer therapeutic targets are needed to boost quality of life and reduce the occurrence of migraine globally.

References

Agostoni EC et al (2019) Current and emerging evidence-based treatment options in chronic migraine: a narrative review. J Headache Pain 20(1):92. https://doi.org/10.1186/s10194-019-1038-4

Amiri P et al (2022) Migraine: a review on its history, global epidemiology, risk factors, and comorbidities. Front Neurol 12:800605. https://doi.org/10.3389/fneur.2021.800605

Andreou AP, Holland PR, Akerman S, Summ O, Fredrick J, Goadsby PJ (2016) Transcranial magnetic stimulation and potential cortical and trigeminothalamic mechanisms in migraine. Brain J Neurol 139(Pt 7):2002–2014. https://doi.org/10.1093/brain/aww118

Cortés JA, Díez L, Cañete FJ et al (2011) Performance analysis of OFDM modulation on indoor broadband PLC channels. EURASIP J Adv Signal Process. https://asp-eurasipjournals.springeropen.com/articles/10.1186/1687-6180-2011-78. Accessed 27 Nov 2023

Crawford J, Liu S, Tao F (2022) Gut microbiota and migraine. Neurobiol Pain 11:100090. https://doi.org/10.1016/j.ynpai.2022.100090

Do TP, Guo S, Ashina M (2019) Therapeutic novelties in migraine: new drugs, new hope? J Headache Pain 20(1):37. https://doi.org/10.1186/s10194-019-0974-3

Drugs.com (2023) What are the new drugs for the treatment of migraines? Drugs.com. https://www.drugs.com/medical-answers/new-drugs-treatment-prevention-migraine-3515053/. Accessed 4 Dec 2023

Gazerani P (2020) Migraine and diet. Nutrients 12(6):1658. https://www.mdpi.com/2072-6643/12/6/1658. Accessed 27 Nov 2023

He Q et al (2023) A causal effects of gut microbiota in the development of migraine. J Headache Pain 24(1):90. https://doi.org/10.1186/s10194-023-01609-x

Headache disorders (2023). https://www.who.int/news-room/fact-sheets/detail/headache-disorders. Accessed 27 Nov 2023

Hindiyeh NA, Zhang N, Farrar M, Banerjee P, Lombard L, Aurora SK (2020) The role of diet and nutrition in migraine triggers and treatment: a systematic literature review. Headache 60(7):1300–1316. https://doi.org/10.1111/head.13836

Hu XH, Markson LE, Lipton RB, Stewart WF, Berger ML (1999) Burden of migraine in the United States: disability and economic costs. Arch Intern Med 159(8):813–818. https://doi.org/10.1001/archinte.159.8.813
Iannone LF, De Cesaris F, Geppetti P (2022) Emerging pharmacological treatments for migraine in the pediatric population. Life (Basel) 12(4):536. https://www.ncbi.nlm.nih.gov/pmc/articles/PMC9031827/. Accessed 23 Nov 2023
Johnson MD et al (2013) Neuromodulation for brain disorders: challenges and opportunities. IEEE Trans Biomed Eng 60(3):610–624. https://doi.org/10.1109/TBME.2013.2244890
Khanal S et al (2022) A systematic review of economic evaluations of pharmacological treatments for adults with chronic migraine. J Headache Pain 23(1):122. https://doi.org/10.1186/s10194-022-01492-y
Migraine diet modifications: what really works? (2023). https://www.everydayhealth.com/migraine/migraine-diet-modifications-what-really-works/. Accessed 27 Nov 2023
Nisar A, Ahmed Z, Yuan H (2023) Novel therapeutic targets for migraine. Biomedicines 11(2):569. https://pubmed.ncbi.nlm.nih.gov/36831105/. Accessed 23 Nov 2023
Ogunlaja O, Karsan N, Goadsby P (2021) Emerging drugs for the prevention of migraine. Expert Opin Emerg Drugs 26(3):271–280. https://www.tandfonline.com/doi/full/10.1080/14728214.2021.1956463. Accessed 27 Nov 2023
Sahithi AST, Muthu T, Saraswathy R (2020) Migraine: update and future perspectives. Int J Nutr Pharmacol Neurol Dis 10(4):179. https://doi.org/10.4103/ijnpnd.ijnpnd_26_20
Schembri E, Barrow M, McKenzie C, Dawson A (2022) The evolving classifications and epidemiological challenges surrounding chronic migraine and medication overuse headache: a review. Korean J Pain 35(1):4–13. https://doi.org/10.3344/kjp.2022.35.1.4
Silberstein SD et al (2017) The emerging role of gammaCore® in the management of cluster headache: expert panel recommendations. Am J Manag Care 23(17 Suppl):S326–S333
Tajti J, Szok D, Csáti A, Szabó Á, Tanaka M, Vécsei L (2023) Exploring novel therapeutic targets in the common pathogenic factors in migraine and neuropathic pain. Int J Mol Sci 24(4):4114. https://doi.org/10.3390/ijms24044114
Tardiolo G, Bramanti P, Mazzon E (2019) Migraine: experimental models and novel therapeutic approaches. Int J Mol Sci 20(12):12. https://doi.org/10.3390/ijms20122932
Tiwari V, Agrawal S (2022) Migraine and neuromodulation: a literature review. Cureus 14(11):e31223. https://www.ncbi.nlm.nih.gov/pmc/articles/PMC9729750/. Accessed 27 Nov 2023
Urits I et al (2020) Peripheral neuromodulation for the management of headache. Anesthesiol Pain Med 10(6):e110515. https://doi.org/10.5812/aapm.110515
Vikelis M, Dermitzakis EV, Spingos KC, Vasiliadis GG, Vlachos GS, Kararizou E (2017) Clinical experience with transcutaneous supraorbital nerve stimulation in patients with refractory migraine or with migraine and intolerance to topiramate: a prospective exploratory clinical study. BMC Neurol 17(1):97. https://doi.org/10.1186/s12883-017-0869-3
Yong D, Lee H, Min HG et al (2023) Altered gut microbiota in individuals with episodic and chronic migraine. Sci Rep 13:626. https://www.nature.com/articles/s41598-023-27586-4. Accessed 23 Nov 2023
Yuan H, Chuang TY (2021) Update of neuromodulation in chronic migraine. Curr Pain Headache Rep 25:71. https://link.springer.com/article/10.1007/s11916-021-00988-7. Accessed 27 Nov 2023

The manufacturer's authorised representative in the EU is Springer
Nature Customer Service Centre GmbH, Europaplatz 3, 69115 Heidelberg,
Germany. If you have any concerns regarding our products, please
contact ProductSafety@springernature.com

Printed and bound by CPI Group (UK) Ltd, Croydon, CR0 4YY
01/10/2026
02225850-0009